AF608510

**ISNM 106:**
**International Series of Numerical Mathematics**
**Internationale Schriftenreihe zur Numerischen Mathematik**
**Série Internationale d'Analyse Numérique**
**Vol. 106**

**Edited by**
**K.-H. Hoffmann, München; H. D. Mittelmann, Tempe;**
**J. Todd, Pasadena**

Springer Basel AG

# Free Boundary Problems in Continuum Mechanics

**International Conference on Free Boundary Problems in Continuum Mechanics, Novosibirsk, July 15–19, 1991**

**Edited by**

**S. N. Antontsev**
**K.-H. Hoffmann**
**A. M. Khludnev**

**Springer Basel AG**

Editors

Prof. S. N. Antontsev
Lavrentyev Institute
of Hydrodynamics
Novosibirsk 630090
Russia

Prof. K.-H. Hoffmann
Institut für Angewandte
Mathematik und Statistik
Dachauer Str. 9 a
D–W–8000 München
Germany

Prof. A. M. Khludnev
Lavrentyev Institute
of Hydrodynamics
Novosibirsk 630090
Russia

A CIP catalogue record for this book is available from the Library of Congress, Washington D.C., USA

**Deutsche Bibliothek Cataloging-in-Publication Data**

**Free boundary problems in continuum mechanics /**
International Conference on Free Boundary Problems in Continuum Mechanics, Novosibirsk, July 15–19, 1991. Ed. by S. N. Antontsev ... – Basel ; Boston ; Berlin : Birkhäuser, 1992
(International series of numerical mathematics ; Vol. 106)
ISBN 978-3-7643-2784-2 ISBN 978-3-0348-8627-7 (eBook)
DOI 10.1007/978-3-0348-8627-7
NE: Antoncev, Stanislav N.; International Conference on Free Boundary Problems in Continuum Mechanics <1991, Novosibirsk>; GT


Originally published by Birkhäuser Verlag Basel in 1992

ISBN 978-3-7643-2784-2

# Contents

# Preface

Progress in different fields of mechanics, such as filtration theory, elastic-plastic problems, crystallization processes, internal and surface waves, etc., is governed to a great extent by the advances in the study of free boundary problems for nonlinear partial differential equations.

Free boundary problems form a scientific area which attracts attention of many specialists in mathematics and mechanics. Increasing interest in the field has given rise to the "International Conferences on Free Boundary Problems and Their Applications" which have convened, since the 1980s, in such countries as England, the United States, Italy, France and Germany.

This book comprises the papers presented at the International Conference "Free Boundary Problems in Continuum Mechanics", organized by the Lavrentyev Institute of Hydrodynamics, Russian Academy of Sciences, July 15-19, 1991, Novosibirsk, Russia.

The scientific committee consisted of:

Co-chairmen: K.-H. Hoffmann, L.V. Ovsiannikov

S. Antontsev (Russia)
M. Frémond (France)
A. Friedman (USA)
K.-H. Hoffmann (Germany)
A. Khludnev (Russia)
V. Monakhov (Russia)
J. Ockendon (UK)
L. Ovsiannikov (Russia)
S. Pokhozhaev (Russia)
M. Primicerio (Italy)
V. Pukhnachov (Russia)
Yu. Shokin (Russia)
V. Teshukov (Russia)

Our thanks are due to the members of the Scientific Committee, all authors, and participants for contributing to the success of the Conference. We would like to express special appreciation to N. Makarenko, J. Mal'tseva and T. Savelieva, Lavrentyev Institute of Hydrodynamics, for their help in preparing this book for publication.

July 1992 S. Antontsev, K.-H. Hoffmann, A. Khludnev

# SOME EXTREMUM AND UNILATERAL BOUNDARY VALUE PROBLEMS IN VISCOUS HYDRODYNAMICS

*G.V.Alekseyev, A.Yu.Chebotarev*
*Institute of Applied Mathematics, Vladivostok 690068, RUSSIA.*

This paper is concerned with investigation of direct and inverse problems for the stationary Stokes system. At first we prove the unique solvability of a direct unilateral boundary value problem and establish some properties of the solution. Then we formulate problems which are inverse to the direct problem and investigate the solvability of one inverse extremum problem.

Key words: Stokes system, unilateral boundary value problem, inverse extremum problem, variational inequality.

## 1.INTRODUCTION.

Simulation of some new processes in fluid mechanics causes the necessity of solving the new boundary value problems for the Navier-Stokes and Euler equations. As an examples of such problems are so-called unilateral boundary value problems and extremum problems arising in viscous and ideal hydrodynamics. Various authors have considered the extremum problems in hydrodynamics. The works most closely related to ours can be found in Lions [1,2], Fursikov [3,4], Chebotarev [5,6]. In all of them some standard linear boundary conditions were posed on the boundaries of domains considered. Unilateral boundary value problems were studied in Lions [7], Kazhikhov [8-9] and others(cf. references to [7], [9]) for the nonstationary Navier-Stokes equations and in Chebotarev [10] for the Euler equations. For an ideal fluid unilateral boundary conditions arise in a natural way in the problems of liquid flowing through a 'tube'. For the first time these problems were considered in stationary case (with classical boundary conditions) in Alekseyev [11-13]. For the Stokes system unilateral boundary value problems connected with the steady viscous fluid flowing through a bounded domain and corresponding extremum problems were studied in Chebotarev [14], where some properties (existence, uniqueness, estimates) of the solutions were derived.

Below we investigate some direct and inverse unilateral boundary value problems for the stationary Stokes system in the space $\mathbb{R}^m$, $m = 2,3$.

## 2. UNILATERAL BOUNDARY VALUE PROBLEM FOR THE STOKES SYSTEM.

Let $\Omega$ be a bounded domain in the space $\mathbb{R}^m$, $m = 2,3$ with a smooth boundary $\Gamma = \Gamma_1 \cup \Gamma_2 \cup \Gamma_0$, $\Gamma \in C^2$. We consider the special unilateral boundary value problem for the Stokes system

$$- \nu \Delta \mathbf{u} + \nabla P = \mathbf{f}, \quad div\, \mathbf{u} = 0, \; x \in \Omega, \tag{1}$$

which describes the stationary flow of the homogeneous viscous incompressible fluid through the domain $\Omega$. In (1) $\mathbf{u}$ denotes the velocity, P is the pressure and $\mathbf{f}$ stands for a given external force. Let $\mathbf{n}$ be the unit outward normal vector to $\Gamma$. We assume that

$$\mathbf{u}_\tau = \mathbf{u} - (\mathbf{u}\cdot\mathbf{n})\mathbf{n} = 0 \quad \text{on} \quad \Gamma \tag{2}$$

and the normal component $u_n = \mathbf{u}\cdot\mathbf{n}$ satisfies the condition

$$u_n = q \leqslant 0 \quad \text{on} \quad \Gamma_1, \; u_n = 0 \quad \text{on} \quad \Gamma_0 \tag{3}$$

and besides

$$u_n \geqslant 0 \;, \quad p \leqslant 0, \quad u_n \cdot p = 0 \quad \text{on} \quad \Gamma_2. \tag{4}$$

The condition (4) is so called unilateral boundary condition. It means in fact that the pressure $P$ equals zero in those points of down-flow part $\Gamma_2$ of the boundary $\Gamma$, where $u_n > 0$. Our purpose is to study the problem 1 (Pr.1), which consists in finding a solution $(\mathbf{u},P)$ of the Stokes system (1) satisfying the boundary conditions (2) - (4).

We need some notations. Let $D(\Omega)$ be the linear space of vector-functions infinitely differentiable and with a compact support in $\Omega$, $H^s(\Omega)$, $s > 0$ be the Sobolev space with the norm $\|\cdot\|_{s,\Omega}$, $H^s_o(\Omega)$ be the subspace of $H^s(\Omega)$ which consists of functions $\mathbf{v}$ vanishing on $\Gamma$, $D'(\Omega)$ denotes the dual space to $D(\Omega)$ usually called the space of distributions on $\Omega$ and $H^{-s}(\Omega)$ be the dual space to $H^s_o(\Omega)$. Similarly, $H^s(\gamma)$ denotes the Sobolev space of vector-functions given on a part $\gamma$ of the boundary $\Gamma$ with the norm $\|\cdot\|_{s,\gamma}$, $H^{-s}(\gamma)$ is the dual space to $H^s(\gamma)$. As in [8] define the following function spaces

$$J^\infty = \left\{ \mathbf{v} \in C^\infty(\Omega):\ div\ \mathbf{v} = 0 \text{ in } \Omega,\ \mathbf{v}_\tau = 0 \text{ on } \Gamma,\ v_n = 0 \text{ on } \Gamma_0 \right\},$$

$$H = \text{closure } J^\infty(\Omega) \text{ in } H^0(\Omega),\ V = \text{closure } J^\infty(\Omega) \text{ in } H^1_0(\Omega),$$

and besides let

$$V_\gamma = \left\{ \mathbf{v} \in V:\ \mathbf{v} = 0 \text{ on } \gamma \subset \Gamma \right\},\ Q = \left\{ q = v_n \big|_{\Gamma_1} :\ \mathbf{v} \in V \right\},$$

$$Q^- = \left\{ q \in Q:\ q \leqslant 0 \right\}.$$

$H$ and $V$ are Hilbert spaces. The scalar products and norms in $H$ and $V$ are denoted by $(\ ,\ )$, $|\cdot|$ and $((\ ,\ ))$, $\|\cdot\|$ respectively. In particular,

$$((\mathbf{u},\mathbf{v})) = \int_\Omega \nabla\mathbf{u}\cdot\nabla\mathbf{v}\ dx \qquad \forall\ \mathbf{u},\mathbf{v} \in V.$$

$Q$ is a Hilbert space with the norm $\|\cdot\|_Q = \|\cdot\|_{1/2,\Gamma_1}$. As usual we identify $H$ with its dual $H^*$ and so we have: $V \subset H \subset V^*$. We shall denote the value of a functional $\mathbf{f} \in V^*$ on an element $\mathbf{v} \in V$ by $\langle \mathbf{f},\mathbf{v} \rangle$, and the same for a functional $p \in H^{-1/2}(\Gamma_2)$ on $v \in H^{1/2}(\Gamma_2)$ by $\int_{\Gamma_2} pv\ ds$.

**LEMMA 1** [8]. Let functions $\mathbf{u}$ and $\mathbf{v}$ satisfy the conditions

$$\mathbf{u} \in V \cap H^2(\Omega),\ \mathbf{v} \in H^1(\Omega),\ \mathbf{v}_\tau = 0 \text{ on } \Gamma.$$

Then

$$-(\Delta\mathbf{u},\mathbf{v}) = ((\mathbf{u},\mathbf{v})) + \int_{\Gamma_1 \cap \Gamma_2} k\mathbf{u}\cdot\mathbf{v}\ ds. \tag{5}$$

Here $k(x)$ is a double mean curvature of the boundary $\Gamma$ in $x \in \Gamma$. Now we present a weak (variational) formulation of the Pr.1. Let

$q \in Q^-$ be fixed. Define a closed convex subset

$$K = K_q = \left\{ \mathbf{v} \in V\colon\ v_n = q \text{ on } \Gamma_1,\ v_n \geqslant 0 \text{ on } \Gamma_2 \right\}.$$

Let $\mathbf{u}, P$ and $\mathbf{f}$ be some smooth functions satisfying (1) - (4). Multiply the first equation in (1) by the function $\mathbf{u} - \mathbf{v}$, where $\mathbf{v} \in K_q$ is a smooth function and integrate it in $\Omega$. Using the Lemma 1 and conditions (2)-(4) we obtain the variational formulation of the Pr.1. It consists in finding such function $\mathbf{u}$ that the following variational inequality holds

$$\mathbf{u} \in K_q\ ,\ A(\mathbf{u},\mathbf{u}-\mathbf{v}) \leqslant \langle \mathbf{f}, \mathbf{u}-\mathbf{v} \rangle \quad \forall\ \mathbf{v} \in K_q\ . \tag{6}$$

Here the bilinear form A is defined by the formula

$$A(\mathbf{u},\mathbf{v}) = \nu((\mathbf{u},\mathbf{v})) + \nu \int_{\Gamma_2} k\mathbf{u}\cdot\mathbf{v}\, ds.$$

One can prove that the opposite takes place as well: if $\mathbf{u}$ satisfies (6),then $\mathbf{u}$ is a solution in a certain sense of the Pr.1.

*Definition 1.* The function $\mathbf{u} \in K_q$, satisfying (6), is called a weak solution of the Pr.1.

**THEOREM 1.** Let the following conditions take place:

*(i)* $\mathbf{f} \in V^*,\ q \in Q^-,\ \Gamma \in C^2$;

*(ii)* the bilinear form A be symmetric and $V$-elliptic, i.e. there exists a constant $\alpha > 0$ such that

$$A(\mathbf{v},\mathbf{v}) \geqslant \alpha \|\mathbf{v}\|^2 \quad \forall\ \mathbf{v} \in V.$$

Then the weak solution $\mathbf{u}$ of the Pr.1 exists and it is unique.

**Proof.** It is well known [1], that the problem of solving the inequality (6) is equivalent to the following minimization problem

$$\nu\|\mathbf{v}\|^2 + \nu \int_{\Gamma_2} k\mathbf{v}^2\, ds - 2\langle \mathbf{f},\mathbf{v} \rangle \longrightarrow \inf,\ \mathbf{v} \in K_q\ . \tag{7}$$

Since $K_q$ is a closed convex set, then it follows from *(ii)*, that the solution of the problem (7) exists and it is unique.

Below the term "the flow $u = u_q$" (corresponding to an input function $q$) will mean the weak solution of the Pr.1. Taking in (6) $v = u + \lambda\varphi$, where $\varphi \in D(\Omega) \cap J^{\infty}$, $\lambda \in \mathbb{R}$ one can show by standard way that the flow $u$ satisfies as well the integral identity

$$\nu((u,\varphi)) = \langle f,\varphi \rangle \quad \forall\ \varphi \in D(\Omega) \cap J^{\infty}. \tag{8}$$

Then there exists a distribution (pressure) $p \in L^2(\Omega)$ such that [16]

$$\nu\Delta u + f = \nabla P \text{ in } V^{*}. \tag{9}$$

If besides $f \in L^2(\Omega)$ then it follows from (9) that

$$\Delta P = div\ f \in H^{-1}(\Omega).$$

Then $P|_{\Gamma} \in H^{-1/2}(\Gamma)$ and besides $P|_{\Gamma_2} \in H^{-1/2}(\Gamma_2)$. Suppose

$\int_{\Gamma_2} Pu_n\, ds = 0$. Using this condition we can deduce from (9) that the flow $u$ and the pressure $P$ satisfy as well the following identity

$$A(u,w) + \int_{\Gamma_2} Pw_n\, ds = \langle f,w \rangle \quad \forall\ w \in V_{\Gamma_1}. \tag{10}$$

Take in (10) $w = u - v$ where $v \in K_q$ and subtract it from (6). We obtain $\int_{\Gamma_2} Pv_n\, ds \leqslant 0$. Thus the flow $u$ and the pressure $P$ satisfy the following analogue of the unilateral boundary condition (4) on $\Gamma_2$:

$$u_n \geqslant 0,\ \int_{\Gamma_2} Pu_n\, ds = 0 \quad \text{and} \quad \int_{\Gamma_2} Pv_n\, ds = 0 \quad \forall\ v \in K_q. \tag{11}$$

**REMARKS.**

1. The V-ellipticity of the form A is obvious, if $k(x) \geqslant 0$ on $\Gamma_2$, (or $|k|$ is sufficiently small at those points of $\Gamma_2$ where $k < 0$).

2. The theorem 1 is valid as well, if the condition *(ii)* takes place only for every $\mathbf{v} \in K_q$.

3. It follows from [14] that if $\mathbf{f} \in L^2(\Omega)$ there exist

$$P\big|_{\Gamma_2} \in H^{1/2}(\Gamma_2), \quad \mathbf{u}\big|_{\Gamma_2} \in H^{3/2}(\Gamma_2)$$

and the following estimates for the solution $(\mathbf{u},P)$ take place

$$\|\mathbf{u}\|^2 \leqslant C_1(\|q\|_Q^2 + |\mathbf{f}|^2), \; \|\mathbf{u}\|_{3/2,\Gamma_2} + \|P\|_{1/2,\Gamma_2} \leqslant C_2. \tag{12}$$

Here $C_1 = C_1(\Omega,\nu)$ - *Const*, constant $C_2$ is independent of $\mathbf{u}$ and $P$ but depends on $\|q\|_Q$, $|\mathbf{f}|$, $\Omega$, $\nu$ and it is bounded if

$$\|q\|_Q \leqslant M < \infty, \quad \int_{\Gamma_1} q\, ds < - \varepsilon < 0.$$

4. At $q = 0$ the set $K_q$ becomes a subspace

$$K_0 = \left\{\mathbf{v} \in V\colon \mathbf{v} = 0 \text{ on } \Gamma\right\}$$

of the space $V$ and the variational inequality (6) transforms to the identity (8). The latter corresponds to the classical boundary problem for the Stokes system (1) with the homogeneous condition $\mathbf{u} = 0$ on $\Gamma$. Thus the last problem can be considered as a particular case of the unilateral Pr.1. We notice also that the second estimate in (12) at $q = 0$ is an obvious consequence of the well known properties: $\mathbf{u} \in H^2(\Gamma)$ and $P \in H^1(\Gamma)$ of the solution $(\mathbf{u},P)$ of this classical problem.

## 3. SOME INVERSE UNILATERAL PROBLEMS FOR THE STOKES SYSTEM.

Let $\mathbf{f} \in L^2(\Omega)$ be fixed. It follows from the Th.1 that there exists the (nonlinear) operator $A\colon L^2(\Gamma_1) \longrightarrow V$ which every input $q \in Q^-$ sets the flow $\mathbf{u}$ so that

$$\mathbf{u} = \mathbf{u}_q = \mathbf{A}q \in V. \tag{13}$$

It is clear that the operator $\mathbf{A}$ is invertible on $R(\mathbf{A}) \subset V$ and its inverse $\mathbf{A}^{-1}: V \to Q$ coincides with the trace operator $\gamma_v|_{\Gamma_1} : V \to Q$, which sets to an every flow $\mathbf{v} \in V$ its normal component $q = v_n$ on $\Gamma_1$.

Using the operator $\mathbf{A}$ one can consider problems which are inverse to the (direct) unilateral Pr.1. To this end we suppose that some information about the flow $\mathbf{u}$ is known and it's necessary to restore an input $q$ on it. For example one can consider the problem of finding an input $q$ in given values $\mathbf{g}$ of the flow $\mathbf{u}$ on some subset $\omega \subset \Omega$. In order to pose the corresponding inverse problem introduce the restriction operator $S_\omega$ to the set $\omega$ defined by the formula $S_\omega \mathbf{u} = \mathbf{u}|_\omega$, $\mathbf{u} \in V$. Then applying the operator $S_\omega$ to (13) and inverting the equation obtained we have

$$S_\omega \mathbf{A}q = \mathbf{g}. \tag{14}$$

Here $\mathbf{g} = S_\omega \mathbf{u}$ is a given function. Thus if the direct unilateral Pr.1 is reduced to finding the flow $\mathbf{u}$ with the help of (13) then solving the inverse unilateral problem is reduced to finding a solution $q \in Q^-$ of the first kind nonlinear operator equation (14).

Besides the inverse problems of the type (14) one can formulate extremum inverse problems. In order to specify such problem choose from the certain (for example physical) considerations a subset $V_0$ in $V$. Consider the problem: find an element $q \in Q^-$ for which the flow $\mathbf{u}_q$ belongs to $V_0$. It is clear that such element $q$ is generally nonunique. So we shall introduce a functional (the cost function)

$$J_\lambda(q) = \lambda \int_\Omega \Phi(\mathbf{u}_q)\, dx + \int_{\Gamma_1} q^2\, ds, \quad \Phi(\mathbf{v}) = \frac{\nu}{2} \sum_{i,j=1} \left( \frac{dv_i}{dx_j} + \frac{dv_j}{dx_i} \right)^2 .$$

Here $\Phi(\mathbf{u})$ is the dissipative function for a viscous incompressible flow $\mathbf{u}$. It is a heat equivalent of the corresponding mechanic energy. Thus, we arrive at the following optimization problem

$$J_\lambda(q) \to \inf\ ,\ q \in Q_0. \tag{15}$$

Here $Q_0$ is an admissible control set which we define as

$$Q_0 = Q_0^-(\varepsilon) = A^{-1}(V_0) \cap \left\{q \in Q^- : \int_{\Gamma_1} q\, ds < -\varepsilon < 0\right\}. \tag{16}$$

**THEOREM 2.** Let the conditions *(i)-(ii)* hold, $f \in L^2(\Omega)$ and

*(iii)* $Q_0 \neq \emptyset$, $V_0$ be a closed convex set;

*(iv)* $\lambda > 0$ (or $\lambda = 0$ but $Q_0$ be a bounded set).

Then there exists at least one solution of the problem (15).

**Proof.** Let $q_m \in Q_0$ be a minimizing sequence for $J_\lambda$ i.e.

$$\lim_{m\to\infty} J_\lambda(q_m) = J_\lambda^* = \inf\left\{J_\lambda(q),\ q \in Q_0\right\}.$$

Let $u_m = Aq_m$ and note that according to *(iv)* and the first Korn inequality [17] the sequences $\{q_m\}$ and $\{u_m\}$ are bounded in $Q$ and $V$ respectively. It follows from the weak compactness of the unit ball in a Hilbert space and the embedding theorems that (by taking subsequences)

$$q_m \to q \in Q^- \text{ weakly in } Q, \qquad u_m \to u \text{ weakly in } V, \tag{17}$$

$$u_m\big|_\Gamma \to u\big|_\Gamma \quad \text{in} \quad L^2(\Gamma) \quad \text{when} \quad m \to \infty. \tag{18}$$

As $u_m \in V_0$ and $V_0$ is a weakly closed set then $u \in V_0$ and so $q \in Q_0$ Moreover it follows from (17), (18) that $u \cdot n = q$ on $\Gamma_1$ so that $u \in K_q$.

Let $P_m$ be the pressure corresponding to the flow $u_m$ such that

$$\int_{\Gamma_2} P_m(u_m \cdot n)\, ds = 0 \quad \text{and} \quad \int_{\Gamma_2} P_m v_n\, ds \leqslant 0 \quad \forall\ v \in K_{q_m}. \tag{19}$$

Then according to the second estimate in (12) (see remark 3) we have ( by taking subsequence ) that

$$P_m\Big|_{\Gamma_2} \longrightarrow p \in H^{1/2}(\Gamma_2) \quad \text{in } L^2(\Gamma_2). \tag{20}$$

It follows from (19)-(20) that

$$\int_{\Gamma_2} pu_n \, ds = 0 \quad \text{and} \quad \int_{\Gamma_2} pv_n \, ds \leqslant 0 \qquad \forall\ \mathbf{v} \in \mathbf{K}_q. \tag{21}$$

Take in (10) $\mathbf{u} = \mathbf{u}_m$, $P = P_m$, $\mathbf{w} = \mathbf{u} - \mathbf{v}$, where $\mathbf{v} \in \mathbf{K}_q$. Then we obtain

$$A(\mathbf{u}_m,\mathbf{u}-\mathbf{v}) + \int_{\Gamma_2} P_m(u_n - v_n)ds = (\mathbf{f},\mathbf{u}-\mathbf{v}) \qquad \forall\ \ \mathbf{v} \in \mathbf{K}_q. \tag{22}$$

In the limit in (22), when $m \to \infty$ we have from (17)-(21) that

$$A(\mathbf{u},\mathbf{u}-\mathbf{v}) = \int_{\Gamma_2} pv_n \, ds + (\mathbf{f},\mathbf{u}-\mathbf{v}) \leqslant (\mathbf{f},\mathbf{u}-\mathbf{v}) \qquad \forall\ \mathbf{v} \in \mathbf{K}_q.$$

This means that $\mathbf{u} = \mathbf{u}_q$ and it follows from obvious inequalities

$$J_\lambda^* \leqslant J_\lambda(q) \leqslant \lim_{m\to\infty} \Big[\lambda\int_{\Gamma_2} \Phi(\mathbf{u}_m)\, dx + \int_{\Gamma_1} q_m^2 \, ds\Big] \leqslant J_\lambda^*,$$

that $q \in Q_0$ is an optimal control. The theorem has been proved.

To conclude we consider one interesting particular case when the variables "control-state" are separated. Let

$$V_0 = \Big\{\mathbf{v} \in V:\ \theta = \theta(\mathbf{v}) = \frac{1}{\text{mes}\ \Gamma_2} \int_{\Gamma_2} v_n \, ds \in [\theta_1,\theta_2]\Big\},$$

where $0 < \theta_1 < \theta_2$. Then the condition $\mathbf{v} \in V_0$ means that the mean output flow belongs to given interval $[\theta_1,\theta_2]$. It is important that such choice of $V_0$ allows us to write the condition on control $q$ directly in terms of the set $V_0$ i.e. through $\theta_1$ and $\theta_2$. Really, as $\operatorname{div} \mathbf{v} = 0$ for any $\mathbf{v} \in V$ we have

$$\theta = \theta(\mathbf{v}) = -\frac{1}{\text{mes}\ \Gamma_2} \int_{\Gamma_1} q \, ds \ \equiv \mu(q).$$

Therefore, we can rewrite the set $Q_0$ as

$$Q_0 = \left\{ q \in Q^- : \mu(q) \in [\theta_1, \theta_2] \right\}. \tag{23}$$

We consider the problem (15),(23)at $\lambda=0$, when the sufficient condition for its solvability is

$$\|q\|_Q \leqslant Const \quad \forall\, q \in Q_0.$$

The process of solving (15),(23) at $\lambda=0$ can be divided into two stages. At first, it is necessary to solve the extremum problem

$$|q|^2_{0,\Gamma_1} = \int_{\Gamma_1} q^2 ds \longrightarrow inf, \quad q \in Q_0. \tag{24}$$

which is generally not connected with the inequality (6). Then, at the second stage, for finding an optimal flow $u_q \in V_0$ it is necessary to solve the variational inequality (6) on $K_q$, where $q$ is a solution of the problem (24).

## REFERENCES

[1] J.L.Lions. Controle optimal de systemes gouvernes par des equations aux derivees (Dunod, Paris, 1968).

[2] J.L.Lions. Controle des systemes distribues singuliers (Gauthier-Villars, Paris, 1983).

[3] A.V. Fursikov. Mathem. Zbornik. V. 115(157). P. 281 (1981).

[4] A.V. Fursikov. Mathem. Zbornik. V. 118(160). P. 323 (1982).

[5] A.Yu. Chebotarev. In "Some problems of analysis and algebra". Izdat Novosibirsk Gos. Univ. P.128. (1984).(in Russian).

[6] A.Yu. Chebotarev. Dinamika sploshnoi sredy. Vyp. 81. Novosibirsk. P.138 (1987) (in Russian).

[7] J.L.Lions. Quelques methodes de resolution les problemes aux limites non lineaires. (Dunod, Paris, 1969).

[8] A.V.Kazhykhov. Dinamika sploshnoi sredy. Vyp.16. Novosibirsk. P.5 (1974) (in Russian)

[9] S.N. Antontzev, V.N. Monakhov and A.V. Kazhikhov. Boundary value problems of the mechanics of nonhomogeneous fluids (Science, Novosibirsk, 1983). (in Russian).

[10] A.Yu. Chebotarev. Dinamika sploshnoi sredy. Vyp. 79. Novosi-

birsk. P.121 (1987) (in Russian).

[11] G.V.Alekseyev. Dinamika sploshnoi sredy. Vyp.10. Novosibirsk. P.5 (1972) (in Russian).

[12] G.V.Alekseyev. Dinamika sploshnoi sredy. Vyp.15. Novosibirsk. P.7 (1973) (in Russian).

[13] G.V.Alekseyev. Prykl. Mech. Techn. Phys. N.2. P.85 (1977) (in Russian).

[14] A.Yu.Chebotarev in " Mathem. Phys. and Mathem. Modelling in Ecology". Vladivostok. P.121 (1990) (in Russian).

[15] J.L.Lions, E.Magenes. Problemes aus limites non homogenes et applications. V.1, (Dunod, Pris, 1986).

[16] R.Temam. Navier-Stokes equations (North-Holland Publishing Company, Amsterdam. New York. Oxford. 1979).

[17] G.Duvaut, J.L.Lions. Les inequations en mecanique et en physique, (Dunod, Paris, 1972).

International Series of Numerical Mathematics, Vol. 106, © 1992 Birkhäuser Verlag Basel 

# ON AXISYMMETRIC MOTION OF THE FLUID WITH A FREE SURFACE

V.K.Andreev
Computing Center, Krasnoyarsk 630036, RUSSIA.

The problem with a free boundary for the equations of the ideal incompressible fluid is considered with the presence of axial symmetry in Lagrangian coordinates $(\eta,\xi)$

$$r_{tt} - \frac{V}{r^3} + \frac{r}{\eta}\left(z_\xi p_\eta - z_\eta p_\xi\right) = 0, \quad z_{tt} - \frac{r}{\eta}\left(r_\xi p_\eta - r_\eta p_\xi\right) = 0, \tag{1}$$

$$r\left(z_\xi r_\eta - r_\xi z_\eta\right) = \eta, \quad (\eta,\xi) \in \Omega; \tag{2}$$

$$p = 2\sigma H, \quad (\eta,\xi) \in \Gamma; \tag{3}$$

$$r = \eta, \quad z = \xi, \quad r_t = u_0(\eta,\xi), \quad z_t = w_0(\eta,\xi), \quad t = 0. \tag{4}$$

Here $V(\eta,\xi)$ is a square of the initial angular momentum of a liquid particle impulse about the axis $z$, $\sigma \geqslant 0$ is a surface tension coefficient, $H$ is an average curvature of the boundary $\Gamma_t$ of the flow area

$$\Omega_t = \left\{r,\ z \;\middle|\; r = r(\eta,\xi,t),\ z = z(\eta,\xi,t),\ (\eta,\xi) \in \Omega\right\}.$$

For the system (1), (2) a problem of new classification groups is solved with respect to the arbitrary element $V(\eta,\xi)$ [1]. The equivalence transformations have a form

$$\bar\eta = a_1\alpha(\eta,\xi), \quad \bar\xi = a_1\beta(\eta,\xi), \quad \bar t = a_2 t, \quad \bar r = a_1 r,$$

$$\bar z = a_1 z, \quad \bar p = a_2^{-2}p + a_3 r^{-2},$$

$$\bar V = a_2^{-2}V(\bar\eta,\bar\xi) - 2a_3 \quad (a_i = const), \tag{5}$$

where functions $\alpha$, $\beta$ satisfy a continuity equation:

$$\alpha_\eta \beta_\xi - \alpha_\xi \beta_\eta = \eta/\alpha.$$

A result of the group classification is represented in Table 1. Lie group appears to be infinite-dimensional also by space coordinates.

The invariance requirement of the initial conditions $r = \eta$, $z = \xi$, $t = 0$ reduces the equivalence group (5), such that $\alpha = \eta$, $\beta = \xi + a_4$. The other variables are transformed according to the same law (5). Here the expansion of functions $V(\eta,\xi)$, which take part in the group classification, occurs, see Table 2.

It is easy to separate a subgroup in every case of 1-9 of Table 2, when boundary condition (3) is also invariant. For example, with $\sigma = 0$ (3) is invariant under the operators $Y_t$, $Y_1$, $Y_2$, $Y_4$ action. As for Table 1,for the case 1 the boundary condition (3) is invariant with respect to operators $\partial_t$, $t\partial_z$, for 2 - with respect to $\partial_t$, $t\partial_z$, $Y_1$, $Y_2$, $\eta^{-1}\varphi_\xi\partial_\eta - \eta^{-1}\varphi_\eta\partial_\xi$, and for 3 - with respect to $\partial_t$, $t\partial_z$, $t\partial_t - 2\eta\partial_\eta + 4\xi\partial_\xi - 2p\partial_p$, $4\eta\partial_\eta - 5\xi\partial_\xi + r\partial_r + z\partial_z + 2p\,\partial_p$, $B(\eta)\partial_\xi$. In the latter case one should take into account, that while classifying a substitution $\eta' = V(\eta,\xi)$, $\xi' = f(\eta,\xi)$, $V_\eta f_\xi - V_\xi f_\eta = \nu/V$ has been made and the initial domain $\Omega$ maps into some domain $\Omega'$ of variables $(\eta',\xi')$.

2. Assume, that $V = V(\eta)$, $V_\eta \neq 0$. This is the case 5 from Table 2 with $\beta = 0$. Take a three-parametrical subgroup $\langle\partial_\xi + \partial_z, t\partial_z, \partial_p\rangle$. There exists a partly invariant solution of rank 2 and of defect 1 of the form $r = r(\eta,t)$, $z = z(\eta,\xi,t)$, $p = p(\eta,\xi,t)$. Then from (1) we find a representation for the solution

$$r = \left[ 2\int \frac{\eta\, d\eta}{a(\eta,t)} + C(t) \right]^{1/2}, \quad z = a(\eta,t)\xi,$$

$$p = \int r_2\left(r_{tt} - \frac{V}{r^3}\right) d\eta - \mu(t)\, a^2(\eta,t)\xi^2/2 + \varphi(t),$$

$$a_{tt} = \mu(t)a, \quad a(\eta,0) = 1, \quad a_t(\eta,0) = a_0(\eta). \tag{6}$$

with arbitrary functions $C(t)$, $\mu(t)$, $V(\eta)$, $a_0(\eta)$, $\varphi(t)$, $C(0) = 0$. This solution can be interpreted as the motion with a conic free boundary, with a free boundary in the form of one-sheet hyperboloid, plane, spheroid, rotation paraboloid or cylindrical boundary. Consider the latter case in detail.

Assume in (6) $a = 1 + Kt$, $K = const$ and consider $C(t) \neq 0$ (with $C = 0$ we obtain a problem on the liquid cylinder expansion [2]). Thus, the solution (6) is specialized to the following

$$r = m(\eta,t)\eta, \quad z = \tau\xi, \quad \tau = 1 + Kt, \quad m = \left[\frac{1}{\tau} + \frac{C}{\eta^2}\right]^{1/2}. \tag{7}$$

In this case a cylindrical layer can be considered as a finite one. Really, let in the initial moment a domain, filled with a fluid, be a cylindrical layer

$$\Omega = \left\{ \eta,\xi \;\middle|\; \eta_1 < \eta < \eta_2, \quad 0 < \xi < h \right\}.$$

Planes $\xi = 0$, $\xi = h$ are inpenetrable walls, and cylindrical surfaces $\eta = \eta_1, \eta_2$ are free boundaries. The initial velocity field on $\Omega$ has a form $u_0 = [C'(0) - Kr^2]/2\eta$, $w_0 = K\xi$, thus a constant $K$ is defined by the initial velocity $W = Kh$ of the solid wall $\xi = h$.

To examine the evolution of the cylindrical layer free boundaries, we write $r_1(t)$ for the inner radius, $r_2(t)$ - for the outer one. From (7) we obtain

$$r_1(t) = m(\eta_1,t)\eta_1, \quad r_2(t) = m(\eta_2,t)\eta_2$$

and hence it follows

$$r_2^2(t) - r_1^2(t) = (\eta_2^2 - \eta_1^2)/\tau. \tag{8}$$

The finite relationship (8) is the conservation law of the cylindrical layer volume. While moving, as follows from (7), (8), the fluid conserves the form of the straight hollow cylinder, moreover, the solid wall $\xi = 0$ remains motionless. The upper wall moves according to the law $z = \tau h$. If $K > 0$, then the cylindrical layer stretches along the axis $z$, with $K < 0$ for the time $t = 1/|K|$ the moving plane meets the motionless one. If $u_{10}$ is the initial velocity of the inner surface, then $C'(0) = 2\eta_1 u_{10} + K\eta_1^2$. Assume, that $C'(0) = 0$, thus the motion is defined only by the constant $K = W/h$.

Let $\sigma_1$, $\sigma_2$ be surface tension coefficients on the inner and outer surfaces. From the dynamic condition on the boundaries $p(r_2(t),t) - p(r_1(t),t) = \sigma_1/r_1(t) + \sigma_2/r_2(t)$ and the equalities (6), (8) we could obtain an ordinary differential equation of the second order for the function $C(t)$. Instead of $C(t)$ it's convenient to introduce a new function $g = 1 + r_1^{-2}\tau C$ and go over to the variable $\mu = \tau^2 = (1 + Kt)^2$ instead of $t$. Then the above mentioned equation with respect to $g(\mu)$ has a form

$$g'' \ln\left(1 + \varepsilon/g\right) + \frac{g'^2}{2}\left[\frac{1}{g+\varepsilon} - \frac{1}{g}\right] + \frac{3\varepsilon}{8\mu^2} + \\ + \mu^{-1/4}\left[\frac{S_1}{\sqrt{g}} + \frac{S_2}{\sqrt{g+\varepsilon}}\right] = 0,$$

$$g(1) = 1, \quad g'(1) = 0, \tag{9}$$

where the following notations of Weber numbers are introduced $\varepsilon = \eta_2^2/\eta_1^2 - 1 > 0$, $S_j = \sigma_j/\eta_1^3 K^2$ $(j = 1, 2)$, (note, that a fluid density is assumed to be equal to unit). By the function $g(\mu)$ radii of inner and outer surfaces are determined by

$$r_1 = \eta_1\mu^{-1/4}\sqrt{g}, \quad r_2 = \eta_1\mu^{-1/4}\sqrt{g+\varepsilon}. \tag{10}$$

It is easily seen from (9), that $g(\mu) \leqslant 1$ for every $\mu \geqslant 1$. It can be shown, that such $\mu_* < \infty$ exists, that

$$g(\mu_*) = 0, \quad g' \sim \gamma[\ln(1 + \varepsilon/g)]^{-1/2}$$

with $\mu \to \mu_*$, $\gamma = const < 0$.

Using (10) we obtain $r_1(t_*) = 0$, $r_2(t_*) = \eta_1\mu_*^{-1/4}\varepsilon^{1/2}$ in the moment $t_* = (\sqrt{\mu_*} - 1)/K$. Using $g'$ behavior, we can easily show, that the inner surface velocity unfinitely increases, and, besides, $r_1' \sim v r_1^{-1}[-\ln(r_1\sqrt{\tau}/\eta_1)]^{-1/2}$ with $r_1 \to 0;$ $v = const.$ This reflects the fact, that in the moment of the space disappearance the hydraulic shock-collapse of the space occurs.

Unfortunately, this motion can't be continued for the time $t > t_* = (\sqrt{\mu_*} - 1)/K$ as a motion of a cylindrical jet. Really, a complete energy of the layer remains a finite one in the collapse moment, however it doesn't coincide with the jet energy in this moment of time. In Fig.1 a plot of a dimensionless collapse time $Kt_*$ is represented, depending on $\varepsilon$, Weber numbers $S_1$, $S_2$. With the increase of $S_1$, $S_2$ the collapse time decreases.

Examine the stability of non-stationary motion of a cylindrical layer (7). A problem on perturbations of such motions has been investigated by the author in [3], therefore, we don't state it here in general statement. For the basic motion (7) it can be specialized to the form

$$\Phi_{\eta\eta} + \frac{\eta^2 - \tau C}{(\eta^2 + \tau C)\eta}\Phi_\eta + \frac{\eta^2}{(\eta^2 - \tau C)^2}\Phi_{\theta\theta} + \frac{\eta^2}{(\eta^2 + \tau C)\tau^3}\Phi_{\xi\xi} = 0 \quad (11)$$

with $(v.\theta,\xi) \in \Omega = \left\{ \eta_1 < \eta < \eta_2, \quad 0 \leqslant \theta \leqslant 2\pi, \quad 0 \leqslant \xi \leqslant h \right\}.$

Boundary conditions lead to two relationships

$$\Phi_t = \left[ \pm \frac{\eta_j m_{jtt}}{\tau m_j} + \frac{\sigma_j}{\tau\eta_j m_j^3}\left[ \frac{\partial^2}{\partial\theta^2} + \frac{\eta_j^2 m_j^2}{\tau^2}\frac{\partial^2}{\partial\xi^2} \right]\right]\left[ s_j \mp \int_0^t \tau^2 m_j^2 \Phi_\eta \, dt \right] = 0$$

$$(\eta = \eta_j, \quad j = 1,\ 2), \ \Phi\Big|_{t=0} = 0, \quad (12)$$

where the upper sign corresponds to $j = 1$ (the inner boundary of the layer), the lower one - to $j = 2$ (the outer boundary), $s_j(\theta,\xi)$ are the initial shifts of free boundaries. As a velocity disturbance is determined by the formula [3] $\vec{U} = M^{*-1}\nabla\Phi$ then inpenetrability condition of the solid walls $\xi = 0$, $\xi = h$ is equivalent to $\Phi_\xi = 0$ with $\xi = 0$, $\xi = h$.

The stability of free boundaries is characterized by the behavior of the functions [3]

$$R_j = \frac{1}{\tau m_j}\left[ s_j \mp \int_0^t t^2 m_j^2 \, \phi_\eta \, dt \right], \quad \eta = \eta_j. \tag{13}$$

Let us take

$$\phi = \sum_{n=0}^{\infty} \sum_{\lambda=0}^{\infty} A_{n\lambda}(\eta, t) e^{i\lambda\theta} \cos \frac{n\pi}{h}\xi, \quad (i^2 = -1),$$

$$s_j = \sum_{n=0}^{\infty} \sum_{\lambda=0}^{\infty} a_{n\lambda,j} e^{i\lambda\theta} \cos \frac{n\pi}{h}\xi,$$

$$N_j = \frac{1}{\eta_1}\left[ a_{n\lambda,j} \mp \frac{1}{K}\int_1^\tau \tau^2 m_j^2 A_\eta \, d\tau \right], \quad \eta = \eta_j. \tag{14}$$

Then from (11), (12), (13) we obtain the system of 4 linear ordinary differential equations

$$\frac{dA_j}{d\tau} = \left[ \pm \frac{\eta_j m_{j\tau\tau}}{\eta_1 \tau m_j} - \frac{\eta_1^2 S_j}{\eta_j^2 \tau m_j^3}\left( \frac{m_j^2 \beta_j^2}{\tau^2} + \lambda^2 - 1 \right) \right] N_j,$$

$$\frac{dN_j}{d\tau} = f_j A_1 + d_j A_2 \quad (A_j(\tau) = A_{n\lambda}(\eta_j, \tau)/K\eta_1^2), \tag{15}$$

$$A_j(1) = 0, \quad N_j(1) = a_j/\eta_1 \quad (j = 1, 2),$$

where $\beta_1 = n\pi\eta_1/h$, $\beta_2 = (1 + \varepsilon)^{1/2}\beta_1$, $f_j$, $d_j$ are the known functions of variable $\tau$. By the known functions $N_1$, $N_2$ the amplitudes of free boundaries disturbances are determined as follows

$$Am\, R_j = \eta_1 N_j(\tau)/\tau m_j. \tag{16}$$

Let $S_1 = S_2 = 0$, coefficients of the system (15) are regular everywhere, except the point $\tau = \tau_* = 1 + Kt_*$, where $g(\tau_*) = 0$. Nevertheless, we can show, that near by this point, i.e. with $\tau \to \tau_*$, which means $r_1 \to 0$, with $t \to t_*$:

$$Am\, R_1 \sim C_1(\tau)\left[-\ln \xi\right]^{1/4} \xi^{-1/2} \quad (\lambda = 0), \; \xi = \eta_1^{-1} r_1 \tau^{1/2}; \tag{17}$$

$$Am\ R_1 \sim C_2(\tau) \left[-\ln \xi\right]^{1/4} \exp\left[i(-2 \ln \xi)^{1/2}\right] \ (\lambda = 1); \tag{18}$$

$$Am\ R_1 \sim C_3(\tau) \left[-\ln \xi\right]^{1/4} \exp\left[i(\lambda - 1)^{1/2} \ln \varepsilon\right] \ (\lambda > 1), \tag{19}$$

where $C_1$, $C_2$, $C_3$ are some finite functions with $\tau \to \tau_*$.

Thus, the inner boundary is always instable at the collapse. As for the outer surface, for every $\lambda \geqslant 0$

$$Am\ R_2 \sim C_4(\tau)\ \xi \left[ -\ln \xi \right]^{3/2}, \tag{20}$$

therefore it is stable while collapsing.

If $S_1$, $S_2 \neq 0$, then asymptotic behavior of some harmonics coincides with (17)-(20). However, for harmonics high enough with the number

$$\lambda \gg \left[ -\xi \ln \xi \right]^{-1/2}$$

the surface tension forces restrict the growth of the disturbances: $|Am\ R_1| < \infty$ with $t \to t_*$.

In Fig.3, 4 the calculation results of the disturbances amplitudes of the cylindrical layer boundaries are represented. Curve 1 corresponds to $Am\ R_1$, and curve 2 corresponds to $Am\ R_2$, curve 3 describes the behavior of the inner radius $r_1(\mu)$.

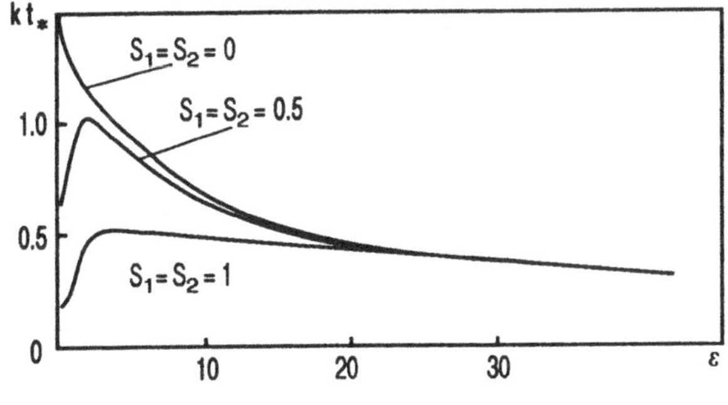


Fig. 1

TABLE 1

| $V(\eta,\xi)\sim$ | Generators |
|---|---|
| 1. Arbitrary | $L=\left\{\partial_t,\ g(t)\partial_z-\ddot{g}(t)z\partial_p,\ \varphi(t)\partial_p\right\}$ |
| 2. 0 | $L,\ \eta\partial_\eta+\xi\partial_\xi+r\partial_r+z\partial_z+2p\partial_p,$ $t\partial_t-2p\partial_p,\ \frac{1}{\eta}\psi_\xi\partial_\eta-\frac{1}{\eta}\psi_\eta\partial_\xi$ |
| 3. $\eta,\ (V_\eta\neq 0)$ | $L,\ t\partial_t-2\eta\partial_\eta+4\xi\partial_\xi-2p\partial_p,$ $4\eta\partial_\eta-5\xi\partial_\xi+r\partial_r+z\partial_z+2p\partial_p,$ $-2\partial_\eta+\frac{2\xi}{\eta}\partial_\xi+r^{-2}\partial_p,\ b(\eta)\partial_\xi$ |

TABLE 2

| $V(\eta,\xi)\sim$ | Generators |
|---|---|
| 1. Arbitrary | $L_0=\left\{Y_m,\ Y_n\right\}$ |
| 2. 0 | $L_0,\ Y_1,\ Y_2,\ Y_4$ |
| 3. $\pm\ln\eta$ | $L_0,\ Y_4,\ 2Y_1+Y_2\pm Y_3/2$ |
| 4. $\eta^\beta$ | $L_0,\ Y_4,\ (4-\beta)Y_1+2Y_2,\ \beta\neq 0$ |
| 5. $\beta\xi+F(\eta)$ | $L_0,\ 2Y_4-\beta Y_3$ |
| 6. $\beta\xi+\eta$ | $L_0,\ 3Y_1+2Y_2,\ -\beta Y_3+2Y_4$ |
| 7. $F(\eta)\exp(\beta\xi)$ | $L_0,\ -\beta Y_1+2Y_4$ |
| 8. $(\eta\xi)^\beta F(\eta/\xi)$ | $L_0,\ (2-\beta)Y_1+Y_2$ |
| 9. $\beta\ln\lvert\eta\xi\rvert+F(\eta/\xi)$ | $L_0,\ 2Y_1+Y_2-\beta Y_3$ |

$$Y_m=m(t)\,\partial_z-\ddot{m}(t)z\partial_p,\quad m(0)=0,\ Y_n=n(t)\partial_p,$$
$$Y_1=t\partial_t-2p\partial_p,\quad Y_2=\eta\partial_\eta+\xi\partial_\xi+r\partial_r+z\partial_z+2p\partial_p,$$
$$Y_3=r^{-2}\partial_p,\ Y_4=\partial_\xi+\partial_z$$

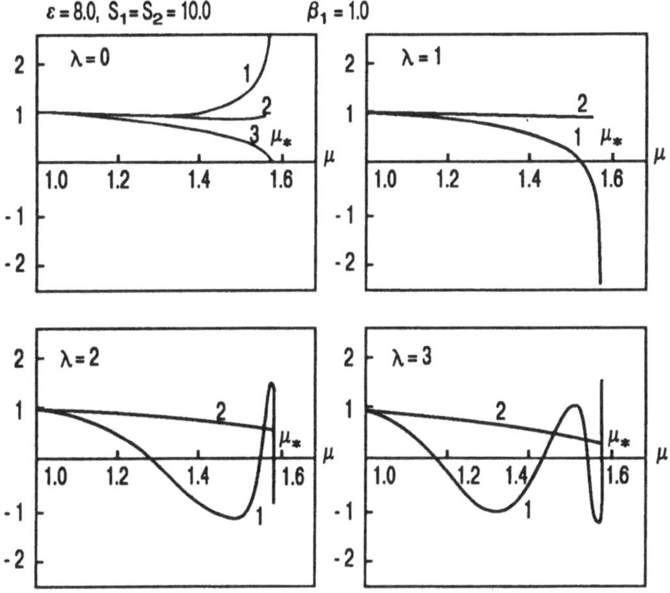


Fig. 2.

## REFERENCES


1. Ovsiannikov L.V. Group analysis of differential equations. - M.: Nauka, 1978, 400 p.
2. Ovsiannikov L.V. General equations and examples.// In book: The problem of nonstationary motion of liquid with a free surface. - Novosibirsk: Nauka, 1967, pp.5-75.
3. Andreev V.K. A small distriburtions of nonstationary motion of liquid under capillary forces action// Dynamics of Continuous Media, Novosibirsk, 1977, vol.32, pp.11-26.

# ON THE OCCURRENCE OF SINGULARITIES IN AXISYMMETRICAL PROBLEMS OF HELE-SHAW TYPE

*D.Andreucci, A.Fasano, M.Primicerio*
*Universita di Firenze*
*Instituto Matematico U.Dini*
*50134 Firenze*

## 1. INTRODUCTION.

It is well known that a Hele-Shaw cell is formed by two parallel plates very close to each other, the space between them being partly filled by a liquid which can be injected or extracted through a horifice. In [3] it is shown that for a sufficiently small spacing $d$ between the plates the problem can be considered two-dimensional and, if we denote by $P(x,y,t)$ the pressure in the fluid, the zone where $P > 0$ is filled with liquid, while the region where $P = 0$ corresponds to a "dry" zone.

In the former, the conservation of mass of the liquid has the form

$$\frac{\partial}{\partial t}(\rho d) + div\ q = 0, \tag{1.1}$$

where $d$ is the spacing between the plates, $\rho$ is the density of the liquid and vector $q$ expresses the mass flux of the liquid. Darcy's law states the proportionality of $q$ and $grad\ P$

$$q = -C\ grad\ P, \tag{1.2}$$

but in general $C$ can depend on $P$ as it is expected if also the spacing $d$ between the plates, and possibly $\rho$, are pressure dependent. In general the dependence of $d$ on $P$ will have a non-local character (a PDE for $d$ having $P$ as a forcing term should be coupled to (1.1), (1.2)), but as a first approach we can consider that $C$ and $\rho d$ are given monotonic functions of $P$, so that we can express all the quantities in terms of the new unknown

$$p(x,y,t) = \int_0^{P(x,y,t)} C(\mu)\, d\mu. \tag{1.3}$$

We have

$$\alpha(p)\, p_t = \Delta p, \tag{1.4}$$

where $\alpha(p)$ is non-negative if we assume that $C \geqslant 0$ and $\rho d$ is non-decreasing.

We consider a problem in radial symmetry, i.e. we prescribe $p$ or $\frac{\partial p}{\partial r}$ on a circle of radius $r_0$ (the horifice) and we look for a solution of

$$\alpha(p)\, p_t = p_{rr} + \frac{1}{r} p_r \tag{1.5}$$

in a region $r \in (r_0, s(t))$ where the free boundary $r = s(t)$ is characterized by the fact that it is a material surface moving according to (1.2) and on which $P$ (and thus $p$) vanishes.

Thus

$$p(s(t),t) = 0, \; -p_r(s(t),t) = (d_0\rho_0)\dot{s}, \tag{1.6}$$

where $d_0$ and $\rho_0$ are the values of $d$ and $\rho$ for $p = 0$.

The case $\alpha(p) \equiv 0$ corresponds to the classical Hele-Shaw problem and we can note that it has the explicit solution

$$p(r,t) = p_0(t) \frac{\ln \frac{s(t)}{r}}{\ln \frac{s(t)}{r_0}}, \tag{1.7}$$

$$s^2(t)\Big[ \ln\frac{s(t)}{r_0} - \frac{1}{2}\Big] = b^2 \Big[\ln \frac{b}{r_0} - \frac{1}{2}\Big] + \frac{2}{d_0\rho_0} \int_0^t p_0(\tau)\, d\tau, \tag{1.8}$$

where $b = s(0)$ and $p_0(t)$ is the prescribed boundary value of $p$ at $r = r_0$.

We note that (1.7) - (1.8) make sense both for $p_0(t) \geqslant 0$

(injected fluid) and for $p_0(t) < 0$ (extracted fluid). But the latter case corresponds to a typical non-equilibrium situation with the wet region at negative pressure; and the solution exists for all times if

$$\frac{2}{L}\int_0^{\infty} p_0(\tau)\, d\tau > \frac{1}{2}\left(b^2 - r_0^2\right) - b^2 \ln\frac{b}{r_0} \equiv B,$$

$B < 0$, otherwise it becomes extinct (i.e. $s = r_0$) for the first time instant $t^*$ such that

$$\frac{2}{L}\int_0^{t^*} p_0(\tau)\, d\tau = B.$$

In such a case it is easy to see that in the neighbourhood of the extinction time we have

$$s(t) - r_0 \simeq \left(\frac{-2}{Lr_0^2}\int_0^{t^*} p_0(\tau)\, d\tau\right)^{1/2}.$$

The case when $\alpha(p)$ takes both positive and zero values corresponds to a nonuniformly parabolic problem which will be the object of a future investigation.

In this paper we consider the simple case $\alpha(p) = 1/\beta > 0$, i.e. we study the following

**Problem ($\mathcal{P}$).** Find a time interval $(0,T)$, a function $s \in C^1(0,T)\cap C^0([0,T])$, and a bounded function $p \in C^{2,1}(D_T)$,

$$D_T \equiv \left\{(r,t)\colon\ 0 < r_0 < r < s(t),\ 0 < t < T\right\},$$

with $p$, $p_r$ continuous up to $r = s(t)$, for $t > 0$, such that the following system is satisfied

$$p_t = \beta\left(p_{rr} + \frac{1}{r}p_r\right) \quad \text{in } D_T, \tag{1.9}$$

$$p(r_0,t) = p_0(t), \quad 0 < t < T, \tag{1.10}$$

$$s(0) = b, \tag{1.11}$$

$$p(r,0) = h(r), \quad r_0 < r < b, \tag{1.12}$$

$$p(s(t),t) = 0, \quad 0 < t < T, \tag{1.13}$$

$$p_r(s(t),t) = -Ls(t), \quad 0 < t < T, \tag{1.14}$$

where $r_0 > 0$, $\beta > 0$, $b > r_0$, $L > 0$ are given positive constants and the boundary and initial data $p_0(t)$, $h(r)$ are bounded piecewise continuous functions. The data in (1.10), (1.12) are taken in a classical pointwise sense.

Problem $(\mathcal{P})$ is nothing but a Stefan problem, whose behaviour is well known when $p_0(t)$ and $h(r)$ are non-negative. When such conditions are not satisfied (non-equilibrium situation), there may be no solutions or solutions exhibiting finite time extinction or blow-up [1].

Stefan problem with supercooling in the planar symmetry case has been investigated e.g. in [6], [7]. The question of possible regularization i.e. modifications of the model eliminating singularities) has been investigated in [4]., [5] in the context of phase change models, and in [8] in the context of planar deformable Hele-Shaw cells.

In [4], [5] it has been stressed that two kinds of blow-up may occur: for some $T^* > 0$,

$$\lim_{t \to T^*_-} s(t) = s(T^*) > 0, \quad \lim_{t \to T^*_-} \dot{s}(t) = -\infty$$

and either

*(i)* the solution cannot be continued (essential blow-up),

*(ii)* the solution can be continued (non-essential blow-up).

An example of the occurrence of *(ii)* was provided.

In this paper we extend the analysis of [8] (including regularization) to Problem $(\mathcal{P})$ and Problem $(\mathcal{P}')$, obtained by replacing (1.10) with

$$p_r(r_0,t) = g(t), \quad 0 < t < T \tag{1.10'}$$

(then $p_r$ is required to be continuous also for $r = r_0$), with $g$ a continuous function in [0,T].

After having summarized some known results (Sec.2), we deal with the question of finite time extinction and of essential blow-up (Sec.3), while regularization is treated briefly, mainly referring to previous papers (Sec.4).

We also show that cases of non-essential blow-up can occurs (Sec.5).

## 2. SUMMARY OF SOME KNOWN RESULTS.

*i). Local existence and uniqueness.*

Local existence and uniqueness are ensured by the sole condition $h(r) > -\beta L$ for $r$ in some interval $(b-\varepsilon,b)$. Such a result has been proved recently in [2].

*ii). Non-existence.*

A sufficient condition for non-existence is that $h(r) \leqslant -\beta L$ in some interval $(b-\varepsilon,b)$ (see [1]).

Both $(i)$ and $(ii)$ apply to problems $(\mathcal{P})$ and $(\mathcal{P}')$, stressing the fact that local existence is essentially determinated by the behaviour of the initial data near the starting point of the free boundary.

It is also useful to recall from [1] a series of results concerning the case of zero flux at $r = r_0$.

*iii). Global existence, extinction or blow-up when* $p_r(r_0,t) = 0$.

Introducing the quantity

$$P = \frac{2}{\beta L}\int_{r_0}^{b} r\left[h(r) + \beta L\right] dr \qquad (2.1)$$

and assuming that the set in $[r_0,b]$ in which $h(r) \geqslant -1$ is an interval, for any solution we have the following a-priori implications

$P > 0 \Leftrightarrow$ global existence, (2.2)

$P = 0 \Leftrightarrow$ finite time extinction, (2.3)

$P < 0 \Leftrightarrow$ essential blow-up. (2.4)

We have already defined what we mean by essential blow-up. By finite time extinction we mean that $s(t) - r_0$ vanishes at some finite time.

*iv). The associated oxygen diffusion-consumption problem.* If $(s,p)$ is a solution of Problem $(\mathcal{P})$, the function

$$c(r,t) = \int_r^{s(t)} \rho^{-1}\, d\rho \int_\rho^{s(t)} y \left[p(y,t) + \beta L\right] dy \tag{2.5}$$

satisfies the free boundary problem

$$c_t - \beta\left(c_{rr} + \frac{1}{2} c_r\right) = -\beta^2 L \quad \text{in } D_T, \tag{2.6}$$

$$s(0) = b > r_0, \tag{2.7}$$

$$c(r,0) = c_0(r) \equiv \int_r^b \rho^{-1}\, d\rho \int_\rho^b y\left[h(y) + \beta L\right] dy, \quad r_0 < y < b, \tag{2.8}$$

$$c(r_0,t) = c_0(r_0) + \beta \int_0^t p_0(\tau)\, d\tau, \quad 0 < t < T, \tag{2.9}$$

$$c(s(t),t) = c_r(s(t),t) = 0, \qquad 0 < t < T. \tag{2.10}$$

If to the system above we add the constraint

$$c \geqslant 0, \tag{2.11}$$

then the solution $c$ to (2.6) - (2.11) solves a variational inequality and it does not necessarily coincide with the function defined by (2.5). Such a remark is at the basis of the regularization theory exposed in [4], [5].

When dealing with problem $(\mathcal{P}')$ the boundary condition (2.9) is replaced by

$$c_r(r_0,t) = c_0'(r_0) + \beta \int_0^t g(\tau)\, d\tau. \tag{2.12}$$

## 3. FINITE TIME EXTINCTION AND BLOW-UP FOR PROBLEMS $(\mathcal{P})$, $(\mathcal{P}')$.

In this section we want to perform a deeper analysis of

possible finite time extinction and blow-up for problems $(\mathcal{P})$, $(\mathcal{P}')$, with the respective conditions $p_0(t) \leqslant 0$, $g(t) \geqslant 0$, concentrating on the influence of the boundary data.

To be specific, we refer to the case

$$h(r) = 0, \quad r_0 < r < b, \tag{3.1}$$

also assuming that $p_0(t)$ or $g(t)$ are not identically zero near the origin.

In the spirit of the physical model considered in [8] this corresponds to a Hele-Shaw cell in which the liquid is initially at rest and at the saturation pressure. A solution always exists, at least locally in time.

We introduce the notation

$$\Lambda(r) = \frac{1}{2}\left(\frac{r}{r_0}\right)^2 \left(\ln\frac{r}{r_0} - \frac{1}{2}\right), \quad r \geqslant r_0, \tag{3.2}$$

remarking that $\Lambda'(r) > 0$ for $r > r_0$.

Let us prove the following result

PROPOSITION 3.1. For Problem $(\mathcal{P})$ with $h(r) \equiv 0$ and $p_0(t) \leqslant 0$ we have the inequality

$$\Lambda(s(t)) > \Lambda(b) + \frac{1}{Lr_0^2}\int_0^t p_0(\tau)d\tau, \quad t > 0, \tag{3.3}$$

and thus there can be extinction only if $t^*$ exists (possibly $t^* = +\infty$) such that

$$\int_0^{t^*} p_0(\tau)\, d\tau = -Lr_0^2 \left[\Lambda(b) - \Lambda(r_0)\right]. \tag{3.4}$$

If in addition

$$p_0(t) \geqslant -\beta L, \quad t > 0, \tag{3.5}$$

then $\dot{s}(t)$ is bounded as long as $s(t) > r_0$ and in this case (3.4) (with $t^* < \infty$) becomes a sufficient condition for finite extinction, the extinction time $t_e$ being the inf of the set of $t^*$ satisfying (3.4).

**PROOF.** Multiplying (1.1) by $r \ln \frac{r}{r_0}$ and integrating over $D_T$, we get

$$\Lambda(s(t)) - \Lambda(b) = -\frac{1}{\beta L r_0^2}\int_{r_0}^{s(t)} p(r,t)\, r \ln\frac{r}{r_0}\, dr + \frac{1}{L r_0^2}\int_0^t p_0(\tau)d\tau, \tag{3.6}$$

and the necessary extinction condition (3.4) follows, together with the inequality (3.3), due to the negativity of $p$.

The last statement is a consequence of the fact that $\dot{s}(t)$ remains bounded as long as no level curve $p = -\beta L$ hits the free boundary [1]. The maximum principle and (3.5) imply that $p(r,t) > -\beta L$ in all inner points, thus excluding of any singularity of $\dot{s}(t)$ before extinction. Moreover, assuming that $s(t_e) > 0$ leads immediately to a contradiction.

For Problem ($\mathcal{P}'$) we have parallel results:

**PROPOSITION 3.2.** For any solution of Problem ($\mathcal{P}'$) with $h \equiv 0$ and $g(t) \geqslant 0$ we have

$$s^2(t) \geqslant b^2 - \frac{2}{L} r_0 \int_0^t g(r)\, dr, \ 0 < t. \tag{3.7}$$

Consequently extinction may take place only if

$$\int_0^{t^*} g(\tau)\, d\tau = \frac{L}{2r_0}\left(b^2 - r_0^2\right) \tag{3.8}$$

for some (possibly unbounded) $t^*$.

If in addition

$$g(t) \leqslant \beta L/(b - r_0), \ 0 < t, \tag{3.9}$$

and if there exists $t^* < +\infty$ satisfying (3.8), extinction occurs for the least possible value of $t^*$.

**PROOF.** The first part of the Proposition is based on the equality

$$s^2(t) = b^2 - \frac{2}{\beta L}\int_{r_0}^{s(t)} r\, p(r,t)\, dr - \frac{2r_0}{L}\int_0^t g(\tau)\, d\tau, \quad t > 0, \tag{3.10}$$

which is obtained by integrating (1.1) multiplied by $r$.

The second part follows by the conclusion that in this case too $p(r,t) > -\beta L$ in $D_T$ as it can be seen by comparing $p(r,t)$ with the function

$$w(r,t) = \beta L(r - b)/(b - r_0). \tag{3.11}$$

We can also describe the behaviour of the free boundary near the extinction point. Lower estimates are easily provided by (3.3) and by (3.7) for Problem $(\mathcal{P})$ and for Problem $(\mathcal{P}')$, respectively:

PROPOSITION 3.3. Let $(s,p)$ be a solution of problem $(\mathcal{P})$ which becomes extinct at time $t^*$. Then near $t^*$ we have

$$s(t) - r_0 > \left[ \frac{2}{L}\int_t^{t^*} |p_0(\tau)|\, d\tau \right]^{1/2}. \tag{3.12}$$

For a solution of Problem $(\mathcal{P}')$ we have instead

$$s(t) - r_0 > \frac{r_0}{s(t)}\,\frac{1}{L}\int_t^{t^*} g(\tau)\, d\tau, \quad t \in (0,T). \tag{3.13}$$

PROOF. The proof of (3.12) is obtained by subtracting $\Lambda(r_0)$ from both sides of (3.3), while (3.13) follows from (3.7) by subtracting $r_0^2$ from both sides.

Upper estimates can be also obtained.

PROPOSITION 3.4. Suppose (3.5) holds in a strict sense, i.e. that $p_0(t) \geqslant -\theta\beta L$, for some $\theta \in (0,1)$. Then near the extinction point we have

$$s(t) - r_0 < \left[ \frac{1}{1-\theta}\,\frac{2}{L}\int_t^{t^*} |p_0(\tau)|\, d\tau \right]^{1/2}. \tag{3.14}$$

for Problem $(\mathcal{P})$.

**PROOF.** In order to prove (3.14) we use the inequality

$$p(r,t) \geqslant -\theta\beta L \ln \frac{b}{r} / \ln \frac{b}{r_0}, \tag{3.15}$$

that gives from (3.6)

$$\Lambda(s(t)) - \Lambda(r_0) < \theta \int_1^{1+\varepsilon} \rho \ln \rho \left(1 - \frac{\ln \rho}{\ln (b/r_0)}\right) d\rho + \frac{1}{Lr_0^2} \int_t^{t^*} |p_0(\tau)|\, d\tau,$$

where $\varepsilon = (s - r_0)/r_0$. It can be checked easily that

$$\Lambda(s(t)) - \Lambda(r_0) \simeq \varepsilon^2/2,$$

and that the first term on the right hand side in (3.6) is less that $\varepsilon^2\theta/2$. Hence (3.14) follows.

An upper estimate of linear type can be obtained for Problem $(\mathcal{P}')$.

**PROPOSITION 3.5.** If $g(t)$ satisfies (3.8), then we can find a constant $A > 0$ such that

$$s(t) - r_0 \leqslant A \int_t^{t^*} g(\tau)\, d\tau, \tag{3.16}$$

for $t$ sufficiently close to the extinction time $t^*$.

**PROOF.** Take $t_0 \in (0,t^*)$ arbitrary. Then we can select $\lambda > 0$ and $\theta \in (0,1)$ such that

$$p(r,t) \geqslant w(r,t) + \lambda \geqslant \theta w(r,t), \tag{3.17}$$

$$r_0 \leqslant r \leqslant s(t), \quad t_0 \leqslant t < t^*,$$

with $w$ given by (3.11). Thus from (3.10) we have

$$s^2(t) - r_0^2 < \theta \int_{r_0}^{s(t)} \frac{b-r}{b-r_0}\, r\, dr + \frac{2r_0}{L} \int_t^{t^*} g(\tau)\, d\tau,$$

whence, setting $s/r_0 = 1+\varepsilon$,

$$(1+\varepsilon)^2 - 1 < 2\theta \int_1^{1+\varepsilon} \rho\, d\rho + \frac{2}{r_0 L} \int_t^{t^*} g(\tau)\, d\tau,$$

i.e., for $t$ near $t^*$,

$$(1-\theta)\varepsilon < \frac{2}{r_0 L} \int_t^{t^*} g(\tau)\, d\tau,$$

which proves (3.16).

Let us conclude this section with some more remarks about extinction and blow-up.

**REMARK 3.6.** We consider those cases in which $t^*$ is not defined uniquely by (3.4) or (3.8), but on the contrary $t^*$ varies in a nonvoid interval $[\underline{t}^*, \overline{t}^*)$, in which $p_0(t)$ or $g(t)$ vanish.

For Problem $(\mathcal{P})$ we have that the associated function $c(r,t)$, defined by (2.5) is such that $c(r_0,t) > 0$ for $t \in (0, \underline{t}^*)$ (note that $c_0(r_0) = \beta L r_0^2 \left[\Lambda(b) - \Lambda(r_0)\right]$), and $c(0,t) = 0$ in $(\underline{t}^*, \overline{t}^*)$. Moreover, if (3.5) holds, we have

$$c_r < 0 < c \quad \text{in } D_{t^*},$$

and assuming $s(\underline{t}^*) > 0$ leads to a contradiction, so that $s(\underline{t}^*) = 0$.

If (3.5) is violated, extinction may occur at some later time in the interval $[\underline{t}^*, \overline{t}^*)$, including $\overline{t}^*$ if it is finite. However, if $\overline{t}^* < +\infty$ and $s(\overline{t}^*) > 0$, necessarily the solution will exhibit essential blow-up at some later time, since a negativity set for $c(r,t)$ will originate from the point $(r_0, \overline{t}^*)$.

We can go through a similar argument for Problem $(\mathcal{P}')$. Indeed from (2.12) and (3.8) we can see that $c_r(r_0,t) > 0$ in $(0, \underline{t}^*)$ and $c_r(r_0,t) = 0$ in $(\underline{t}^*, \overline{t}^*)$. Therefore, if $\overline{t}^* < +\infty$ and $s(\overline{t}^*) > 0$ we have $c > 0$ in $D_{\overline{t}^*}$ and blow-up will necessarily occur at some time greater than $\bar{t}^*$. We can also conclude that

$$\int_0^\infty g(t)\,dt < \frac{L}{2r_0}(b^2 - r_0^2)$$

is a necessary and sufficient condition for global existence.

## 4. REGULARIZATION.

Regularization is a procedure which prevents blow-up by letting a new free boundary appear according to some criterion inspired by the physical problem we want to describe.

In [4], [5] a regularization procedure was discussed based on a "nuclation principle" and applicable to charge of phase processes with supercooling.

In [8] a regularization "by cavitation" was discussed, referring to the case of the deformable Hele-Shaw cell.

Also in this case, we have that the value $p = -\beta L$ corresponds to a physical limit of the model. Indeed, in the simple case we are considering $\rho d$ is assumed to depend linearly on $p$ and thus

$$\rho d = \rho_0 d_0 + \frac{p}{\alpha},$$

and $p = -\beta L = -\alpha\rho_0 d_0$ corresponds to the vanishing of the quantity expressing the liquid content for unit surface of the cell.

There is no particular difficulty in extending the theory in the above mentioned papers to the case under consideration.

It has to be noticed that whenever $p$ takes the value $-\beta L$, this must happen for the first time at $r = r_0$ (both for Problems $(\mathcal{P})$ and $(\mathcal{P}')$).

If $t_0 > 0$ is such a time, for $t > t_0$ we regularize by solving a problem with two free boundaries $x = \sigma(t)$, $x = s(t)$ the newly appeared free boundary satisfying the initial condition $\sigma(t_0) = r_0$ and bearing $p(\sigma(t),t) = 0$, $L\dot{\sigma} = -p_r(\sigma(t),t)$.

The initial values for the regularized problem are provided by the limit of $p(r,t)$ for $t \to t_0^-$ and by

$$s_0 = \lim_{t \to t_0^-} s(t).$$

One can easily realize that $\dot{\sigma}(t) > 0$, $\dot{s}(t) < 0$ and also evaluate the limits of $\sigma$ and $s$ as $t \to \infty$ in a way similar to the one shown in [8].

It is worth observing that a regularization of the type proposed in [4], [5] is not suitable to the present case, since it does not prevent $\rho d$ from taking negative values.

## 5. NON-ESSENTIAL BLOW-UP.

It is some interest to look for the axisymmetrical analog of the example given in [4] about the actual occurrence of non-essential blow-up for a one-phase one-dimensional supercooled Stefan problem.

We consider Problem $(\mathcal{P}')$ with $g \equiv 0$ and we introduce the following one-parameter family of initial data:

$$h_N(r) = -N, \quad r_0 < a_1 < r < a_2 < b, \tag{5.1}$$

$$h_N(r) = 0 \quad \text{otherwise},$$

where $N \in \mathbb{R}_+$ and $a_1$, $a_2$ are to be chosen. We refer to this problem as problem $(\mathcal{P}'_N)$.

The corresponding value $P_N$ of the quantity defined by (2.1) is

$$P_N = - N \frac{a_2^2 - a_1^2}{\beta L} + b^2 - r_0^2, \tag{5.2}$$

while the initial value of the "oxygen concentration" (2.8) at $r = r_0$ is

$$c_0(r_0) = \frac{\beta L}{2} \left[ b^2 \ln \frac{b}{r_0} - \frac{1}{2} (b^2 - r_0^2) \right] - \tag{5.3}$$

$$- \frac{N}{2} \left[ a_2^2 \ln \frac{a_2}{r_0} - a_1^2 \ln \frac{a_1}{r_0} - \frac{1}{2} (a_2^2 - a_1^2) \right].$$

For $N/\beta L < 1$ we have global existence, since $p_N(r,t) > > -\beta L$ and $P_N > 0$ and consequently we can have no singularity for the free boundary, nor finite time extinction.

Therefore there is an interval for $N/\beta L$ in which the corresponding free boundary problem has global existence. It is known that, as long as the associated function $c_N$ is non-negative everywhere, the family $S_N$ of free boundaries are monotonically decreasing for $N$ increasing.

We want to show that, if $a_1$, $a_2$ are appropriate, we can let $N$ vary in an interval such that we have at the same time $P_N > 0$ (excluding finite time extinction) and $c_0(r_0) < 0$, implying essential blow-up. In other words, we can satisfy the system

$$\frac{N}{\beta L} < \frac{b^2 r_0^2}{a_2^2 - a_1^2}, \tag{5.4}$$

$$\frac{N}{\beta L} > \frac{b^2 \ln(b/r_0) - (b^2 - r_0^2)/2}{a_2^2 \ln(a_2/r_0) - a_1^2 \ln(a_1/r_0) - (a_2^2 - a_1^2)/2}. \tag{5.5}$$

Indeed the inequalities above are consistent if $a_1$, $a_2$ are such that

$$\frac{\rho^2 \ln \rho}{\rho^2 - 1} < \frac{\pi^2 \ln \pi}{\pi^2 - 1} + \ln \frac{a_1}{r_0}, \tag{5.6}$$

where we have set $\rho = b/r_0$ and $\pi = a_2/a_1$, $1 < \pi < \rho$.

It is now clear that

$$\ln \frac{a_1}{r_0} = -\ln \pi + \ln \frac{a_2}{r_0} = -\ln \pi + \ln \rho - \ln \frac{b}{a_2}.$$

Hence (5.6) reduces to

$$F(\rho) < F(\pi) - \ln \frac{b}{a_2}, \tag{5.7}$$

with

$$F(\pi) = \frac{\pi^2 \ln \pi}{\pi^2 - 1} - \ln \pi = \frac{\ln \pi}{\pi^2 - 1}.$$

It is easy to check that $F(\pi) > F(\rho)$ for $1 < \pi < \rho$.

Therefore for any $\pi$ given in $(1,\rho)$ we can select $b/a_2$ sufficiently close to 1 so that (5.7) is satisfied. Thus, when $a_1$, $a_2$ are chosen in this way there is an interval for $N/\beta L$ in which both (5.4) and (5.5) are valid and the corresponding solutions have essential blow-up.

Moreover, we can realize that the r.h.s. of (5.5) is greater than 1, since it can be written $\dfrac{G(\rho) - G(1)}{G(\rho_2) - G(\rho_1)}$ with

$$G(\rho) = \rho^2 \ln \rho - \rho^2/2, \ \rho_i = a_i/r_0 \ \text{ and } \ G'(\rho) > 0.$$

We can conclude that there exists $N^* \in \mathbb{R}^+$,

$$N^* = \sup \left\{ N > 0 \;\middle|\; (\mathcal{P}'_n) \text{ has a global solution for } 0 < n < N \right\} > \beta L.$$

For $N \to N^*$ the solutions to $(\mathcal{P}'_N)$ tend to a solution still defined for any $T$ (because the associated function is non-negative, being the monotone limit of non-negative functions). Such a limit function will be characterized by the fact that the level curve $p = -\beta L$ connecting the points $(a_1, 0)$, $(a_2, 0)$ touches the free boundary in just one point, where we have a singularity.

**REMARK 5.1.** Even in the case $h(r) \equiv 0$, one can exhibit an example of essential blow-up caused by a suitable Dirichlet datum prescribed on the fixed boundary $r = r_0$. This can be done following the lines of the proof in the Appendix of [8]. For the sake of brevity, we omit the details.

## REFERENCES.

1. D.Andreucci. Continuation of the solution of a free boundary problem in cylindrical symmetry. Meccanica 19 (1984), p.91-97.
2. D.Andreucci, A.Fasano, M.Primicerio. Titolo. (to appear).
3. C.M.Elliot, J.R.Ockendon. Weak and variational methods for moving boundary problems. - Research Notes in Mathematics, 59, Pitman, London, 1982.
4. A.Fasano, S.D.Howison, J.R.Ockendon, M.Primicerio. On the singularities of one-dimensional Stefan problems with super cooling. - Math Models in Phase Change Problems (J.F.Rodrigues ed.). - International series of Numerical Mathematics, 88, Birkhäuser Verlag, Basel, 1989.
5. A.Fasano, S.D.Howison, J.R.Ockendon, M.Primicerio. Some remarks on the regularization of supercooled one-phase Stefan problem in one dimension. - Quart. Appl. Math 48(1990), pp.153-168.
6. A.Fasano, M.Primicerio. New results on some classical parabolic free boundary problems. - Quart. Appl. Math 38(1981), pp.439-459.
7. A.Fasano, M.Primicerio. A critical case for the solvability of Stefan-like problems - Math. Meth. in the Appl. Sci. 5 (1983), pp.84-96.
8. A.Fasano, M.Primicerio. Blow-up and regularization for the Hele-Shaw problem. - IMA Minneapolis (to appear).

Work partially supported by the Italian MURST Project "Equazioni di evoluzione..."

International Series of Numerical Mathematics, Vol. 106, © 1992 Birkhäuser Verlag Basel 

# NEW ASYMPTOTIC METHOD FOR SOLVING OF MIXED BOUNDARY VALUE PROBLEM

*I.V. Andrianov, A.O. Ivankov*
*Dnepropetrovsk 320005, UKRAINE*

The basic idea of method presented may be descibed as follows. Parameter $\varepsilon$ is introduced into boundary condnitions in such a way that $\varepsilon = 0$ case corresponds to the simple boundary problems and $\varepsilon = 1$ case corresponds to the common problem under consideration. Then the $\varepsilon$-expansion of the solution is obtained. As a rule, just at point $\varepsilon = 1$ the expansion of the solution is divergent. PA may be used to remove divergence.

Key words: mixed boundary value problem; perturbation procedure; Pade approximation.

## INTRODUCTION.

Strain-stress state analysis on plates and shells under mixed boundary conditions is of significant practical value: a lot of problems arising in machine design, civil engineering, etc. are reduced to similar ones. The problems mentioned are usually solved using numerical methods such as finite element procedure. Nevertheless, numerical approach does not adequately meet the requirements of optimal structural design ideology. Approximate analytical expression, accurate enough, will be of great practical advantage for these needs. Effective analytical approach combining boundary conditions perturbation technique and Pade approximants (PA) of perturbation series is presented in this paper.

## 1. STABILITY ANALYSIS.

Let us consider the application of the approach presented to the stability analysis of rectangular plate $(-0.5a \leqslant x \leqslant 0.5a;$ $-0.5b \leqslant y \leqslant 0.5b)$ uniformly compressed in $x$-direction. We suppose that in-plane boundary conditions provide uniformity of prebuckling state. The plate is simply supported along the sides $x = \pm a/2$ and subjected to mixed boundary conditions along the sides $y = \pm b/2$, symmetrical with respect to $x$.

Governing differential equation may be written as follows

$$\nabla^4 W + NW_{xx} = 0, \tag{1.1}$$

Here

$$D = Eh^3 \frac{1}{12(1 - \nu^2)}, \quad \nabla^2 = \frac{\partial^2}{\partial x^2} + \frac{\partial^2}{\partial y^2},$$

$W = \overline{W}/b$, $N = \overline{N}/b$, $x = \overline{x}/b$, $y = \overline{y}/b$, $k = a/b$, $\overline{W}$, $\overline{N}$, $\overline{y}$, $\overline{x}$ -initial value of variable.

Boundary condition may be formalized as

$$W = 0, \quad W_{xx} = 0 \quad \text{when } x = \pm k/2, \tag{1.2}$$

$$W = 0, \quad \left[1 - \overline{H}(x)\right] W_{yy} \pm \overline{H}(x) W_y = 0 \quad \text{when } y = \pm 1/2. \tag{1.3}$$

where $\overline{H}(x) = H(x-\mu k) + H(-x-\mu k)$, $H(x)$ - Heaviside function.

Introducing the parameter $\varepsilon$ into the boundary condition according to the above-mentioned procedure, one obtains

$$W = 0, \quad W_{yy} = -\varepsilon \overline{H}(x)(W_{yy} \pm W_y) \quad \text{when } y = \pm 1/2. \tag{1.4}$$

The case $\varepsilon = 0$ brings us the plate simply supported along the boundary; the case $\varepsilon = 1$ corresponds to the problem under consideration (1.1)-(1.3).

The intermediate values of $\varepsilon$ are related to mixed conditions of "simple support - elastic clamping" kind with elastic support coefficient $u = \varepsilon/(1 - \varepsilon)$.

Let us apply the perturbation technique to the equation (1.1) and boundary conditions (1.2), (1.4) representing $N$, $W$ as $\varepsilon$ - expansions

$$N = \sum_{i=0}^{\infty} N_i \varepsilon^i, \quad W = \sum_{i=0}^{\infty} W_i \varepsilon^i. \tag{1.5}$$

Substituting (1.5) into equation (1.1) and boundary conditions (1.2),(1.4) and splitting it into the powers of $\varepsilon$, one obtains the recurrent sequence of eigenvalue problems

$$\varepsilon^0: \quad \nabla^4 W_0 + N W_{0xx} = 0, \tag{1.6}$$

$$W_0 = 0, \quad W_{0xx} = 0 \quad \text{when } x = \pm k/2,$$
$$W_0 = 0, \quad W_{0yy} = 0 \quad \text{when } y = \pm 1/2. \tag{1.7}$$

$$\varepsilon^j: \nabla^4 W_j + N_0 W_{jxx} = -\sum_{i=0}^{j-1} N_{j-1} W_{ixx}, \tag{1.8}$$

$$W_j = 0, \quad W_{jxx} = 0, \quad \text{when } x = \pm\, k/2,$$
$$W_j = 0, \quad W_{jyy} = \mp \bar{H}(x) \sum_{i=0}^{j-1} W_{jy} \quad \text{when } y = \pm 1/2. \tag{1.9}$$

Expressions (1.6), (1.7) describe the stability of simply supported plate subject to uniform compression in one direction.

Perturbation procedure leads to the following solution of system (1.6)-(1.9):

$$N = \pi^2 \frac{k^2}{m^2} \delta^2 + 4k^2 \frac{n^2}{m} \gamma_{mm} \varepsilon + \frac{k^2}{\pi^2 m^2} \left\{ 4\pi^2 n^2 \gamma_{mm} - 2\delta^{-1} \gamma_{mm} \left[ \frac{\alpha}{2} \, th \, \frac{\alpha}{2} - 1 \right] - \right.$$
$$\left. - 4\pi^2 n^2 \sum_{\substack{i=1,3,5. \\ i \neq m}}^{\infty} \gamma_{im} \left( \alpha_i \, th \frac{\alpha_i}{2} - A_i - 2\delta^{-2} \left( n^2 - k^2/m^2 \right) \gamma_{mm} \right\} \varepsilon^2 + \ldots,$$

$$\alpha = \pi \sqrt{2m^2/k^2 + n^2}, \quad \delta = n^2 + m^2/k^2.$$

$$\alpha_i = \pi \sqrt{i \left( \frac{i+m}{k^2} + \frac{n^2}{m} \right)}, \quad \beta_i = \pi \sqrt{i \left( \frac{i+m}{k^2} - \frac{n^2}{m} \right)},$$

$$\gamma_i = \pi \sqrt{i \left( \frac{n^2}{m} - \frac{i-m}{k^2} \right)},$$

$$A_i = \begin{cases} \beta_i \, cth \, (-1)^m \beta_i/2, & m(i - m) > n^2k^2, \\ -\gamma_i \, ctg \, (-1)^m \gamma_i/2, & m(i - m) < n^2k^2, \end{cases}$$

$$\gamma_{im} = \begin{cases} 2(0.5-\mu) - \frac{1}{m\pi} \sin 2\pi\mu m, & i = m, \\ \frac{4}{\pi} (m^2-i^2)^{-1} \left[ i \sin \pi\mu i \cos \pi\mu m - m \sin \pi\mu \cos\pi\mu i \right], & i \neq m, \end{cases}$$

$m(n)$ - wave number in $x(y)$ direction.

For the plate shown in Fig. 2, solution may be obtained similarly. In this case $\gamma_{mm}$ can be written in the following form

$$\gamma_{mm} = 2\mu - \frac{(-1)^m}{4\pi m} \sin 2\pi\mu m.$$

Let us determine the error of approximate solutions (1.10) for the case when sides $y = \pm 0.5$ are completely clamped ($\varepsilon = 1$, $\mu = 0.5$). Exact solutions are $N = 8.6044 \, \pi^2$ for $k = 1$, $N = 7.6913 \, \pi^2$ for $k = 2$; approximate solutions (1.10) give $N = 4.7757 \, \pi^2$ for $k = 1$ (error - (-44.5%)) and $N = 6.4456 \, \pi^2$ (error - (16.1%)) for $k = 2$. Then, expressions (1.10) are a poor approximation to the true value of $N$, and we can use Pade approximations technique to eliminate this drawback.

Let us produce the PA-definition [2]. For expansion given by

$$F(\varepsilon) = \sum_{i=0}^{\infty} c_i \varepsilon^i, \tag{1.11}$$

the fractional-rational function $F(\varepsilon)[m/n]$

$$F(\varepsilon)[m/n] = \left( \sum_{i=0}^{m} a_i \varepsilon^i \right) \left( \sum_{i=0}^{m} b_i \varepsilon^i \right)^{-1} \tag{1.12}$$

represents PA of expansion (1.11) if McLoran series of $F(\varepsilon)$ expression shows the coincidence of its coefficients with corresponding ones of (1.11) up to the terms of $(m+n+1)$-th order. The features of PA are the following: it possesses uniqueness while $m$

and $n$ are chosen; it performs meromorphic continuation of function; for its definition from the source expansion (1.11) the linear algebraic problem arises [2].

We have in our case for truncated series (1.10)

$$N_{[1/1]}(\varepsilon) = (a_0 + a_1\varepsilon)(b_0 + b_1\varepsilon)^{-1}, \tag{1.13}$$

where

$$a_0 = N_0,\ a_1 = N_1 + b_1N_0,\ b_0 = 1,\ b_1 = -N_2/N_1.$$

Then error of formula (1.13) in comparison with exact solutions is +1.27% for $k = 1$ and -0.31% for $k = 2$.

The numerical results calculated by the above-mentioned method (formula (1.13)) are compared with results obtained by the R-function method (authors thank DSc L.V. Kurpa for these results); in [3], [4]). The discrepancy of critical loads does not exceed 5% which confirms the acceptable accuracy of the method presented.

## 2. FREE OSCILLATION.

Mixed boundary problem is to be considered:

$$\nabla^4 W - \lambda W = 0, \tag{2.1}$$

where $\lambda = \omega\rho h b^4 D^{-1}$, $\omega$ - natural frequency, and boundary conditions (1.2), (1.4).

Eigenvalue $\lambda$ and eigenfunction $W$ are presented by $\varepsilon$-based expansions:

$$\lambda = \sum_{i=0}^{\infty} \lambda_i \varepsilon^i,\quad W = \sum_{i=0}^{\infty} W_i \varepsilon^i. \tag{2.2}$$

Substituting series (2.2) into boundary problem, governing relations (2.1), (1.2), (1.4) and splitting it with respect to the power of $\varepsilon$, anyone obtains the recurrent sequence of boundary problems:

$$\varepsilon^0:\ \nabla^4 W_0 - \lambda_0 W_0 = 0,$$

$$W_0 = 0, \quad W_{0xx} = 0 \quad \text{when } x = \pm k/2,$$

$$W_0 = 0, \quad W_{0yy} = 0 \quad \text{when } y = \pm 1/2.$$

$$\varepsilon^j: \ \nabla^4 W_j - \lambda_0 W_j = \sum_{i=1}^{j} \lambda_j W_{j-i},$$

$$W_j = 0, \quad W_{jxx} = 0, \quad \text{when } x = \pm\, k/2,$$

$$W_j = 0, \quad W_{jyy} = \pm\, \bar{H}(x) \sum_{i=0}^{j-1} (-W_{jy}) \quad \text{while } y = \pm 1/2, \quad j = 1,\ 2,\ 3,..$$

Eliminating the nonuniformity of asymptotic expansions, we obtain the expression for eigenvalue in the form of truncated perturbation expansion:

$$\lambda = \pi^4\alpha^2 + 4\pi^2 n^2 \gamma_{mm}\varepsilon + \tag{2.3}$$

$$+ \left\{ 4\pi^2 n^2 \gamma_{mm}\left(1 - \frac{\gamma_{mm}}{\pi^2\alpha}\left[\frac{\pi\beta_1}{2}\, cth\, (-1)^i\, \frac{\pi\beta_1}{2} + \frac{n^2}{\alpha_m} - \frac{3}{2}\right]\right) - 2 - \right.$$

$$\left. - \frac{n^2}{2} \sum_{\substack{i=1,3,5..\\ i=2,4,6,.\\ i=m}}^{\infty} \gamma_{im}^2 \left[\alpha_{1i}\, cth(-1)^i\, \frac{\alpha_{1i}}{2} + \left\{\begin{matrix} \varphi_{1i}\ cth((-1)^i\varphi_{1i}/2) \\ \beta_{1i} cth((-1)^i\beta_{1i}/2) \end{matrix}\right\}\right]\right\} \varepsilon^2,$$

$$m,\ n = 1,\ 2,\ 3,\ldots \quad \left\{\begin{matrix} i^2 > m^2 + n^2/k^2 \\ i^2 < m^2 + n^2/k^2 \end{matrix}\right\},$$

$$\alpha_m = n^2 + \frac{m^2}{k^2}, \quad \beta_1 = \sqrt{2m^2/k^2 + n^2}, \quad \beta_2 = \sqrt{m^2/k^2 + 2n^2},$$

$$\alpha_{1i} = \pi\sqrt{\frac{i^2+m^2}{k^2} + n^2}, \quad \beta_{1i} = \pi\sqrt{\frac{i^2+m^2}{k^2} + n^2}, \quad \varphi_{1i} = \pi\sqrt{\frac{i^2-m^2}{k^2} - n^2},$$

$$\gamma_{im} = \begin{cases} 2(0.5-\mu) - \dfrac{(-1)^m}{\pi m} \sin 2\pi\mu m, & i = m, \\ \dfrac{4}{\pi}\dfrac{1}{m^2 - i^2}\left[\begin{Bmatrix} i \\ m \end{Bmatrix} \sin \pi\mu i \cos \pi\mu m - \right. \\ \left. \qquad - \begin{Bmatrix} i \\ m \end{Bmatrix} \sin \pi\mu m \cos \pi\mu i \right], & i \neq m. \end{cases}$$

Then eigenfunction $W$ may be obtained easily (this expression is very complicated, that's why we can't write it here).

Then we'll use our method for the first eigenvalue of the eigenvalue problem (2.1), (1.2), (1.4) because it depends on boundary condition most of all.

For the case $\mu = 0$ the exact eigenvalue is known [5] (for $k = m = n = 1$ we have $\lambda = (1.7050p)^4$).

For the basic case $\varepsilon = 1$ Pade transformation yields $F(1)_{[1/1]} = (1.7081p)^4$ showing less than 0.2% discrepancy.

**REFERENCES.**

[1] Dorodnitsyn A.A. The use of perturbation method for numerical analysis of mathematical physics equations. Numerical method of solution of elastic media problems of mechanics. Moscow, 1969, pp.85-101 (in Russian).

[2] Baker G.A., Graves-Morris P.R. Pade approximants. Part 1: Basic theory. Part 2: Extension and application. Addison-Wesley Publ. Company, 1981, 325 p.

[3] Keer L.M., Stahl B. Eigenvalue problems of rectangular plates with mixed edge conditions. ASME J. of Appl. Mech., vol.39 No.2 (1972), pp.513-520.

[4] Hamada M., Inoue Y., Hashimoto H. Buckling of simply supported partially clamped recatangular plates uniformly compressed in one direction. Bull. of J.S.M.E, vol.10, No.37 (1967), pp.35-40.

[5] Timoshenko S.P., Yong D.H., Weaver W. Vibration problems in engineering. New York, John Wiley & Sons, 1974.

# SOME RESULTS ON THE THERMISTOR PROBLEM

*S.N.Antontsev*
*Lavrentyev Institute of Hydrodynamics*
*Novosibirsk 630090, RUSSIA*

*M.Chipot*
*Universite de Metz*
*Departement de Mathematiques*
*Ile de Saulcy, 57045 Metz-Cedex 01*
*France*

## 1. INTRODUCTION.

We would like to consider here the so called Thermistor problem. The heat produced is a conductor by an electric current leads to the system:

$$\begin{cases} u_t - \nabla\cdot k(u)\nabla u = \sigma(u)|\nabla\varphi|^2 \quad \text{in} \quad \Omega\times(0,T), & (1.1.1) \\ u = 0 \quad \text{on} \quad \Gamma\times(0,T), \quad u(\cdot,0) = u_0, & (1.1.2) \\ \nabla\cdot\sigma(u)\nabla\varphi = 0 \quad \text{in} \quad \Omega\times(0,T), & (1.1.3) \\ \varphi = \varphi_0 \quad \text{on} \quad \Gamma\times(0,T). & (1.1.4) \end{cases} \qquad (1.1)$$

Here, $\Omega$ is a smooth bounded open set of $R^n$, $\Gamma$ denotes its boundary, $T$ is some positive given number, $\varphi$ is the electrical potential, $u$ the temperature inside the conductor, $k(u) > 0$ the thermal conductivity and $\sigma(u) > 0$ the electrical conductivity. Of course the physical situation is when $n = 3$ and $\Omega$ is the spatial domain occupied by the body that we consider and which is assumed to conduct both heat and electricity. However, the mathematical results are worth to be considered for any $n \geqslant 1$.

If $\mathcal{I}$ denotes the current density and $\mathcal{Q}$ the vector of heat flow then the Ohm law and the Fourier law read respectively

$$\mathcal{I} = -\sigma(u)\ \nabla\varphi, \qquad (1.2)$$

$$\mathcal{Q} = -k(u)\ \nabla u. \qquad (1.3)$$

The equations (1.1.1) and (1.1.3) follow then from the conservation laws

$$\nabla\cdot\mathcal{I} = 0, \quad \frac{\partial u}{\partial t} + \nabla\cdot Q = \mathcal{I}\cdot\mathcal{E}, \tag{1.4}$$

where $\mathcal{E}$ denotes the electric field (see also [C.i], [C.P.], [H.R.S.]. [Ko]).

**Remark 1:** Due to (1.1.3), (1.1.1) reads also

$$u_t = \nabla\cdot(k(u)\nabla u + \sigma(u)\varphi\nabla\varphi) \text{ in } \Omega\times(0,T).$$

It should be noticed the similarity with the two phase filtration problem. Indeed, if $u$ is the concentration and $\varphi$ the pressure then the equations of two phase filtration read

$$u_t = \nabla\cdot(k(u)\nabla u + b(u)\nabla\varphi) \text{ in } \Omega\times(0,T),$$

$$\nabla\cdot\sigma(u)\nabla\varphi = 0 \text{ in } \Omega\times(0,T).$$

We refer the reader to [A.K.M.] for details.

Instead of (1.1.2) one will also consider the boundary condition

$$\frac{\partial u}{\partial n} = 0 \text{ on } \Gamma\times(0,T), \quad u(\cdot,0) = u_0, \tag{1.1.2'}$$

where $\frac{\partial u}{\partial n}$ denotes the outward normal derivative of $u$.

We will assume all along that

$$k(u) = 1 \tag{1.5}$$

so that (1.1.1) will be

$$u_t - \Delta u = \sigma(u)|\nabla\varphi|^2 \text{ in } \Omega\times(0,T). \tag{1.6}$$

We refer for instance to [C.D.K.] for some other cases.

The paper is divided as follows. In section 2 we will show existence of a solution to (1.1). In section 3 we will focus on the question of uniqueness and in a final section on the problem of global existence or blow up.

## 2. EXISTENCE OF A WEAK SOLUTION.

Let $V$ be a subspace of $H^1(\Omega)$ containing $H_0^1(\Omega)$, $V'$ its dual (see for instance [B.L.],[J.L.L.] or [G.T.] for the definition and the properties of the Sobolev spaces). Recall first the following result of J.L.Lions (see [B.L.]):

Assume

$$u_0 \in L^2(\Omega). \tag{2.1}$$

Then:

**THEOREM 0.** If $f \in L^2(0,T;V')$, there exists a unique $u$ such that

$$u \in L^2(0,T;V), \quad u_t \in L^2(0,T;V'), \tag{2.2}$$

$$< \frac{d}{dt} u,v > + \int_\Omega \nabla u \cdot \nabla v \, dx = <f,v> \quad a.e. \ t \in (0,T), \quad \forall \ v \in V, \tag{2.3}$$

$$u(0) = u_0, \quad u \in C([0,T];L^2(\Omega)). \tag{2.4}$$

Moreover one has the estimate

$$\frac{1}{2} |u(t)|_2^2 + \int_0^t \|\nabla u(t)\|_2^2 dt = \frac{1}{2} |u(0)|_2^2 + \int_0^t <f,u(t)> dt$$

$$a.e. \ t \in (0,T) \tag{2.5}$$

($< \ >$ is the duality brackets between $V'$, $V$, $|\cdot|_p$ the usual $L^p$ norm, $|\nabla u|$ the Euclidean norm of the gradient of $u$).

We will assume that

$$\varphi_0 \in L^\infty(0,T;H^1(\Omega) \cap L^\infty(\Omega)), \tag{2.6}$$

$$\sigma \text{ continuous, } 0 < \sigma_1 \leqslant \sigma \leqslant \sigma_2 \tag{2.7}$$

where $\sigma_1$, $\sigma_2$ are two positive constants. Then we can prove:

**THEOREM 1.** If (2.1), (2.6), (2.7) hold then there exists a weak solution to (1.1) with the boundary conditions (1.1.2) or (1.1.2').

**Proof.** First remark that by (1.1.3) the second side of (1.1.1) can be written

$$\sigma(u)|\nabla\varphi|^2 = \nabla\cdot(\sigma(u)\varphi\nabla\varphi). \tag{2.8}$$

For any $w \in L^2(0,T;L^2(\Omega))$ consider $\varphi$ the solution to

$$\nabla\cdot\sigma(w)\nabla\varphi = 0 \quad \text{in} \quad \Omega\times(0,T), \quad \varphi = \varphi_0 \quad \text{on} \quad \Gamma\times(0,T).$$

If $\varphi_0$ satisfies (2.6) one can derive easily from the maximum principle (see [P.W.]) that

$$\varphi \in L^\infty(0,T;H^1(\Omega)\cap L^\infty(\Omega))$$

and for almost every $t$ one has

$$\int_\Omega |\nabla\varphi(t,x)|^2 \, dx \leqslant C(\sigma_1,\sigma_2,\varphi_0). \tag{2.9}$$

It is clear then that

$$<\nabla\cdot(\sigma(w)\varphi\nabla\varphi),v> = \int_\Omega \sigma(w)\varphi\nabla\varphi\times\nabla v \, dx \quad \forall \, v \in V$$

defines a element of $L^2(0,T;V')$. According to Theorem 0 there exists a unique $u$ satisfying (2.2) - (2.4). Using then (2.5) one can derive estimates showing that

$$w \to u \tag{2.10}$$

maps some ball of $L^2(0,T;L^2(\Omega))$ compactly into itself. A simple application of the Schauder fixed point theorem shows then that (2.10) has a fixed point which is a solution to (1.1). We refer the reader to [A.C.] for details.

**Remark 2.** Under some smoothness assumptions on the data, $(u,\varphi)$ is smooth. We refer to [A.C.]. See also [Ch.C.].

## 3. UNIQUENESS OF THE WEAK SOLUTION.

In this section we will assume that (2.6), (2.7) hold and that $\sigma$ is Lipsschitz continuous i.e. that for some constant $K$

$$|\sigma(u_1) - \sigma(u_2)| \leqslant K|u_1 - u_2| \quad \forall \, u_1, \, u_2 \in R. \tag{3.1}$$

Then we have

**THEOREM 2.** There exists at most one weak solution to (1.1) such that

$$\nabla\varphi \in L^{\frac{2q}{q-n}}(0,T;L^q(\Omega)), \quad q > n \vee 2, \tag{3.2}$$

where $\vee$ denotes the maximum of 2 and $n$.

**Proof.** Consider $(u_1,\varphi_1)$, $(u_2,\varphi_2)$ two weak solutions of (1.1) such that (3.2) holds. Subtracting the equation satisfied by $u_2$ from the one satisfied by $u_1$ we obtain if we set $w = u_1 - u_2$, $\varphi = \varphi_1 - \varphi_2$,

$$\begin{aligned} w_t &= \Delta w + \sigma(u_1)|\nabla\varphi_1|^2 - \sigma(u_2)|\nabla\varphi_2|^2 = \\ &= \Delta w + (\sigma(u_1) - \sigma(u_2))|\nabla\varphi_1|^2 + \sigma(u_2)\nabla\varphi\cdot(\nabla\varphi_1 + \nabla\varphi_2). \end{aligned} \tag{3.3}$$

If we multiply by $w$ and integrate over $\Omega$ we get

$$\frac{1}{2}\frac{d}{dt}|w|_2^2 + \|\nabla w\|_2^2 = I_1 + I_2, \tag{3.4}$$

where

$$\begin{aligned} I_1 &= \int_\Omega (\sigma(u_1) - \sigma(u_2))|\nabla\varphi_1|^2 w \, dx, \\ I_2 &= \int_\Omega \sigma(u_2)\nabla\varphi\cdot(\nabla\varphi_1 + \nabla\varphi_2) w \, dx. \end{aligned} \tag{3.5}$$

Using the Lipschitz continuity of $\sigma$ we have by Hölder's Inequality

$$|I_1| \leqslant K \int_\Omega |\nabla\varphi_1|^2 |w|^2 \, dx \leqslant K\|\nabla\varphi_1\|_q^2 \; |w|^2_{\frac{2q}{q-2}} . \tag{3.6}$$

From the Gagliardo-Nirenberg interpolation inequality one has for some constant $C$

$$|w|_{\frac{2q}{q-2}} \leqslant C \; |w|_2^{1-n/q}\left(|w|_2^2 + \|\nabla w\|_2^2\right)^{n/2q} \quad \forall \; w \in H^1(\Omega). \tag{3.7}$$

Hence (3.6) becomes

$$|I_1| \leqslant C\ \|\nabla\varphi_1\|_q^2\ |w|_2^{2(1-n/q)}\Big(|w|_2^2 + \|\nabla w\|_2^2\Big)^{n/q}. \tag{3.8}$$

To estimate $I_2$ we first use the Holder Inequality to get since $\sigma$ is bounded by $\sigma_2$

$$|I_1| \leqslant \sigma_2\|\nabla\varphi\|_1\|\nabla(\varphi_1 + \varphi_2)\|_q|w|_{\frac{2q}{q-2}}. \tag{3.9}$$

Next, to estimate $\varphi$ we use the equation satisfied by $\varphi_1$ and $\varphi_2$ to get

$$-\nabla\cdot(\sigma(u_2)\nabla\varphi) = -\nabla\cdot\Big(\Big[\sigma(u_2) - \sigma(u_1)\Big]\nabla\varphi_1\Big). \tag{3.10}$$

Multiplying this equation by $\varphi$ and integrating over $\Omega$ leads to

$$\sigma_1\|\nabla\varphi\|_2^2 \leqslant \int_\Omega \sigma(u_2)|\nabla\varphi|^2dx = \int_\Omega(\sigma(u_2) - \sigma(u_1))\nabla\varphi_1\cdot\nabla\varphi\ dx \leqslant$$

$$\leqslant K\int_\Omega|w||\nabla\varphi_1||\nabla\varphi|\ dx \leqslant K\|\nabla\varphi\|_2\left[\int_\Omega|w|^2|\nabla\varphi_1|^2dx\right]^{1/2}. \tag{3.11}$$

Hence

$$\|\nabla\varphi\|_2^2 \leqslant C\int_\Omega|w|^2|\nabla\varphi_1|^2dx .$$

Following (3.6),(3,7) we obtain

$$\|\nabla\varphi\|_2 \leqslant C\ \|\nabla\varphi_1\|_q|w|_2^{1-n/q}\Big(|w|_2^2 + \|\nabla w\|_2^2\Big)^{n/2q}. \tag{3.12}$$

Combining (3.7),(3.9), (3.12) we get

$$|I_2| \leqslant C\ \Big(\|\nabla\varphi_1\|_q + \|\nabla\varphi_2\|_q\Big)^2|w|_2^{2(1-n/q)}\Big(|w|_2^2 + \|\nabla w\|_2^2\Big)^{n/2q}. \tag{3.13}$$

By (3.4), (3.8), (3.13) we thus obtain

$$\frac{1}{2}\frac{d}{dt}|w|_2^2 + \|\nabla w\|_2^2 \leqslant$$

$$\leqslant C \left(\|\nabla\varphi_1\|_q^2 + \|\nabla\varphi_2\|_q^2\right)^2 |w|_2^{2(1-n/q)}\left(|w|_2^2 + \|\nabla w\|_2^2\right)^{n/q}. \qquad (3.14)$$

Applying the Young Inequality

$$ab \leqslant \varepsilon a^{q/n} + C_\varepsilon b^{\frac{q}{q-n}},$$

it follows that for any $\varepsilon > 0$

$$\frac{1}{2}\frac{d}{dt}|w|_2^2 + \|\nabla w\|_2^2 \leqslant$$

$$\leqslant 2\varepsilon \left(|w|_2^2 + \|\nabla w\|_2^2\right) + C_\varepsilon \left( \|\nabla\varphi_1\|_q^{\frac{2q}{q-n}} + \|\nabla\varphi_2\|_q^{\frac{2q}{q-n}} \right)|w|_2^2 \qquad (3.15)$$

where $C_\varepsilon$ is some constant depending on $\varepsilon$. Hence, choosing $\varepsilon < 1/2$,

$$\frac{d}{dt}|w|_2^2 \leqslant C \left(1 + \|\nabla\varphi_1\|_q^{\frac{2q}{q-n}} + \|\nabla\varphi_2\|_q^{\frac{2q}{q-n}} \right)|w|_2^2.$$

But now by our assumptions we have

$$\left(1 + \|\nabla\varphi_1\|_q^{\frac{2q}{q-n}} + \|\nabla\varphi_2\|_q^{\frac{2q}{q-n}} \right) \in L^1(0,T),$$

and the result follows from the Gronwall Inequality since $|w(0)|_2^2 = 0$.

**Remark 3.** This result improves preceding results of [Ch.C.]. Note that (3.2) holds automatically when $n = 1$, see [ Ch.C.].

## 4. A BLOW UP RESULT.

Let us consider $(u(x,t),\varphi(x,t))$ a local solution to

$$\begin{cases} u_t = \Delta u + \sigma(u)|\nabla\varphi|^2 \quad \text{in} \quad \Omega\times(0,T), \\ \dfrac{\partial u}{\partial n} = 0 \quad x \in \Gamma,\ t > 0, \\ u(x,0) = u_0(x) \quad x \in \Omega, \\ \nabla\cdot\sigma(u)\nabla\varphi = 0 \quad \text{in} \quad \Omega\times(0,T), \\ \varphi = \varphi_0 \quad \text{on} \quad \Gamma\times(0,T). \end{cases} \tag{4.1}$$

Let us assume that

$$u_0(x) \geqslant 0 \quad x \in \Omega, \tag{4.2}$$

$0 < \sigma(s) < +\infty \quad \forall\ s \geqslant 0,\quad \sigma$ differentiable, $\sigma'(s) \geqslant 0 \quad \forall\ s \geqslant 0.$

Then we can proof the following:

**THEOREM 3.** Assume that

$$\int_\Gamma \varphi_0(x)n(x)\ d\sigma(x) \neq 0, \tag{4.5}$$

then (4.1) cannot have a smooth global solution ($n$ is the outward normal to $\Gamma$, $d\sigma(x)$ the superficial measure on $\Gamma$).

**Proof.** Let us assume that (4.1) has a smooth global solution. First from (4.5) it is clear that

$$0 < \delta = \left| \int_\Gamma \varphi_0(x)n(x)\ d\sigma(x) \right|, \tag{4.6}$$

Define

$$Y(t) = \int_\Omega \left( \int_{u(x,t)}^{+\infty} \frac{ds}{\sigma(s)} \right) dx. \tag{4.7}$$

From (4.2) and the maximum principle (see [F.]) it is clear that

$$u(x,t) \geqslant 0, \quad x \in \Omega, \quad t > 0, \tag{4.8}$$

and thus $Y(t)$ makes sense and is nonnegative (see (4.4)).

Differentiating we obtain using (4.1)

$$\frac{dY(t)}{dt} = -\int_\Omega \frac{u_1}{\sigma(u)}\, dx = -\int_\Omega \frac{\Delta u + \sigma(u)|\nabla\varphi|^2}{\sigma(u)}\, dx =$$

$$= -\int_\Omega \Delta u \cdot \frac{1}{\sigma(u)}\, dx - \int_\Omega |\nabla\varphi|^2\, dx. \tag{4.9}$$

Integrating by parts we have since $\partial u/\partial n = 0$ on $\Gamma$ and by (4.3)

$$-\int_\Omega \Delta u \cdot \frac{1}{\sigma(u)}\, dx = -\int_\Omega \frac{\sigma'(u)}{\sigma^2(u)} |\nabla u|^2\, dx \leqslant 0. \tag{4.10}$$

Hence

$$\frac{dY(t)}{dt} \leqslant -\int_\Omega |\nabla\varphi|^2\, dx. \tag{4.11}$$

Applying now the divergence formula and taking into account (4.1) we have

$$\delta = \left| \int_\Gamma \varphi_0(x)\cdot n(x)\, d\sigma(x) \right| = \left|\int_\Omega \nabla\varphi\, dx\right| \leqslant$$

$$\leqslant \int_\Omega |\nabla\varphi|\, dx \leqslant |\Omega|^{1/2}\left( \int_\Omega |\nabla\varphi|^2 dx \right)^{1/2}. \tag{4.12}$$

It then follows from (4.11) that

$$\frac{dY(t)}{dt} \leqslant -\frac{\delta^2}{|\Omega|}.$$

Hence

$$0 \leqslant Y(t) \leqslant Y(0) - \frac{\delta^2}{|\Omega|}\, t. \tag{4.13}$$

which is impossible for $t$ large.

**REMARK 4.** It is clear from (4.13) that the blow up time is bounded from above by

$$t^* = \frac{|\Omega|}{\delta^2} \int_\Omega \left[ \int_{u_0(x)}^{+\infty} \frac{ds}{\sigma(s)} \right] dx. \tag{4.14}$$

We do not know if this estimate is sharp.

**REMARK 5.** It is proved that in the case of one spatial variable the solution nesessary blows-up at almost every $x \in \Omega$.

**REMARK 6.** If

$$\varphi_0(x) = C = const, \tag{4.15}$$

then clearly $\varphi = C$ and (4.1) has a global solution. Of course in this case (4.5) fails.

**Acknowledgements.** This work has been done when both authors were visiting the Institute for Mathematics and its Applications in Minneapolis. We thank this institution for its support.

**REFERENCES.**

[A.C.] S.N.Antontsev, M.Chipot. Existence, Stability, Blow up of Solution for Thermistor Problem. Dokl. Akad. Nauk. Russian (to appear).

[A.K.M.] S.N.Antontsev, A.V.Kazhikov, V.N.Monakhov. Boundary Value Problems in Mechanics of Nonhomogeneous Fluids. Studies in Mathematics and its Applications # 22, (1990), North Holland.

[B.L.] A.Bensoussan, L.J.Lions. Aplications des inequations variationnelles en controle stochastique, (1978), Dunod, Paris.

[C.D.K.] M.Chipot, J.I.Diaz, R.Kersner. Existence and uniqueness results for the Thermistor problem with temperature dependent conductivity. (To appear)

[Ch.C.] M.Chipot, G.Gimatti. A uniqueness result for the Thermistor problem. European J. of Applied Math., 2, (1991),p.97-103.

[C.1] G.Gimatti. Existence of weak solutions for the nonstationary problem of the Jouhle heating of a conductor. Preprint, Universita di Pisa. (to appear)

[C.2] G.Gimatti. A bound for the temperature in the thermistor problem. J. of Applied Math.,40, (1988), p.15-22.

[C.3] G.Gimatti. Remark on existence and uniqueness for the thermistor problem, Quaterly of Applied Math., 47, (1989), p.117-121.

[C.P.] G.Gimatti, G.Prodi. Existence results for a nonlinear elliptic system modelling a temperature dependent electrical resistor, Ann.Mat.Pura Apl.152, (1989), p.227-236.

[F.] A.Friedman. Partial Differential equations of parabolic type, Prentice Hall, (1984).

[G.T.] D.Gilbarg, N.S.Trudinger. Elliptic partial differential equations of second order. Springer Verlag, (1985).

[H.R.S.] S.D.Howison, J.F.Rodriques, M.Shillor. Stationary solution to the Thermistor problem. J.Math.Anal.Apl. (To apear).

[Ko.] F.Kohlrausch. Uber den stationaren Temperatur-zustand eines electrisch geheizten Leiters. Ann.Physics 1, (1990), p.132-158.

[J.L.L.] J.L.Lions. Quelques methodes de resolution des problemes aux limites non lineaires, Dunod, (1969).

[P.W.] M.H.Protter, H.F.Weinberger. Maximum principles in differential equations, Prentice Hall, (1967).

International Series of Numerical Mathematics, Vol. 106, © 1992 Birkhäuser Verlag Basel


# NEW APPLICATIONS OF ENERGY METHODS TO PARABOLIC AND ELLIPTIC FREE BOUNDARY PROBLEMS

*S.N. Antontsev,*
*Lavrentiev Institute of Hydrodynamics, Novosibirsk, RUSSIA,*

*J.I. Diaz*
*Universidad Complutense de Madrid, Madrid, SPAIN,*

*S.I. Shmarev,*
*Lavrentiev Institute of Hydrodynamics, Novosibirsk, RUSSIA.*

We present some recent results on the application of different energy methods for the study of free boundary problems. Such methods offer an alternative way when the maximum principle fails. So they are of special interest for the study of systems of equations and higher order equations. They are also useful for single equations with complicated structure making difficult the construction of super and subsolutions: this is the case, for instance, when there are unbounded data; or the nonlinearities depend on $x$ and $t$; when there are first order differential terms in the equation, and so on. A monograph [1] (in preparation) collects many results in this direction. We present here several different applications.

Key words: degenerate elliptic and parabolic equations, local energy method, vanishing properties of solutions.

## 1. FREE BOUNDARY PROBLEMS IN STATIONARY GAS DYNAMICS.

Let us consider two-dimensional flow of barotropic gas. Let $\vec{v} = (v_1, v_2)$, $\rho \equiv \rho(q)$, $(q^2 = v_1^2 + v_2^2)$, be, correspondingly, the velocity and the density of a gas. Then on the plane of complex potential $\Omega = \left\{(\phi,\psi)\colon\ 0 < \phi < \infty,\ 0 < \psi < 1\right\}$ the function

$$u(\phi,\psi) = \int_q^{q_s} \frac{\rho(\tau)}{\tau}\, d\tau$$

satisfies equation

$$\frac{\partial}{\partial\phi}\left[ K(u)\,\frac{\partial u}{\partial\phi}\right] + \frac{\partial^2 u}{\partial\psi^2} = 0$$

where

$$K(u) = \frac{1-M^2}{\rho^2} = \frac{1}{\rho}\,\frac{d}{dq}\,(\rho q),$$

$M^2(q)$ is the Mach number, $q_s$ is the sonic speed, $u(q_s)=0$, $M^2(q_s)=1$, $K(0) = 0$. For equation (1) we consider the boundary-value problem

$$u(\phi,1) = h(\phi) \geqslant 0,\ u_\psi(\phi,0) = 0,\ 0 < \phi < \infty;\ u(0,\psi) = u_0(\psi). \tag{2}$$

Problem (1)-(2) describes the motion of the plane gas jet moving along the given straight boundary (being the image of the line $\psi = 0$). The unknown (free) boundary of the jet, (the image of the line $\psi = 1$), is defined by the prescribed distribution of the speed modulus. It is assumed that the given functions satisfy inequality $0 \leqslant (h,u_0)$, implying, due to the maximum principle, that $0 \leqslant u(\phi,\psi)$, $q = |\vec{v}| \leqslant q_s$, $K(u)= \dfrac{1-M^2}{\rho^2} \geqslant 0$. We study local properties of weak solutions $u(\phi,\psi)$ of the problem (1)-(2) such that

$$0 \leqslant u(\phi,\psi) \leqslant M_0,\quad E(\phi) = \int_\phi^\infty \int_0^1 (Ku_\phi^2 + u_\psi^2)\,d\phi d\psi \leqslant M_1,\quad \phi > 0. \tag{3}$$

The existence of such solutions is proved in [2].

Let the functions $K(u)$, $h(\phi)$ satisfy the conditions

$$0 \leqslant K(u) \leqslant K_0 u^\alpha, \qquad \forall\ u \geqslant 0,\ \alpha > 0. \tag{4}$$

$$0 \leqslant h(\phi) \leqslant h_0 \equiv \varepsilon\left[1 - \frac{\phi}{T}\right]_+^\sigma,\ u_+ = max\ (u,0). \tag{5}$$

$$\int_\phi^\infty h_\phi^2\ d\phi \leqslant \varepsilon^2 C\left[1 - \frac{\phi}{T}\right]_+^{2\sigma-1},\ \sigma > 1/2. \tag{6}$$

Here $K_0$, $\alpha$, $\varepsilon$, $\sigma$, $T$, $C(\sigma,T)$ are some positive constants. Let us remark that (6) is valid for $h_0$ with $C = \dfrac{\sigma^2}{T(2\sigma-1)}$.

**THEOREM 1.** *(waiting distance).* Let $u(\phi,\psi)$ be a weak solution of the problem (1)-(2) and the conditions (3)-(6) hold with $\sigma > max(1/2, 2/\alpha)$. Then

$$u(\phi,\psi) \equiv 0 \quad \text{as} \quad \phi > T \text{ for any } 0 < \psi < 1$$

if $\varepsilon$ and $M_1$ are small enough.

**Proof.** Any weak solution $u(\phi,\psi)$ of the problem (1)-(2) possesses the following energy equality

$$E(\phi) = \int_\phi^\infty \int_0^1 (Ku_\phi^2 + u_\psi^2) d\phi d\psi = I_1 + I_2 \tag{7}$$

where

$$I_1 = -\int_0^1 Ku_\phi (u-h) d\psi, \qquad I_2 = \int_\phi^\infty \int_0^1 Ku_\phi h_\phi \, d\psi d\phi.$$

Using (3)-(6) and relations

$$(u-h)^2 = \left[\int_1^\psi u_\psi(\phi,\sigma)d\sigma\right]^2 \leqslant -E', \; E' = \frac{dE}{d\phi} \leqslant 0, \; u(\phi,\psi) \leqslant 2(h^2-E')$$

one may estimate $I_1$, $I_2$ as follows

$$|I_1| \leqslant \left[\int_0^1 Ku_\phi^2 \, d\psi\right]^{1/2} \left[\int_0^1 |Ku^{-\alpha}| u^\alpha (u-h)^2 d\psi\right]^{1/2} \leqslant$$

$$\leqslant 2^{\alpha/4} K_0^{1/2} (-E')^{1/2} (h^2-E')^{\alpha/4} (-E')^{1/2} \leqslant C(h^2-E')^{\alpha/4}. \tag{8}$$

$$|I_2| \leqslant 2^{\alpha/4} K_0^{1/2} E^{1/2} (h^2-E')^{\alpha/4} \left[\int_\phi^\infty h_\phi^2 d\phi\right]^{1/2} \leqslant$$

$$\leqslant \delta E^{(4+\alpha)/8}\Big[\int_{\phi}^{\infty} h_{\phi}^{2} d\phi\Big]^{(4+\alpha)/8} + C_{\delta}(h^{2}-E')^{(\alpha+4)/4}, \ \delta > 0. \qquad (9)$$

If $\alpha \geqslant 4$ then, applying (3)-(6), choosing $\delta$ so that

$$\delta E^{\frac{4+\alpha}{8}-1}\Big[\int_{\phi}^{\infty} h_{\phi}^{2} d\phi\Big]^{(4+\alpha)/8} \leqslant 1/2$$

and using (7)-(9), we get inequality

$$E \leqslant C\ (h^{2}-E')^{(\alpha+4)/4}.$$

Here and elsewhere later we denote by $C$ different constants depending only on $\alpha$, $T$, $M_1$, $K_0$. If $\alpha < 4$ then we obtain, applying Young inequality to the first addend in the right-side part of (9):

$$|I_2| \leqslant E/2 + C\Big[(h^{2}-E')^{(\alpha+4)/4} + \Big(\int_{\phi}^{\infty} h_{\phi}^{2} d\phi\Big)^{(4+\alpha)/(4-\alpha)}\Big]$$

and, respectively,

$$E \leqslant C\Big[(h^{2}-E')^{(\alpha+4)/4} + \Big(\int_{\phi}^{\infty} h_{\phi}^{2}\ d\phi\ \Big)^{(4+\alpha)/(4-\alpha)}\Big].$$

Hence, in the both cases the energy function E satisfies ordinary differential inequality

$$E' + CE^{\alpha/(4+\alpha)} \leqslant C\Big[\ h^{2} + \iota(\alpha)\Big(\int h_{\phi}^{2}\ d\phi\ \Big)^{4/(4-\alpha)}\Big] \leqslant H_0,$$

where $H_0 = C\varepsilon^2\Big[1 - \frac{\phi}{T}\Big]_+^{4/\alpha}$, $\iota(\alpha) = 0$ if $\alpha \geqslant 4$, $\iota(\alpha) = 1$ if $\alpha < 4$.

By (3) $E(T) = 0$ if only $M_1$ and $\varepsilon$ are small enough. Thus, $u_{\psi}(\phi,\psi) = 0$, $u(\phi,\psi) = 0$ as $\phi \geqslant T$ and the Theorem 1 is proven.

## 2. THE FLOW OF IMMISCIBLE FLUIDS THROUGH A POROUS MEDIUM.

Consider the system of equations

$$\begin{cases} \phi(x)\dfrac{\partial s}{\partial t} - div(k_0(x)a(s)\nabla s) = div(k_0(x)b(s)\nabla p) + f(x,t), \\ \\ div(k_0(x)d(s)\nabla p)=0 , \end{cases} \tag{10}$$

under the structural assumptions: $0 < C_1 \leqslant f(x) \leqslant C_2$,

$$C_3|\xi|^2 \leqslant (k_0(x)\xi,\xi) < C_4|\xi|^2, \ \forall\ \xi \in \mathbb{R}^N-\{0\},\ 0 < C_5 \leqslant d(s)$$

and $C_6 s^\alpha (1-s)^\beta \leqslant a(s) \leqslant < C_7 s^\alpha (1-s)^\alpha$. This system arises in the study of immiscible fluids flow through a porous medium. References on the physical derivation of the system and on the basic theory of the existence of weak solutions can be found in [3]. We make emphasis in the absence of the maximum principle for the system (10). To illustrate the application of energy methods we concentrate our attention in the degenerate case $\alpha > 0$.

**THEOREM 2.** *(finite speed of propagation).* Let $(s,p)$ be any local weak solution of (10) such that $p \in L^\infty(0,T;W^{1,q}(B_{r_1}(x_0)))$ for some $q > 2$, $r_1 > 0$, $x_0 \in \mathbb{R}^N$. Assume $\alpha > 0$ and $(b')^2 \leqslant Ms^\alpha$. Let $s(x,0)$ and $f(x,t)$ vanishing on $B_{r_1}(x_0)$ and $B_{r_1}(x_0)\times(0,T)$ respectively. Then there exist $t_0\in(0,T)$ and $0 < r(t) < r_1$ such that $s(x,t) = 0$ on $B_{r(t)}(x_0)$ for any $t \in [0,t_0]$.

**THEOREM 3.** *(waiting time).* Assume (for simplicity) $f\equiv 0$ and the assumptions of Theorem 1. If in addition

$$\int_{B_r(x_0)} |s(x,0)|^2 dx \leqslant C(r-r_1)_+^q$$

for any $r \in [0,r_2]$, $r_2 > r_1$ and a suitable $q>0$ then there exists $t^* > 0$ such that $s(x,t) = 0$ on $B_r(x_0)$ for any $t \in [0,t^*]$.

The proofs as well as other qualitative properties for the case $\alpha \in (-1,0)$ can be found in [3].

## 3. ON THE BOUNDARY LAYER FOR DILATANT FLUIDS.

The study of the boundary layer for a dilatant fluid of viscosity n>0 leads (after the formulation as a Prandtl' system in von Mises new unknowns) to the problem

$$\begin{cases} \dfrac{\nu}{2^{n-1}} \sqrt{w} \dfrac{\partial}{\partial \phi}\left(\left|\dfrac{\partial w}{\partial \psi}\right|^{n-1} \dfrac{\partial w}{\partial \psi}\right) - \dfrac{\partial w}{\partial x} - v_0(x)\dfrac{\partial w}{\partial \psi} + 2UU_x = 0, \quad 0<x<X, \quad 0<\psi<\infty, \\ w(0,\psi)=w_0(\psi) \quad w(x,0)=0 \; , \; w(x,\psi) \to U^2(x) \text{ as } \psi \to \infty, \end{cases}$$

where $U$, $w_0$ and $v_0$ are given functions satisfying $v_0(x) < 0$, $U_x > 0$, $U(x) > 0$, $w_0(0) = 0$, $w_0(\psi) > 0$ if $\psi > 0$. The case $n = 1$ has been studied by O.A. OLEINIK in a series of important works. Here we assume n>1 and use some technical results that allow us to assume that $0 < C_1 < w(x,\psi) < C_2$ for any $x \in (0,X)$ and $\psi \geq \psi_0$, for some $\psi_0 > 0$ (see [4]). The application of energy methods allows one to improve the results of [4]on the localization of the coincidence set where $W(x,\psi) = U^2(x)$:

**THEOREM** 4. Assume $w_0(\psi) = U^2(0)$ for any $\psi \geq \psi_1$, for some $\psi_1 \geq \psi_0$. Then there exist $C > 0$, and $\alpha > 0$ such that

$$w(x,\psi) = U^2(x) \text{ for any } \psi \geq \psi_1 + Cx^\alpha \text{ and any } x \in [0,X].$$

**THEOREM** 5 *(waiting distance)*. Assume in addition

$$\int_\psi^\infty (U^2(0)-w_0(\sigma))^2 d\sigma \leq C(\psi_0 - \psi)_+^q$$

for some suitable q>0 and any $\psi \in (\psi_3,+\infty)$ for some $\psi_0 \leq \psi_3 \leq \psi_1$. Then there exists $x^* > 0$ such that $w(x,\psi) = U^2(x)$ for any $\psi \in [\psi_1,\infty)$ and any $x \in [0,x^*]$.

## 4. FORMATION OF "DEAD CORES" IN REACTION-DIFFUSION EQUATIONS UNDER STRONG ABSORPTION.

By introducing of new domains of integration in the definition of the energy functions it is possible to consider not vanishing initial data in the study of the formation of "dead cores" ([5,6]). Consider equation

$$\frac{\partial}{\partial t}\left(|u|^{\alpha-1}u\right) - div\,\vec{A}(x,t,u,\nabla u) + B(x,t,u) = 0 \tag{11}$$

where $\alpha > 0$, $(\vec{\xi},\vec{A}(x,t,u,\vec{\xi}) \geq C_0|\vec{\xi}|^p$, $(\vec{\xi},\vec{A}(x,t,u,\vec{\xi}) \leq C_1|\vec{\xi}|^p$, $p > 1$ and $B(x,t,u)u \geq C_2|u|^{\gamma}$.

**THEOREM 6.** Assume $(p-1)/\alpha \geq 1 > \gamma/\alpha$. Let

$$u \in C^0(B_{r_1}(x_0) \times [0,\infty)) \cap L^\infty(B_{r_1}(x_0) \times [0,\infty))$$

be any local weak solution of (11). Then there exist $0 \leq T_0 < \infty$ and $r:[T_0,\infty) \to \mathbb{R}^+$, $r(T_0) = 0$, such that $u(x,t) \equiv 0$ on $B_{r(t)}(x_0)$ $\forall$ $t > T_0$.

Rigorous proofs of this assertion with different functions r(t) are given in [6,7].

### REFERENCES

[1] S.N. Antontsev, J.I. Diaz: Book in Birkhuser, 1992.

[2] S.N. Antontsev, Dokl.Akad nauk SSSR, 216, 1974, n.3, p.p.473-476. (Translation in Soviet Math. Doklady, 15, 1974, n.3, p.p.803-807).

[3] S.N. Antontsev, J.I. Diaz: Nonlinear Analysis, Theory, Methods Applications, 16, 4, pp.299-313.

[4] V.N. Samokhin, Trudy Seminara imeni I.G. Petrovskogo N$^0$14, pp. 89-108 (1989) (English translation) pp.2358-2373, 1990.

[5] J.I. Diaz, J. Hernandez: In Trends in theory and practice of Nonlinear differential equations, V. Lakshmikantham ed. Marcel Dekker, 1984, pp. 149-156.

[6] S.I. Shmarev, Dinamika Sploshnoy Sredi. Novosibirsk 1990, V. 95,97.

[7] S.N.Antontsev, S.I.Shmarev Dokl. Akad. Nauk SSSR, 318, 1991, n.4, p.p.777-781. (Translation in Soviet Math. Doklady, v.43, 1991).

# A LOCALIZED FINITE ELEMENT METHOD FOR NONLINEAR WATER WAVE PROBLEMS

*Kwang June Bai & Jang Whan Kim*
*Department of Naval Architecture*
*Seoul National University, Seoul, KOREA*

This paper describes an application of the localized finite element method to nonlinear water wave problems. A mathematical formulation of the nonlinear free-surface flow phenomena of incompressible ideal fluid can be described by variational principles based on the classical Hamilton's principle. In the numerical computations a modified variational functional is defined by subdividing the original fluid domain into three subdomains: the fully nonlinear subdomain including the source of disturbance, the linear infinite subdomain, and the nonlinear-to-linear transition buffer subdomain between the above two. In the buffer subdomain the free-surface boundary and its nonlinear boundary condition are artificially and gradually reduced to the mean free surface and the linear condition. In the numerical computations, the truncated infinite linear subdomain is excluded by representing the linear solutions in the subdomain with appropriate matching conditions along the juncture boundary between this subdomain and the buffer subdomain.

Key words: water waves, numerical method, finite-element method, nonlinear waves.

## 1. INTRODUCTION.

A free surface flow problem has been one of the most important research areas in ship hydrodynamics and ocean engineering. The main difficulty arises from the presence of the free surface boundary which is not known a priori. This is a free boundary problem. Due to the presence of a free surface, water waves are generated when the fluid boundary is disturbed by a solid body or a pressure disribution on the free surface.

In the past the linearized problem has been extensively treated. However, there has been a growing interest in the prediction of the nonlinear phenomena, often observed in the real physical problems. Thus the development of an efficient computational method for nonlinear water wave problems has been one of the most challenging research topics. The main difficulties arising in the solution procedure are from the nonlinear free surface condition and the luck of an appropriate radiation condition to be imposed in the computational domain. It is also very desirable to reduce the computational domain as small as possible to obtain the final matrix equation to be a manageable size.

There are several computational methods specifically developed for the nonlinear water-wave problems with restricted applications, for example, Vinje and Brevig (1981) and Dommermuth and Yue (1988). In the present paper we give an application of the finite-element method, based on the variational principle, to nonlinear free-surface flow problems. As the applications of the present numerical method, computations are successfully made for several flow cases, i.e. two- and three-dimensional problems for both steady and unsteady motions.

## 2.MATHEMATICAL FORMULATION.

Let $Oxyz$ be a coordinate system with Oz opposing the direction of gravity and $z = 0$ coincides the undisturbed free surface. We assume that the fluid is inviscid and incompressible and its motion is irrotational such that the velocity field of the fluid $\mathbf{u}$ can be defined as

$$\mathbf{u}(x,y,z,t) = \nabla\phi(x,y,z,t), \tag{1}$$

where $\phi$ is the velocity potential and satisfies the Laplace equation

$$\nabla^2\phi(x,y,z,t) = 0 \tag{2}$$

in fluid domain $D$.

If the free surfaces is represented by $z = \zeta(x,y,t)$ the kinematic and dynamic boundary conditions on the free surface $S_F$ can be given as

$$\zeta_t = \frac{1}{n_z} \phi_n, \tag{3.a}$$

$$\phi_t = -\frac{1}{2} |\nabla\phi|^2 - g\zeta - \frac{p}{\rho}, \tag{3.b}$$

where $g$ and $\rho$ denote the gravitational constant and the density of fluid and $p = p(x,y,t)$ is taken zero when the pressure distribution is absent.

By assuming that the fluid is initially at rest, the initial condition may be given as

$$\phi = \phi_t = 0 \quad \text{at} \quad t = 0 \tag{4}$$

and the resulting radiation condition is given as

$$\phi \to 0 \quad \text{as} \quad x^2 + y^2 \to \infty. \tag{5}$$

It should be noted that this radiation condition is replaced by an appropriate numerical radiation conditions in the computations.

## 3. LOCALIZED FINITE-ELEMENT METHOD.

This method has been mainly applied to the free-surface wave problems in the ship hydrodynamics by Bai and Yeung (1974) and Bai (1977, 1978). At the early stage this method successfully applied to the linear problems. However, recently the method has been extended to treat nonlinear problems. Recent investigations in the nonlinear problems can be found in Bai, Kim & Kim (1989), Lee (1990), Kim & Bai (1991), Han (1991) and Kim (1991). A typical numerical procedure of this method is described below:

The localized finite element method for the nonlinear free-surface flow problem is based on a variational principle equivalent to Luke's variational principle. In this formulation, a notable modification to the Luke's formulation is introduced for the treatment of the numerical radiation condition. The Lagrangian $L$ for this formulation can be written in terms of the velocity potential $\phi(x,y,z,t)$ as follows:

$$L = \iint \Phi\, \zeta_t \, dx\, dy - \frac{1}{2} \iiint_{-h}^{\varepsilon\zeta} \nabla\phi\cdot\nabla\phi \, dz\, dx\, dy - \frac{g}{2} \iint \zeta^2 \, dx\, dy -$$

$$- \frac{1}{\rho} \iint p(x,y,t)\zeta \, dx\, dy, \tag{6}$$

where $\Phi$ denotes the velocity potential on the free surface, i.e., $\Phi(x,y,t) = \phi(x,y,\zeta,t)$. Here $h$ is the water depth, $p(x,y,t)$ is the pressure on the free surface and $\varepsilon(x,y)$ is a locally-linearizing parameter. Taking variation of the above functional, we can recover Laplace equation in fluid domain and other boundary conditions including the following modified nonlinear free-surface conditions.

$$\zeta_t = \frac{\phi_n}{n_z} \quad \text{on} \quad z = \varepsilon\zeta, \tag{7}$$

$$\phi_t + \frac{\varepsilon}{2} \nabla\phi\cdot\nabla\phi + g\zeta = - \frac{p(x,y)}{\rho} \quad \text{on} \quad z = \varepsilon\zeta. \tag{8}$$

In the computations the original fluid domain is subdivided into three subdomains; i.e. the fully nonlinear subdomain including the source of disturbance, the truncated linear infinite subdomain, and the nonlinear-to-linear transition buffer subdomain between the above two subdomains. The value of $\varepsilon(x,y)$ is taken to be one and zero in the nonlinear and linear subdomains, respectively. However, in the buffer subdomain the value of $\varepsilon$ changes from one to zero gradually and smoothly to match the adjacent subdomains. In the solutions procedures a drastic saving in the computation time is achieved by utilizing the modal analysis. A more detail can be found in Kim (1991). In the present method, the main equations to solve, for example, in a unsteady problem, are the sets of the nonlinear ordinary differential equations on the free surface. These are integrated by 4th order Runge-Kutta method while the Laplace equation in the fluid domain is reduced to a linear matrix equation which plays a role of the constraint. This numerical scheme has been applied to the unsteady and steady water wave problems.

## 4. NUMERICAL RESULTS.

The present numerical method has been applied to several problems: two- and three-dimensional sloshing, two-dimensional hydrofoil and numerical simulations of towing tank experiment. Here we present only two computed results. Fig. 1 shows the pressure distribution on a two dimensional Joukowski symmetric hydrofoil (12 % thickness) moving with a constant speed and $5^0$ angle of attack. Also shown is the generated wave elevation due to the hydrofoil. The present nonlinear results agree much better than the linear results with those of the experimental measurements in Parkin et al (1955). Fig. 2 shows the computed free surface elevations in the numerical towing tank test. From the comparisons between the present computed results and other existing results, we can conclude that the present numerical method can be applied to various nonlinear water wave problems.

## ACKNOWLEDGEMENT.

This work has been partially supported by the Korean Science and Engineering Foundation under the Nonlinear Ship Hydrodynamics Program, Grant Number 87020703.

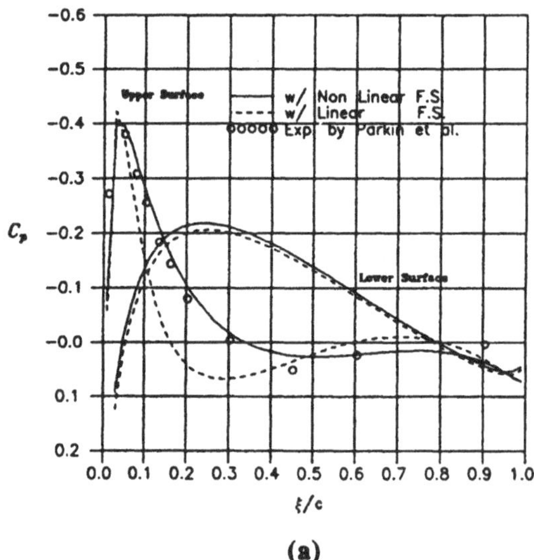


(a)

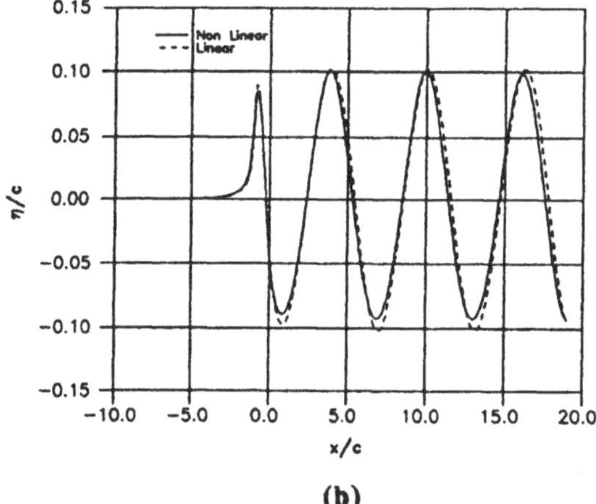


(b)

**Fig.1** Pressure distributions and wave profiles due to a 12% thickness Joukowski symmetric hydrofoil. The speed and the submergence of the hydrofoil and water depth is $0.989\sqrt{gc}$, 0.2c and 6c, respectively. Here c is the chord length of the hydrofoil.

(a) Pressure distributions

(b) Wave profiles

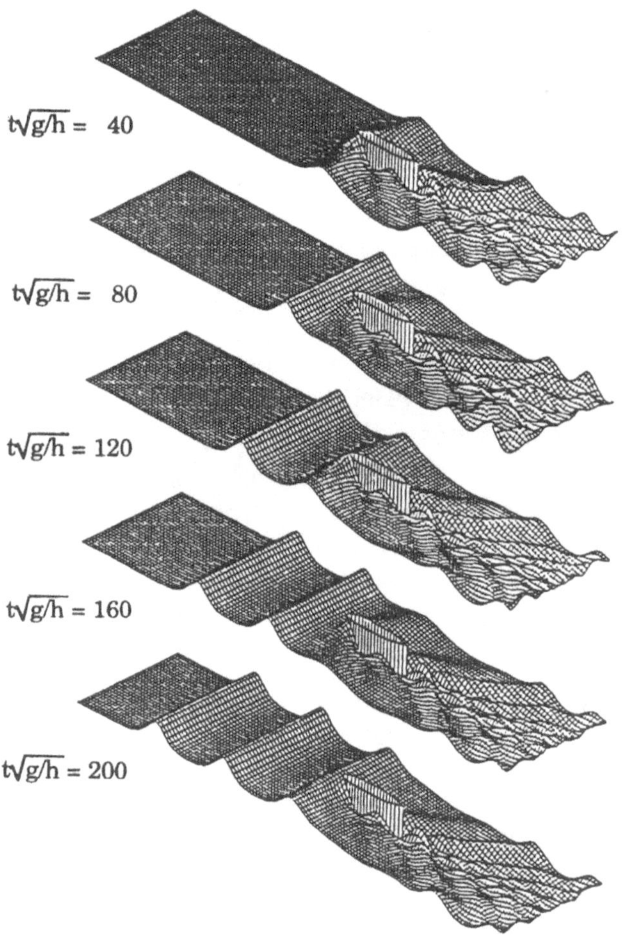


**Fig.2** Wave profiles generated by a series 60 ship model ( $C_b = 0.8$).
The speed of ship model, U, and width of towing tank, W, are given as $\frac{U}{\sqrt{gh}} = 1.1$, W/h = 8.

## REFERENCES.


Bai K.J. & Yeung R. Numerical solutions to free-surface flow problems. 10th Symposium on Naval Hydrodynamics, Office of Naval Research, MIT, Cambridge, Mass., USA, pp. 609-633, 1974.

Bai K.J. A localized finite-element method for steady three-dimensional free-surface flow problems, 2nd Int. Conf. on Numerical Ship Hydrodynamics, Univ. of Calif. Berkeley, USA, 1977.

Bai K.J. A localized finite-element method for two-dimensional steady potential flows with a free surface. J.Ship Research, 22(4), pp. 216-230, 1978.

Bai K.J.,Kim J.W. & Kim Y.H. Numerical computations for a nonlinear free surface flow problem, Proc. 5th Int. Conf. on Num. Ship Hydro., Hiroshima, Japan, 1989.

Dommermuth D.G. & Yue D.K.P. The nonlinear three-dimensional waves generated by a moving surface disturbance. 17th ONR Symposium on Naval Hydrodynamics, Hague, 1988.

Han J.H. A numerical analysis of nonlinear free-surface waves due to a two-dimensional hydrofoil, M.S.Thesis, Seoul Nat'l Univ. (in Korean), 1991( also submitted to J. of Ship Research).

Kim J.W. & Bai K.J. A note on Hamilton's principle for a free surface flow problem. J.Soc. of Naval Architects of Korea, vol. 27, No.3, pp. 19-30 (in Korean), 1991.

Kim J.W. A numerical method for nonlinear wave-making phenomena, Ph. D. Thesis, Seoul Nat'l Univ. (In Korean), 1991.

Lee H.S. A numerical analysis of two-dimensional free surface flow problem. M.S.Thesis, Seoul Nat'l Univ (in Korean), 1990.

Miles J.W. On Hamilton's principle for surface waves. J.Fluid Mech. 83, pp. 153-158, 1977.

Parkin B.R., Perry B. & Wu T.Y. Pressure distribution on a hydrofoil running near the water surface, Hydrodynamic Laboratory.

Report No. 47-2, Calif. Institute of Tech., 1955.

Vinje T. & Brevig P. Nonlinear ship motions, Proc. 3rd Int. Conf. on Numerical Ship Hydrodynamics, Paris, 1981.

International Series of Numerical Mathematics, Vol. 106, © 1992 Birkhäuser Verlag Basel 

# APPROXIMATE METHOD OF INVESTIGATION OF NORMAL OSCILLATIONS OF VISCOUS INCOMPRESSIBLE LIQUID IN CONTAINER

*M.Ya.Barnyak*
*Institute of Mathematics, Kiev 252601, UKRAINE*

Investigation of small normal oscillations of viscous incompressible liquid partially filling a vessel is connected with construction of solutions for a spectral boundary problem of the form

$$-\frac{1}{H}\Delta V+\nabla p=\lambda V,\ div\ V=0 \text{ in } \Omega,\ V=0 \text{ on } S,\ \int_\Sigma h\,dS=0,$$
$$V_z=-\lambda h,\ \frac{\partial V_x}{\partial z}+\frac{\partial V_z}{\partial x}=0,\ \frac{\partial V_y}{\partial z}+\frac{\partial V_z}{\partial y}=0,\ \frac{2}{H}\frac{\partial V_z}{\partial z}-p+h=0 \text{ on } \Sigma, \tag{1}$$

where $\Omega$ is the domain filled with liquid, $S$ is the solid wall of the container, $\Sigma$ is the non-perturbed free surface of the liquid, $V(x,y,z)$ is a velocity vector for particles of the liquid, $p(x,y,z)$ is pressure inside of the liquid, $h(x,y)$ is deviation with respect to the vertical of the free surface of the liquid, $\lambda$ is the frequency parameter, $H=g^{1/2}L^{3/2}\nu^{-1}$ is the Galilei number, $g$ is the free fall acceleration, $L$ is the characterictic length of the domain, $\nu$ is the coefficient of kinematical viscosity, $(x,y,z)$ is the Gartesian coordinate system, the $z$ axis being directed upward vertically and $\Sigma$ lying on the plane $z=0$.

The spectral problem described above was investigated earlier by means of functional analysis technique in [1,2]. It had been shown in these papers that the problem (1) has a discrete spectrum. All eigenvalues of the problem but finite number of pairs of non-real eigenvalues are real and are located on the positive part of the real line with two points of closeness - the zero and the infinity. When small values of $L$ or large values of $\nu$ correspond to small values of $H$ all eigenvalues of the problem (1) are real.

Following [1] let us consider the Sobolev space $W_2^1(\Omega)$ of the solenoid vector-functions which are integrable with their square and first derivatives, and vanishing on $S$. The norm in $W_2^1(\Omega)$ can be defined in the following form [1]:

$$\|\mathbf{V}\|^2_{W_2^1(\Omega)} = E(\mathbf{V},\mathbf{V}) = \frac{1}{2} \sum_{i,j=1}^{3} \int_\Omega \left| \frac{\partial V_i}{\partial x_j} + \frac{\partial V_j}{\partial x_i} \right|^2 d\Omega. \tag{2}$$

Let us consider on $\Sigma$ the Hilbert space $L_2(\Sigma)$ of scalar functions $h(x,y)$ satisfying the condition $\int_\Sigma h\, dS = 0$. The direct sum of the Hilbert spaces $W_2^1(\Omega)$ and $L_2(\Sigma)$ i.e. the set of all possible pairs $w(x,y,z) = \langle \mathbf{V}(x,y,z), h(x,y)\rangle$ ($\mathbf{V} \in W_2^1(\Omega)$, $h \in L_2(\Sigma)$) will be designated as $G_1(\Omega) = W_2^1(\Omega)\oplus L_2(\Sigma)$. We shall designate $G_1$ the direct sum of the vector-functions space $\mathbf{V} \in L_2(\Omega)$ and scalar functions space $h \in L_2(\Sigma)$ as $G$. Scalar products in $G$ and $G_1$ take the forms

$$(w_1,w_2)_G = \int_\Omega (\mathbf{V}_1,\mathbf{V}_2)d\Omega + \int_\Sigma h_1\bar{h}_2 dS,$$

$$(w_1,w_2)_{G_1} = E(\mathbf{V}_1,\mathbf{V}_2) + \int_\Sigma h_1\bar{h}_2\, dS,$$

where

$$w_1 = \langle \mathbf{V}_1, h_1\rangle, \quad w_2 = \langle \mathbf{V}_2, h_2\rangle.$$

Along with the above described definite scalar products in $G$ and $G_1$ we shall consider indefinite scalar products of the following form

$$\begin{aligned} [w_1,w_2]_G &= (\mathbf{V}_1,\mathbf{V}_2)_{L_2(\Omega)} - (h_1,h_2)_{L_2(\Sigma)}, \\ [w_1,w_2]_{G_1} &= E(\mathbf{V}_1,\mathbf{V}_2) + H\,(V_{1,z},h_2)_{L_2(\Sigma)} + H(h_1,V_{1,z})_{L_2(\Sigma)}, \end{aligned} \tag{3}$$

where $V_z$ is the normal component of the vector-function $\mathbf{V}$ on $\Sigma$.

A pair $\lambda$ and $\langle \mathbf{V},h\rangle \in G_1$ is called the generalized solution of problem (1) if for any arbitrary pair of functions

$\langle U,h\rangle \in G_1$ the following equation is satisfied:

$$\frac{1}{H}[\langle V,h\rangle,\langle U,\eta\rangle]_{G_1} - \lambda[\langle V,h\rangle,\langle U,\eta\rangle]_G = 0. \tag{4}$$

Every classical solution of the problem (1) satisfies the relation (4) and conversely, when the generalized solution of the problem (1) is smooth enough it can be shown that this solution satisfies the equation and the boundary conditions in the usual sense. Taking into account (3) we can rewrite (4) in the form

$$\frac{1}{H}E(V,U) + \int_\Sigma (V_z\bar{\eta} + h\bar{U}_z)dS - \lambda\left[\int_\Omega (V,U)d\Omega - \int_\Sigma h\bar{\eta}\,dS\right] = 0. \tag{5}$$

Restricting ourselves to consideration only of the pairs of functions $\langle V,h\rangle \in G_1$ and $\langle U,\eta\rangle \in G$ for which $V_z = -\lambda h$ and $U_z = -\lambda\eta$ on $\Sigma$ we come to relation

$$\lambda^2 T(V,U) - \frac{\lambda}{H}E(V,U) + \Pi(V,U) = 0, \tag{6}$$

where

$$T(V,U) = \int_\Omega (V,U)d\Omega, \qquad \Pi(V,U) = \int_\Sigma V_z\bar{U}_z dS.$$

If the above relation is satisfied for arbitrary $U \in W_2^1(\Omega)$ then as shown in [2] it is possible to define a generalized solution of the problem (1) and to realize on this basis the projection method for construction of approximate solutions of the problem (1).

Another projection method for investigation of normal oscillations of viscous incompressible liquid in a container can be formulated on the basis of consideration of Rayleigh functional of the following form [3]:

$$F_k(V) = \frac{E(V,V) + (-1)^k\sqrt{E^2(V,V) - 4H^2T(V,V)\Pi(V,V)}}{2H\Pi(V,V)}, \tag{7}$$

where $k = 1,2$.

The following theorem is valid [3]:

**THEOREM 1.** Let one of the functionals (7) or both simultaneously take a stationary value for a vector-function $V_i \in W_2^1(\Omega)$. Then $V_i$ is a generalized eigenvalue of the problem (1) and the value of this functional when $V = V_i$ is equal to the corresponding eigenvalue of the problem (1), h being equal to

$$h = -V_{i,z}/\lambda_i \quad \text{on } \Sigma .$$

The proof of this theorem is based on the stationary condition of the functional $F_k(V)$

$$\frac{\partial}{\partial\varepsilon} F_k(V_i + \varepsilon U)\Big|_{\varepsilon=0} = 0 \quad \forall\, U \in W_2^1(\Omega).$$

The set of solutions of the problem (1) can be divided into two subclasses with respect to taking stationary values by the corresponding functional $F_1(V)$ or $F_2(V)$:

$$\lambda_{j,1} = F_1(V_{j,1}), \quad \lambda_{j,2} = F_2(V_{j,2}). \tag{8}$$

The functional $F_k(V)$ can take both real and non-real values. When the Galilei number $H$ is small enough the functional $F_k(V)$ takes only real values, i.e. the bundle is strongly damped. This statement known from [2] can be confirmed by the fact that the functional

$$\Phi(V) = \frac{E^2(V,V)}{\Pi(V,V)\ T(V,V)}, \tag{9}$$

defined on the class of vector-functions $V \in W_2^1(\Omega)$ is bounded below.

The following theorem is valid [3]:

**THEOREM 2.** Let the functional $\Phi(V)$ for $V = V_i^* \in W_2^1(\Omega)$ take a stationary value. Then the functionals $F_1(V)$ and $F_2(V)$ for $V = V_i^*$ and when $H = H_i^* = 1/2\ \Phi(V_i^*)$ also take stationary values i.e. the vector-function $V_i^*$ is a generalized solution of the problem (1),

$$\lambda_{i.1} = \lambda_{i.2} = \frac{E(V_i^*, V_i^*)}{2\ H_i^*\ T(V_i^*, V_i^*)} \tag{10}$$

It is interesting to determine the critical values of the parameter $H = H_i$ in going over which in the problem (1) a pair of non-real eigenvalues can appear or disappear. For this purpose the following theorem is useful [3]:

**THEOREM 3.** If the functional $\Phi(\mathbf{V})$ takes stationary value for the vector-function $\mathbf{V}_i^*$ then for the same function the following functional takes a stationary value:

$$\Phi_{\varkappa}(\mathbf{V}) = \frac{\varkappa^2 E(\mathbf{V},\mathbf{V}) - \varkappa^4 T(\mathbf{V},\mathbf{V})}{\Pi(\mathbf{V},\mathbf{V})}, \tag{11}$$

here the parameter $\varkappa$ being varied along with $\mathbf{V}$.

The following inequalities were proved in [2] on the base of Sobolev embedding theorems:

$$E(\mathbf{V},\mathbf{V}) \geqslant \chi_1 T(\mathbf{V},\mathbf{V}),\ E(\mathbf{V},\mathbf{V}) \geqslant \sigma_1 \Pi(\mathbf{V},\mathbf{V})\ (\chi_1 > 0,\ \sigma_1 > 0) \tag{12}$$

These inequalities allow to conclude that when the values of $\varkappa < \sqrt{\chi_1}$ are small enough the functional $\Phi_{\varkappa}(\mathbf{V})$ is positive and its stationary values are defined as eigenvalues $\omega$ of the following spectral problem:

$$\begin{gathered} -\Delta\mathbf{V} + \nabla\varphi = \varkappa^2\mathbf{V},\ div\,\mathbf{V} = 0\ in\ \Omega,\ \mathbf{V} = 0 \text{ on } S,\ \int_{\Sigma} V_z dS = 0, \\ \frac{\partial V_x}{\partial z} + \frac{\partial V_z}{\partial x} = 0,\ \frac{\partial V_y}{\partial z} + \frac{\partial V_z}{\partial y} = 0,\ \varkappa^2\left[2\frac{\partial V_z}{\partial z} - \varphi\right] = \omega V_z \text{ on } \Sigma . \end{gathered} \tag{13}$$

Let us note that the problem (12) coincides with the problem (1) if we put in (1) $h = -1/\lambda\, V_z$, $H^2 = \omega$, $p = \varphi/H$.

Stationary values of the functionals $\Phi(\mathbf{V})$ and $\Phi_{\varkappa}(\mathbf{V})$ as it follows from theorems 2 and 3 are taken on vector-functions $\mathbf{V}$ satisfying the following system of differential equations:

$$-\Delta\mathbf{V} + \nabla p = \varkappa^2\mathbf{V}\ ,\ div\,\mathbf{V} = 0 \text{ in } \Omega . \tag{14}$$

Therefore it makes sense to choose vector-functions of comparison on the class of solutions of the system (14). In this way calculation of the functional $\Phi_{\varkappa}(\mathbf{V})$ reduces to calculation of surface integrals over the surfaces $S$ and $\Sigma$ what is much more economical comparing with calculation of the functional $\Phi(\mathbf{V})$. When $0 < æ < \sqrt{\chi_1}$ there exists the positive minimum of the functional $\Phi_æ(\mathbf{V})$. On this very interval the first stationary value of the functional $\Phi_æ(\mathbf{V})$ is reached.

Every solution of the system (14) can be represented in the form

$$\mathbf{V} = \nabla\varphi + rot(k\,\psi_1) + rot\;rot(k\,\psi_2), \quad p = æ^2\varphi, \tag{15}$$

where $\mathbf{k}$ is the unit vector of the axis $z$,

$$\Delta\varphi = 0, \quad \Delta\psi_i + æ^2\psi_i = 0 \text{ in } \Omega \quad (i = 1,2).$$

Let $\{w_k\}_{k=1}^{\infty}$ and $\{f_k\}_{k=1}^{\infty}$ be systems of coordinate functions satisfying correspondingly the Laplace and the Helmholtz equations and possessing the required properties of smoothness and completeness. Let us divide these systems of functions with respect to evenness on the variable $z$ into the subsystems of even $\{w_k^{(2)}\}$ and odd $\{w_k^{(1)}\}$ functions on $z$. Then according to the formula (15) we consider the following subsystems of vector-functions satisfying the system of differential equations (14)

$$\mathbf{V}_k^{(1)} = \nabla w_k^{(1)}, \; \mathbf{V}_k^{(2)} = \nabla w_k^{(2)}, \; p_k^{(i)} = æ^2 w_k^{(i)} \quad (i = 1,2), \; \mathbf{V}_k^{(3)} = rot\,(\mathbf{k}f_k^{(1)}),$$

$$\mathbf{V}_k^{(4)} = rot(\mathbf{k}f_k^{(2)}), \; \mathbf{V}_k^{(5)} = rot\;rot\;(\mathbf{k}f_k^{(1)}), \; \mathbf{V}_k^{(6)} = rot\;rot\;(\mathbf{k}f_k^{(2)}),$$

$$p_k^{(i)} = 0 \;\; (i \geqslant 3).$$

The functions $\mathbf{V}_k^{(2)}$, $\mathbf{V}_k^{(4)}$ and $\mathbf{V}_k^{(5)}$ satisfy the first two of boundary conditions on $\Sigma$ of the problems (1) and (2)

$$\frac{\partial V_z}{\partial x} + \frac{\partial V_x}{\partial z} = 0, \quad \frac{\partial V_z}{\partial y} + \frac{\partial V_y}{\partial z} = 0 \text{ on } \Sigma. \tag{16}$$

On the basis of vector-functions $\mathbf{V}_k^{(1)}$, $\mathbf{V}_k^{(3)}$ and $\mathbf{V}_k^{(6)}$ we construct one more system of vector-functions

$$\mathbf{V}_k^{(7)} = \mathbf{V}_k^{(1)} - \sum_{j=1}^{N_3} C_{j,k}^{(3)}\mathbf{V}_j^{(3)} - \sum_{j=1}^{N_6} C_{j,k}^{(6)}\mathbf{V}_j^{(6)}, \quad P_k^{(7)} = æ^2\, w_k^{(1)}.$$

satisfying the conditions (16) approximately where we determine the coefficients $C^{(3)}_{j,k}$ and $C^{(6)}_{j,k}$ from the boundary conditions (16) by means of some approximate method. After that using the constructed systems of coordinate functions we find solutions of the system of differential equations (13) satisfying approximately the condition $V = 0$ on $S$

$$V^{(8)}_k = V^{(1)}_k - \sum_{i=2}^{6} \sum_{j=1}^{N_i} C^{(i)}_{j,k} V^{(i)}_j , \quad p^{(8)}_k = \text{æ}^2 [w^{(1)}_k - \sum_{j=1}^{N_2} C^{(2)}_{j,k} w^{(2)}_j ] .$$

The coefficients $C^{(i)}_{j,k}$ $i = 2,4,5$ are defined with the least-square technique.

On the class of functions satisfying the boundary conditions of the problem (13) on $S$, the equations (14) and the boundary conditions (16) functional $\Phi_{\text{æ}}(V)$ takes the following form:

$$\Phi_{\text{æ}}(V) = \frac{\text{æ}^2 \int_{\Sigma} (2 \frac{\partial V_z}{\partial z} - p) V_z dS}{\int_{\Sigma} |V_z|^2 dS}$$

We construct its stationary solutions by means of the Ritz method approximating the searched solution of the problem (13) by finite series of the form.

$$V = \sum_{k=1}^{N_1} a_k V^{(8)}_k , \quad p_k = \sum_{k=1}^{N_1} a_k p^{(8)}_k .$$

In the particular case of an infinite cylinder horizontal domain and of lateral oscillations of liquid i.e. when $V_x = 0$ (the axis $OZ$ is parallel to the cylinder's element) a solution of the problem (13) can be represented in the form [3]

$$V = \nabla\varphi(y,z) + rot(i\psi(y,z)), \quad \Delta\varphi = 0, \ \Delta\psi + \text{æ}^2\psi = 0 \text{ in } G.$$

For the case of the cross-section of $G$ being the segment of the unit circle with the height $h_k = 0{,}5$ and $h_k = 0{,}75$ in [4] the following values of $H_i = \sqrt{\omega_i}$ were obtained, depending on the value of the parameter æ:

| | $h_k = 0.5$ | | $h_k = 0.75$ | |
|---|---|---|---|---|
| $æ$ | $H_1$ | $H_2$ | $H_1$ | $H_2$ |
| 0.5 | 1.9497 | 2.2829 | 1.6135 | 1.9743 |
| 1.0 | 3.8661 | 4.5393 | 3.1659 | 3.9150 |
| 1.5 | 5.7116 | 6.7415 | 4.5856 | 5.7891 |
| 2.0 | 7.4378 | 8.8598 | 5.7785 | 7.5638 |
| 2.5 | 8.9764 | 10.8640 | 6.6082 | 9.2076 |
| 3.0 | 10.2247 | 12.7265 | 6.8554 | 10.6869 |
| 3.5 | 11.0236 | 14.4254 | 6.0815 | 11.9617 |
| 4.0 | 11.1164 | 15.9439 | 2.3966 | 12.9756 |
| 4.5 | 10.0283 | 17.2629 | — | 13.6389 |
| 5.0 | 6.3511 | 18.3518 | — | 13.7836 |
| 5.5 | — | 19.1621 | — | 12.9943 |
| 6.0 | — | 19.6198 | — | 8.9821 |

Stationary values of the functional $\Phi_{æ}(V)$ ( the maximal in these cases ) are taken with the following values of $æ_1^*$ и $æ_2^*$:
for $h_k = 0.5$ $æ_1^* = 3.789$, $H_1^* = 11.169$; $æ_2^* = 6.240$, $H_2^* = 19.647$,
for $h_k = 0.75$ $æ_1^* = 2.871$, $H_1^* = 6.872$; $æ_2^* = 4.829$, $H_2^* = 13.811$.

## REFERENCES

[1] Krein S.G. On Oscillations of Viscous Liquid in a Container // Dokl AN USSR, 1964, vol.159, N 2, pp.262-265 (in Russian).

[2] Askerov N.K., Krein S.G., Laptev G.I. Problem of Oscillations of Viscous Liquid and Operator Equations Related to it //Functional Analysis and its Applications, 1968, vol.2, N 2, pp. 21-32.

[3] Barnyak M.Ya. Small Oscillations of Viscous Incompressible liquid in a Container, Kiev, 1989 (Preprint/Acad. Sci. Ukr. SSR. Institute of Mathematics; 89.48), 60 p.

[4] Barnyak M.Ya. Investigation of Normal Oscillations of Viscous Liquid in Horizontal Channel by Means of Projection Technique // Modelling of Dynamic Processes. in Systems of Bodies with Liquid. - Kiev:Institute of Mathematics, Acad.Sci.Ukr.SSR, 1990. - pp.34-41.

# THE CLASSICAL STEFAN PROBLEM AS THE LIMIT CASE OF THE STEFAN PROBLEM WITH A KINETIC CONDITION AT THE FREE BOUNDARY

*B.V.Bazaliy, S.P.Degtyarev*
*Institute of Applied Mathematics and Mechanics*
*Donetsk 340114, UKRAINE*

In some papers mathematical models of solidification have been grounded as the limit of the boundary problems of the phase-field equations (see e.g. [1]). At the same time with the classical Stefan problem, which contains a constant-temperature condition at the interface (this temperature is an equilibrium melting temperature), there arises a modified Stefan problem, in which the phase-change temperature depends on the velocity and curvature of the interface (classical Stefan condition is replaced by the kinetic condition).

In our paper [2] the question of the classical solvability of the modified Stefan problem has been discussed. In this report we are studying the question of the limit transition at the set of classical solutions of the modified Stefan problem when the kinetic condition at the interface is reduced to the classical one. In the one-dimensional case the Stefan problem with a kinetic condition has been considered in [3,4].

Key words: Stefan problem, kinetic condition.

1. Let $\Omega \subset R^3$ be a given domain with a boundary consisting of disjoint components $\Gamma^+$ and $\Gamma^-$, where $\Gamma^+$ lies inside the bounded domain, the boundary of which is $\Gamma^-$. Let surface $\Gamma \subset \Omega$ divide $\Omega$ into two connected subdomains $\Omega^\pm$ so that $\partial\Omega^\pm = \Gamma \cup \Gamma^\pm$. For the points on the surface $\Gamma$ we introduce coordinates $\omega = (\omega_1,\omega_2)$. We also denote by $y(\omega) \in \Gamma$ the corresponding points in $R^3$ and let $\vec{\nu}(\omega)$ be the unit normal to $\Gamma$ directed into $\Omega^+$. Let $\gamma_0$ be a given positive number, such that the surfaces $\{y = y(\omega) \pm 2\vec{\nu}(\omega)\gamma,\ 0 < \gamma < \gamma_0\}$ have no self intersection and do not intersect $\Gamma$ and $\Gamma^\pm$. We denote by $\Gamma_T = \Gamma\times[0,T]$, $\Gamma_T^\pm = \Gamma^\pm\times[0,T]$, $\Omega_T^\pm = \Omega^\pm\times(0,T)$, $\Omega^\pm_{\rho,T}$ - the region bounded by the

planes $\tau = 0$ and $\tau = T$ and the surfaces $\Gamma_T^{\pm}$ and

$$\Gamma_{\rho,T} = \left\{(y,\tau):\ y = y(\omega)+\vec{\nu}(\omega)\rho(\omega,\tau),\ \tau \in [0,T]\right\},$$

where $\rho(\omega,\tau)$ is a sufficiently smooth function such that $\rho(\omega,0) = 0$ and $|\rho(\omega,\tau)| < \gamma_0$. In general the Stefan problem consists in finding the temperature distribution $u^{\pm}(y,\tau)$ and function $\rho(\omega,\tau)$ defined on a priory unknown surface $\Gamma_{\rho,T}$ on the basis of the conditions:

$$\begin{aligned}
&\frac{\partial u^{\pm}}{\partial \tau} - a^{\pm}\nabla_y^2 u^{\pm} = 0, \quad (y,\tau) \in \Omega^{+}_{\rho,T};\\
&æV = (a^{-}\nabla_y u^{-} - a^{+}\nabla_y u^{+})\vec{n}(y,\tau),\\
&u^{+} = u^{-} = \varepsilon k(y,\tau) - \varepsilon V, \quad (y,\tau) \in \Gamma_{\rho,T};\\
&u^{\pm} = b^{\pm}(y,\tau), \quad (y,\tau) \in \Gamma_T^{\pm};\\
&u^{\pm}(y,0) = u_0^{\pm}(y), \quad \rho(\omega,0) = 0,
\end{aligned} \tag{1}$$

where $a^{\pm}$, $æ$ are given positive numbers, $k(y,\tau)$ is the sum of principal curvatures at the point on the interface $\Gamma_{\rho,T}$, $V$ is the velocity of the free boundary in the direction of the normal $\vec{n}(y,\tau)$ which is directed into $\Omega^{+}_{\rho,T}$, $b^{\pm}(y,\tau)$, $u_0^{\pm}(y)$ are given functions,

$$\nabla_y = \left(\partial/\partial y_1, \partial/\partial y_2, \partial/\partial y_3\right).$$

The modified Stefan problem (1) (for $\varepsilon > 0$) is reduced to the classical one for $\varepsilon = 0$.

For $\varepsilon = 0$ the classical solvability of problem (1) in small with respect to time has been proved by various methods in [5-8]. For $\varepsilon > 0$ in [2] we have proved

**THEOREM 1.** Assume that for problem (1) the compatibility conditions of first order are fulfilled, $\Gamma \in H^{5+\alpha}$, $\Gamma^{\pm} \in H^{5+\alpha}$, $b^{\pm}(y,\tau) \in H^{4+\alpha,(4+\alpha)/2}$, $u_0^{\pm}(y) \in H^{4+\alpha}$. Then there is such $T_0 > 0$ depending on data of the problem that the solution

$$u^{\pm}(y,\tau) \in H^{2+\alpha,(2+\alpha)/2}(\overline{\Omega}^{\pm}_{\rho,T}),\ \rho(\omega,\tau) \in H^{3+\alpha,(3+\alpha)/2}(\Gamma_T)$$

exists for $0 < T \leqslant T_0$.

The definition of Hölder spaces $H^{l,l/2}$ may be founded in [9]. We introduce Banach spaces:

$$\hat{H}^{k+\alpha}(\Gamma_T)=\left\{\rho \in H^{k+\alpha,(k+\alpha)/2}:\ \|\rho\|_{\Gamma_T}^{(k+\alpha)}\equiv |\rho|_{\Gamma_T}^{(k+\alpha)} + |\rho|_{\Gamma_T}^{(k-1+\alpha)} < \infty\right\},$$

$$k = 1,\ 2,$$

$\hat{H}^{\alpha}(\Gamma_T)$ is the subspace of the space $H^{\alpha,\alpha/2}$ consisting of functions $f(\omega,t)$, which have a weak derivative with respect to $t$ and

$$\frac{\partial f}{\partial t}(\omega,t) = a(\omega,t) + \nabla_\omega \vec{b}(\omega,t),$$

where $a,\ b_i,\ i = 1,\ 2,$ belong to $H^{\alpha,\alpha/2}(\Gamma_T)$, and the norm in this space is defined as follows:

$$\|f\|_{\Gamma_T}^{(\alpha)} = |f|_{\Gamma_T}^{(\alpha)} + \inf_{a,b}\left(|a|_{\Gamma_T}^{(\alpha)} + |\vec{b}|_{\Gamma_T}^{(\alpha)}\right).$$

It appears that solutions of problem (1) have estimates which do not depend on $\varepsilon$. This permits to perform a limit transition as $\varepsilon \to 0$ in corresponding spaces.

**THEOREM 2.** Let conditions of theorem 1 be fulfilled and $\rho_0(\omega,\tau)$, $u_0^{\pm}(y,\tau)$ be a solution of problem (1) for $\varepsilon = 0$, $\rho_\varepsilon(\omega,\tau)$, $u_\varepsilon^{\pm}(y,\tau)$ is the sequence of solutions of problem (1) for $\varepsilon > 0$. Then $\rho_\varepsilon(\omega,\tau)$ converges to $\rho_0(\omega,\tau)$ in the space $\hat{H}^{2+\alpha'}(\Gamma_{T_0})$, $u_\varepsilon^{\pm}(y,\tau)$ converges to $u_0^{\pm}(y,\tau)$ in the space $H^{2+\alpha',(2+\alpha')/2}$, where $\alpha' < \alpha$, on some interval $[0,T_0]$, which does not depend on $\varepsilon \geqslant 0$.

2. We denote

$$\mathcal{H}_\psi \equiv \overset{0}{H}{}^{2+\alpha,(2+\alpha)/2}(\overline{\Omega}_T^+)\times \overset{0}{H}{}^{2+\alpha,(2+\alpha)/2}(\overline{\Omega}_T^-)\times \overset{0}{\hat{H}}{}^{2+\alpha}(\Gamma_T),$$

$$\mathcal{H}_\mathcal{F} \equiv \overset{0}{H}{}^{\alpha,\alpha/2}(\overline{\Omega}_T^+)\times \overset{0}{H}{}^{\alpha,\alpha/2}(\overline{\Omega}_T^-)\times \overset{0}{H}{}^{1+\alpha,(1+\alpha)/2}(\Gamma_T)\times \overset{0}{H}{}^{2+\alpha,(2+\alpha)/2}(\Gamma_T^+)\times$$

$$\times \overset{0}{H}{}^{2+\alpha,(2+\alpha)/2}(\Gamma_T^-)\times {}^{\wedge}\overset{0}{H}{}^{\alpha}(\Gamma_T)\times \overset{0}{\hat{H}}{}^{\alpha}(\Gamma_T).$$

Problem (1) can be transformed to some problem in the fixed region and after linearization it can be rewritten in the following form [8,9]:

$$A(\varepsilon)\psi = \mathcal{F}(\varepsilon,\psi), \quad \psi = (\theta^{+}(x,t),\theta^{-}(x,t),\sigma(\omega,t)), \tag{2}$$

where the linear operator $A(\varepsilon)$ and the nonlinear operator $\mathcal{F}(\varepsilon,\psi)$ are acting from $\mathcal{H}_{\psi}$ to $\mathcal{H}_{\mathcal{F}}$ and for $\mathcal{F}(\varepsilon,\psi)$ we have

$$\begin{gathered} \|\mathcal{F}(\varepsilon,0)\|_{\mathcal{H}_{\mathcal{F}}} \leqslant cT^{\alpha/2}, \\ \|\mathcal{F}(\varepsilon,\psi_2) - \mathcal{F}(\varepsilon,\psi_1)\|_{\mathcal{H}_{\mathcal{F}}} \leqslant c\left[T^{\alpha/2} + g(r)\right]\|\psi_1 - \psi_2\|_{\mathcal{H}_{\mathcal{F}}} \end{gathered} \tag{3}$$

for $\psi_1$, $\psi_2 \in B_r$, $R_r$ is a ball the center of which is zero of $\mathcal{H}_{\psi}$, $g(r) \to 0$ as $r \to 0$, constant $c$ does not depend on $\varepsilon > 0$.

If the operator $A(\varepsilon)$ has inverse $A^{-1}(\varepsilon)$ with the norm independent of $\varepsilon$, estimates (3) and the theorem about the fixed point of contracting mappings lead to the proof of Theorem 2.

The proof of the invertability of the operator $A(\varepsilon)$ is reduced to the study of some model problem (see below) and building of some regularizator similar to that given in the boundary problems theory for parabolic equations (see [9,ch. IV]).

3. Let us consider in the domains

$$R_T^{\pm} = \left\{(z,t):\ z = (z',z_3),\ z' \in R^2,\ \pm z_3 > 0,\ t \in (0,T)\right\},$$

$$R_T' = \left\{(z',t):\ z' \in R^2,\ t \in (0,T)\right\}$$

the next boundary problem (model problem) for unknown functions $u^{\pm}(z,t)$, $\rho(z',t)$

$$\begin{gathered} \frac{\partial u^{\pm}}{\partial t} - a^{\pm}\Delta u^{\pm} = 0, \quad (z,t) \in R_T^{\pm}; \\ \frac{\partial \rho}{\partial t} + k^{+}\frac{\partial u^{+}}{\partial z_3} - k^{-}\frac{\partial u^{-}}{\partial z_3} + \sum_{i=1}^{2} b_i \rho_{z_i} = f_1(z',t), \\ u^{\pm} + \alpha^{\pm}\rho - \varepsilon\sum_{i,j=1}^{2}\mu_{ij}\rho_{z_i z_j} + \varepsilon\rho_t = \varepsilon f_3(z',t), \quad z_3 = 0; \end{gathered} \tag{4}$$

$$u^{\pm} \in \underset{0}{H}^{2+\alpha,(2+\alpha)/2}(R_T^{\pm}), \quad \rho \in \underset{0}{\hat{H}}^{2+\alpha}(R_T'),$$

where $k^{\pm}$, $a^{\pm}$, $d^{\pm}$ are positive constants, $\vec{b} = (b_1, b_2) \in R^2$, the constants $\mu_{ij}$ satisfy the condition

$$\nu|\xi|^2 \leqslant \sum_{i,j} \mu_{ij}\xi_i\xi_j \leqslant \nu^{-1}|\xi|^2, \quad \nu = const,$$

$$\varepsilon \in [0,1].$$

The functions $f_1$, $f_3$ are finite and

$$f_1 \in \underset{0}{H}^{1+\alpha,(1+\alpha)/2}, \quad f_3 \in \underset{0}{\hat{H}}^{\alpha}.$$

We denote by $\tilde{f}(\lambda, z_3, p)$ the Fourier transformation with respect to the variables $z'$ and the Laplace transformation with respect to $t$ of the functions $f(z', z_3, t)$:

$$\tilde{f}(\lambda, z_3, p) = \int_{R'} dz' \int_0^{\infty} f(z', z_3, t)\, e^{-i(z',\lambda)-pt} dt.$$

Applying the indicated integral transformation to the first equation in (4) and solving the ordinary differential equation with respect to the variable $z_3$, we obtain

$$u^{\pm}(\lambda, z_3, p) = \tilde{M}^{\pm}(\lambda, p)\, exp\left[\mp \sqrt{\frac{p + a^{\pm}\lambda^2}{a^{\pm}}}\; z_3\right],$$

where

$$M^{\pm}(z', t) = u^{\pm}(z', 0, t).$$

Now, applying the integral transformation to the remaining relation in (4) and using the obtained representation for $u^{\pm}(\lambda, z_3, p)$ we find that

$$\tilde{\rho}(\lambda, p) = \tilde{K}_0(\lambda, p)\tilde{f}_1(\lambda, p) + \tilde{K}_1(\lambda, p)\tilde{f}_3(\lambda, p),$$

$$\tilde{K}_0(\lambda, p) = \left\{\varepsilon\left[(\mu\lambda, \lambda) + p\right]\left( k^{+}\sqrt{\frac{p + a^{+}\lambda^2}{a^{+}}} + k^{-}\sqrt{\frac{p + a^{-}\lambda^2}{a^{-}}} + \right.\right.$$

$$+ d^+k^+\sqrt{\frac{p+a^+\lambda^2}{a^+}} + d^-k^-\sqrt{\frac{p+a^-\lambda^2}{a^-}} + p - i(b,\lambda)\Big\}^{-1},$$

$$\tilde{\mathcal{K}}_1(\lambda,p) = \varepsilon\left[ k^+\sqrt{\frac{p+a^+\lambda^2}{a^+}} + k^-\sqrt{\frac{p+a^-\lambda^2}{a^-}} \right] \tilde{\mathcal{K}}_0(\lambda,p).$$

To prove $\rho \in \underset{0}{\hat{H}}{}^{2+\alpha}(R_T')$ in the model problem we shall use the results of [10].

**THEOREM 3.** (Mogilevski J.Ch., Solonnikov V.A.) Let the function $f(z',t) \in \underset{0}{H}{}^{1,1/2}$ and the function $u(z',t)$ with zero initial conditions be connected with $f(z',t)$ by the condition

$$\tilde{u}(\lambda,p) = \tilde{\mathcal{K}}(\lambda,p)\tilde{f}(\lambda,p)$$

and

$$|\tilde{\mathcal{K}}(\lambda,p)| \leqslant c_1|r|^{-\beta}, \quad \left|\frac{\partial\tilde{\mathcal{K}}}{\partial\xi_0}\right| \leqslant c_2|r|^{-\beta-2},$$

$$\left|\frac{\partial}{\partial\lambda_j}\tilde{\mathcal{K}}(\lambda,p)\right| \leqslant c_3|r|^{-\beta-1}, \quad \left|\frac{\partial^2}{\partial\xi_0\partial\lambda_j}\tilde{\mathcal{K}}(\lambda,p)\right| \leqslant c_4|r|^{-\beta-3}, \tag{5}$$

$$\left|\frac{\partial^2}{\partial\lambda_i\partial\lambda_j}\tilde{\mathcal{K}}(\lambda,p)\right| \leqslant c_5|r|^{-\beta-2}, \quad \left|\frac{\partial^3}{\partial\xi_0\partial\lambda_i\partial\lambda_j}\tilde{\mathcal{K}}(\lambda,p)\right| \leqslant c_6|r|^{-\beta-4},$$

for $p = a + i\xi_0$, $a > 0$, $\beta \geqslant 0$, $r = \sqrt{p + \lambda^2}$, $i, j = 1, 2$.

Then the following inequality holds:

$$\langle u\rangle_{R_T'}^{(1+\beta,(1+\beta)/2)} \leqslant c\, e^{aT}\langle f\rangle_{R_T'}^{(1,1/2)},$$

where $c$ depends on $c_i$, $i = \overline{1,6}$, $a$, $l$ and $\beta$.

We denote

$$z_1 = \varepsilon\big(p + (\mu\lambda,\lambda)\big)\left[k^+\sqrt{\frac{p+a^+\lambda^2}{a^+}} + k^-\sqrt{\frac{p+a^-\lambda^2}{a^-}}\right], \quad z_2 = p,$$

$$z_3 = d^+k^+\sqrt{\frac{p+a^+\lambda^2}{a^+}} + d^-k^-\sqrt{\frac{p+a^-\lambda^2}{a^-}}.$$

**LEMMA 1.** There exist constants $a > 0$, $\nu_0 > 0$, depending only on $\mu$, $k^{\pm}$, $a^{\pm}$, $d^{\pm}$, $\vec{b}$ and not depending on $\varepsilon$, such that for $p = a + i\xi_0$, $\xi_0 \in R$

$$|z_1 + z_2 + z_3 - i(b,\lambda)| \geqslant \nu_0\left(\varepsilon|r|^3 + |p| + |r|\right).$$

With the help of this result it is easy to prove

**LEMMA 2.** The functions $\tilde{\mathcal{K}}_i(\lambda,p)$ satisfy the conditions (5) of Theorem 3 with $\beta = \beta_i = 1 + i$, $i = 0, 1$.

**LEMMA 3.** Problem (4) has a unique solution under the above assumptions and

$$|u^{\pm}|^{(2+\alpha)}_{R^{\pm}_T} + \|\rho\|^{(2+\alpha)}_{R'_T} \leqslant c\left[|f_1|^{(2+\alpha)}_{R'_T} + \|f_3\|^{(\alpha)}_{R'_T}\right],$$

where the constant $c$ does not depend on $\varepsilon$.

In lemma 3 it is the most important fact to prove the existence of the bounded inverse operator $A^{-1}(\varepsilon)$ from relation [2], the norm of which $\|A^{-1}(\varepsilon)\|_{\mathcal{H}_{\mathcal{F}}\to\mathcal{H}_{\psi}}$ does not depend on $\varepsilon$.

To prove theorem 2 we rewrite now relation (2) in the form

$$\psi = A^{-1}(\varepsilon)\mathcal{F}(\varepsilon,\psi) \equiv G(\varepsilon,\psi). \tag{6}$$

The estimates from (3) show that the operator $G(\varepsilon,\psi)$ is contractive in the ball $B_r$. Hence equation (6) has a unique solution in $B_r$ for any $\varepsilon > 0$, moreover the number $T$ in the definition of $B_r$ does not depend on $\varepsilon \geqslant 0$. Taking into consideration that the space $H^{l+\alpha',(l+\alpha')/2}$ is embedded compactly into space $H^{l+\alpha,(l+\alpha)/2}$ for $\alpha' < \alpha$ we obtain the statement of Theorem 2.

**REFERENCES.**

[1] Caginalp G., Stefan and Hele-Show type model as asymptotic limits of the phase-field equations , Physical Rev. - (1989), 39,N 11, pp.5887-5896.

[2] Bazaliy B.V., Degtyarev S.P., Classical solvability of the Stefan problem with a kinetic condition at the free boundary, Nonlinear boundary problems, Kiev, Naukova Dumka (to be published).

[3] Visintin A., The Stefan problem with a kinetic condition at the free boundary, Ann.mat.pure ed appl., (1987), 144, N6, pp.97-122.

[4] Xie W.The Stefan problem with a kinetic condition at the free boundary, SIAM, J.Math.Anal., (1990), 21, N 2, pp.362-373.

[5] Hanzava E.I. Classical solutions of the Stefan problem, Tohoku Math.J., (1981), 33, pp.297-335.

[6] Meirmanov A.M., On the classical solution of the multidimensional Stefan problem for quasilinear parabolic equation, Mat.Sb., (1980), 100(142), pp. 170-192.

[7] Bazaliy B.V., The Stefan problem, Dokl.Akad.Nauk Ukrain.SSR., Ser.A, (1986), N 11, pp.3-7.

[8] Bazaliy B.V., Degtyarev S.P., On classical solvability of the multidimensional Stefan problem for convective motion of a viscous incompressible fluid, Math.Sb., (1987), 132(174), N 1, pp.3-19.

[9] Ladyzhenskaya O.A., Solonnikov V.A., Ural'tseva N.N., Linear and quasilinear equations of parabolic type, Moscow, Nauka, (1967).

[10] Mogilevskiy J.Ch., Solonnikov V.A. On the solution of some uncoercitive initial boundary value problem for the Stokes system in Holder spaces (case of the half-space), Zeit Anal. und Anwendungen, (1988), 8, N 4, pp.329-347.

# A MATHEMATICAL MODEL OF OSCILLATIONS ENERGY DISSIPATION OF VISCOUS LIQUID IN A TANK

*I.B. Bogoryad*
*Research Institute of Applied Mathematics and Mechanics, Tomsk 634050, RUSSIA*

Up to present time an experiment [1,2] rather then a computational method gives the most reliable results on determination of damping factors of viscous liquid oscillations. The reason lies in mathematical difficulties of solving of hydrodynamic problems in setting up which should reflect the physics of the process sufficiently completely.

Nevertheless the development of a new mathematical model and method of their numerical analysis is urgent if only it is impossible to create a complete dynamical and geometrical likeness of the experiment with real conditions of the scheme motion.

Below a mathematical model of dissipation of energy of capillary liquid of small viscosity partly filling the tank differing from the traditional mathematical model is suggested.

I. The traditional mathematical model of oscillations in tank of viscous liquid with the free surface essentially rests upon two assumptions [3,4].

1. Oscillations of liquid occur at the Reynolds numbers $Re \gg 1$ ($Re = \omega R_0^2 \nu^{-1}$, $\omega$ is the frequency of sloshing, $R_0$ is the typical size, e.g. the radius of the free surface, $\nu$ is the kinematic coefficient of viscosity).

2. The amplitude of oscillations of the free surface $S_0$ is small ($S_0/R_0 \ll 1$) so that the Navier-Stokes equations and boundary conditions can be linearized with respect to the small parameter $\varepsilon \sim S_0/R_0$.

The first assumption makes it possible to construct asymptotic approximations according to powers $Re^{-n/2}$ $(n = 0,1,\ldots)$ and with $n = 1$ to obtain an effective theory of calculating damping in approximation of a laminar boundary layer.

The second assumption reduces the boundary-value problem with the boundary unknown in advance - the free surface $\Sigma$ to the problem with the fixed boundary $\Sigma_0 + S$ ($\Sigma_0$ is an undisturbed free surface, $S$ is wetted surface). Besides this evident simplification the second assumption combined with the first one leads to the consequence of fundamental importance: the free surface with the accuracy to the values of the order $\varepsilon^2$ does not affect the friction stress in the boundary layer.

From literature other approaches to the solution on the basis of Navier-Stokes complete equations [5-7] are known as well. However, on the whole results obtained there have a character of model calculations for the numbers $Re \sim 10^2 \div 10^4$.

The comparison of the experimental data on determination of the coefficient of damping oscillation $\beta$ [1,2] with the calcu lated ones [3,4] shows that the traditional model gives an underestimated value $\beta$. For agreement of the experimental and calculated data the empirical correction $\sqrt{2}$ is introduced in [1]. It is evident that this correction does not provide an adequate description of a process, therefore one has to correct the mathematical model.

2. In the suggested mathematical model which is also calculated for the flows with $Re \gg 1$ two new features evident from the point of view of the physics of the process are introduced.

Firstly, it is taken into account that the dissipation of the energy of liquid oscillations due to its viscosity occurs not only under the level of the undisturbed free surface $\Sigma_0$, but above $\Sigma_0$ in a thin liquid film which remains on the surface of the tank after the preceding cycle of oscillations (flow up and flow down of liquid over the film).

Secondly, it is taken into account that at some $Re > Re^*$ turbulization of the laminar boundary layer is possible.

Thus on $S$ three characteristic zones deep down the liquid stand out in each of which dissipation mechanism of its own works:

(a) The zone of the film at the wall. The thickness of the film and the friction stress in it are determined by the formulae [7,8]

$$h \approx c\, Re^{-2/3}\left[1 + \frac{R_0}{\langle S_0\rangle Fr}\right]^{-1/2}\left(B_0\frac{\langle S_0\rangle}{R_0}\right)^{1/6},$$

$$\tau = \rho\sqrt{\frac{\nu}{\pi}}\int_{-\infty}^{t}\frac{dw}{d\xi}\,\theta_4\left(0,\ exp\left[-\frac{h^2}{\nu(t-\xi)}\right]\right)\frac{d\xi}{\sqrt{t-\xi}}\,, \tag{1}$$

where $c \sim 1$, $B_0 = \rho\omega^2 R_0^2\gamma^{-1}$ is the Bond number, $\langle S_0\rangle$ is the average amplitude during the period of oscillations, $\gamma$ is the coefficient of the surface tension, $\theta_4$ is elliptic theta-function,

$$\frac{dw}{dt} = \frac{du}{dt} + \frac{1}{\rho}\frac{\partial p}{\partial x} + g.$$

Let us note that we do not make distinction between stresses at flow down and flow up.

(b) The zone of a laminar boundary layer of the thickness

$$h_L \sim Re^{-1/2}.$$

In this zone tangential stresses arise [4]

$$\tau_L = \rho\sqrt{\frac{\nu}{\pi}}\int_{-\infty}^{t}\frac{dw}{dt}\frac{d\xi}{\sqrt{t-\xi}}\,. \tag{2}$$

It is seen that

$$\tau_L = \lim_{\nu \to 0} \tau \tag{3}$$

(c) The zone of turbulized sublayer. Estimates [9] take place here

$$h_T \sim \left(\frac{\omega\,\langle S_0\rangle}{\nu_T\, R_0}\right)^{-1/2},\quad \tau_T \sim \rho l^2\left|\frac{\partial u}{\partial y}\right|\frac{\partial u}{\partial y}\,, \tag{4}$$

where $l$ is the length of turbulence, $\nu_T$ is the kinematic coefficient of turbulent exchange.

3. It is known that for the plate the transition point of the laminar form of flow into the turbulent one corresponds to the Reynolds number

$$Re_1^* = \frac{\omega \langle S_0 \rangle h_1}{\nu} \approx 10^3.$$

Here $h_1 \approx 0.3h_L$ the displacement thickness is connected with the current coordinate $x$

by relationship

$$h_1 = c\, Re^{-1/2} \left(\frac{x}{\langle S_0 \rangle}\right)^{-1/2}, \quad c \sim 1$$

with $S_0/R_0 \sim 0.1$, $Re^* \approx 10^7$ and the transition point $x^*/R_0 \sim 0.1$ corresponds to the limiting $Re_1^* \approx 10^3$.

It means that with $Re > 10^7$ the turbulent flow may arise at the distance $\sim S_0$ under the surface $\Sigma_0$. At this depth the velocity of liquid at the flow core ($u_\infty$) is about $(0.8 \div 0.9)\omega S_0$.

The turbulent flow may arise both at $Re < 10^7$ and $x < 0.1R_0$ on the rough walls of the tank. At the height of roughness $k \sim h_1$ [9]

$$(Re_1^*)_{rough} \sim 10^{-1} (Re_1^*)_{smooth}$$

Evidently acoustic vibrations of the tank walls may be the reason of much earlier beginning of turbulization. In this case in the boundary layer as well as at roughness the components of the velocity of liquid v normal to $S$ are excited.

The estimates show that with $\omega/\omega_s \sim 10^{-2}$ ($\omega_s$ is the frequency of oscillations of the wetted surface) the component of the liquid velocity [10]

$$v = v_s \exp\left[-\frac{y}{h_L}\left(\frac{\omega_s}{\omega}\right)^{1/2}\right]$$

normal to $S$ is essential at the distance $y \sim h_L$ from the tank walls. Thus on $S$ $(y = 0)$ $v = v_s$; with $y = 0.3h_L$, $v \approx 0.1v_s$; with $y = h_L$, $v \approx 10^{-3}v_s$.

4. We shall obtain an approximation expression for the damping coefficient $\beta$ of oscillations of liquid in the tank. Let us use the relationship for the derivative from the total energy

$$\frac{dE}{dt} = -\int_S u\,\tau\,ds$$

and from (1),(2),(4) find for each of the considered zones the following estimates

$$\frac{dE_a}{dt} \sim 2\rho\nu^{1/2}\omega^{5/2}\langle S_0\rangle^2 R^2{}_0 ,$$

$$\frac{dE_b}{dt} \sim 2\rho\nu^{1/2}\omega^{5/2}\langle S_0\rangle^{5/2}R_0^{3/2}\left(\frac{\gamma}{\rho\nu\omega\,\langle S_0\rangle}\right)^{1/6},$$

$$\frac{dE_c}{dt} \sim 2\rho\nu_T^{1/2}\omega^{5/2}\langle S_0\rangle^{5/2}R_0^{1/2}\Delta x^* ,$$

where $\Delta x^*$ is the characteristic extent of the zone of turbulent boundary layer.

As far as

$$\beta = \frac{1}{2E}\,\frac{dE}{dt} = \frac{1}{2E}\left(\frac{dE_a}{dt} + \frac{dE_b}{dt} + \frac{dE_c}{dt}\right)$$

and

$$E \sim \rho\omega^2\,\langle S_0\rangle R_0^3 ,$$

then

$$\beta = c_1\omega\, Re^{-1/2}\left[1 + c_\gamma\left(\frac{S_0}{R_0}\right)^{1/3} + c_\tau\left(\frac{\nu_T}{\nu}\right)^{1/2}\left(\frac{S_0}{R_0}\right)^{3/2}\right] \qquad (5)$$

Here $c_\gamma = c_S\left(\frac{\alpha}{h_L}\right)^{1/3}$, $\alpha = \left(\frac{\gamma}{\rho g}\right)^{1/2}$ is the capillary constant, $c_1 \sim c_S \sim c_T \sim 1$.

Let us consider in more detail the obtained dependence.

With $Re < Re^*$ the Reynolds stresses in (5) are absent. In [7,8] it is shown: the contribution to $\beta$ of viscous-capillary stresses in the film above $\Sigma_0$ (the second addend) reaches 30÷50% and improves the agreement of experimental and calculated data over the range of numbers $Re \sim 10^4 \div 10^6$. As $Re$ decreases (for example, in connection with the fact that $g \to 0$) the contribution of this factor increases and with $Re \sim 10^2$ it becomes a decisive factor.

On the contrary with the increase of $Re$ the contribution of viscous-capillary stresses decreases as $Re^{-1/3}$ and with $Re \sim 10^7 \div 10^8$ it is not more than $10\%$. In this range of $Re$ numbers the main role in the formation of the value $\beta$ belongs to the Reynolds component of stress: with $Re \sim 10^8$ it approximately three times exceeds the component generated by stresses in the laminar boundary layer.

## REFERENCES


[1] Mikishev,G.N. Experimental methods in dynamics of space apparatuses. Moscow, Mashinostroenie, 1978.

[2] Miles,J.W. Ring damping of free surface oscillations in a circular tank // J.Appl.Mech., 1958, v.25, N6.

[3] Chernousko,F.L. Motion of a solid body with cavities containing viscous liquid. Computer Center of A.S. USSR, 1968.

[4] Rabinovich,B,I. On the theory of small oscillations of solid body with cavity partly filled by viscous incompressible liquid. // Appl.Mech., 1968, v.5, Issue 9.

[5] Harlow,F.H., Welch,I.E. Numerical calculation of timedependent incompressible flow of fluid with free surface // Phys. of Fluids, 1965, v.8, N12.

[6] Hirt,C.W. Cook,J.L. Butler,T.D. A lagrangian method for calculation the dynamics in incompressible fluid with free surface // J. of Comput. Phys., 1970, v.5, N2.

[7] Bogoryad,I.B. Dynamics of viscous liquid with free surface. 1980, Tomsk University Press.

[8] Bogoryad,I.B. On a damping coefficient due to a liquid with free surface in moving container. // Appl.Mech., 1990, v.26, N4.

[9] Schlichting,H. Grenzschicht-Theorie. 1964. Verlag G.Brawn Karlsruhe.

[10] Lamb,H. Hydrodynamics. 1932, Cambridge University Press.

International Series of Numerical Mathematics, Vol. 106, © 1992 Birkhäuser Verlag Basel


# EXISTENCE OF THE CLASSIC SOLUTION OF A TWO PHASE MULTIDIMENSIONAL STEFAN PROBLEM ON ANY FINITE TIME INTERVAL

*M.A.Borodin*
*Donetsk State University,*
*Donetsk 340055, UKRAINE*

A method introduced in this paper makes possible to prove the existence of the classic solution in a two phase multidimensional Stefan problem on any finite time interval and to establish the smoothness of free (unknown) boundary.

Key words: two phase Stefan problem, existence of the classic solution, any finite time interval.

This paper is dedicated to studying a two phase multidimensional Stefan problem. There were published many papers on this problem for last years. Resume of them is published in [1]-[3]. We mark some of them [3]-[5]. In those papers the existence of the classic solution on a small time interval was proved. In last years the idea that was proposed by C.Baiocchi [6] became very popular. By using this idea the Stefan problem can be driven variational inequalities in a fixed domain. The existence of the classic solution of single-phase time-dependent and quasi-time-independent Stefan problems was proved just using this method [7]-[9]. However, in case of two or more phases it was possible only to prove the existence of the general solutions by means of variational inequalities.

A method introduced in this paper makes possible to prove the existence of the classical solution of the two phase multidimensional Stefan problem on any finite time interval and to establish the smoothness of free (unknown) boundary. The idea of the method is: firstly, a sequence of elliptic differential-difference approximate problems should be constructed, secondly, uniform estimates should be established, and thirdly, the pass to the limit should be performed. For the first time this idea was used by the author in dedicated to studying the Stefan problem in the case of two space variables [10].

## 1. INTRODUCTION.

Let

$$D = \left\{ x \in R^3 : R_1 < |x| < R_2,\ R_1 > 0 \right\},\ D_T = D\times(0,T),$$

$$B_{R_i} = \left\{ x \in R^3 : |x| < R_i,\ i = 1,\ 2 \right\},$$

$\Omega_0$ is a simply connected domain.

It is required to find a triple $\{u(x,t),\ \Omega_T,\ G_T\}$ which satisfies

$$\Delta u - a(u)\frac{\partial u}{\partial t} = 0 \quad \forall\ (x,t) \in \Omega_T \cup G_T, \tag{1.1}$$

where

$$\Omega_T = \left\{(x,t) \in D_T:\ u(x,t) < 1\right\},\ G_T = \left\{(x,t) \in D_T:\ u(x,t) > 1\right\}$$

on the given boundary

$$\begin{aligned} &u(x,t) = 0 \quad \forall\ (x,t) \in \partial B_{R_1}\times(0,T),\\ &u(x,t) = \varphi(x,t) \quad \forall\ (x,t) \in \partial B_{R_2}\times(0,T) \end{aligned} \tag{1.2}$$

on the unknown (free) boundary

$$\gamma_T = \left( \bar{\Omega}_T \cap D_T \right) \cap \left( \bar{G}_T \cap D_T \right) \tag{1.3}$$

$$u(x,t) = 1,\quad \sum_{i=1}^{3}\left[\frac{\partial u}{\partial x_k}\right]\cos(n,x_k) + \lambda\cos(n,t) = 0.$$

Initial conditions are

$$u(x,0) = \psi(x) \quad \forall\ x \in D,\quad \Omega_0 = \left\{x \in D:\ 0 < \psi(x) < 1\right\},$$

$$\gamma_0 = \left\{x \in D:\ \psi(x) = 1\right\}, \tag{1.4}$$

$$\psi(x) = 0 \ \text{ on }\ \partial B_{R_1},\quad \psi(x) = \varphi(x,0)\ \text{ on }\ \partial B_{R_2}.$$

Here $a(u)$ is a piecewise constant function equal to $a_1$ in $\Omega_T$ and $a_2$ in $G_T$, $\varphi(x,t)$, $\psi(x)$ are given functions, $n$ is a normal to the surface $\gamma_T$, directed towards increasing of $u(x,t)$. $[u_{x_k}(x,t)]$ is the difference between the limit values on $\gamma_T$ from the domain $\Omega_T$ and $G_T$ respectively, $\lambda$, $a_1$, $a_2$, $R_1$, $R_2$ are positive constants.

Denote by $(\rho,\theta_1,\theta_2)$ spherical coordinates of a point

$$x \in D,\ \rho = |x|,\ \theta = (\theta_1,\theta_2) = \arg x,\ -\pi \leqslant \theta_1 \leqslant \pi,\ 0 \leqslant \theta_2 \leqslant \pi,$$

$$\Pi = \left\{(\theta_1,\theta_2):\ -\pi \leqslant \theta_1 \leqslant \pi,\ 0 \leqslant \theta_2 \leqslant \pi\right\}.$$

In spherical coordinates Laplacian may be represented as

$$\Delta\varphi = \rho^{-2}\left(\rho^2\varphi_\rho\right)_\rho + \rho^{-2}A\varphi,$$

$$A\varphi = \left(a_{ij}\varphi_{\theta_j}\right)_{\theta_i},\quad a_{ij} = a_{ji},\quad i,\ j = 1,\ 2.$$

in the second order operator on $\theta_1$, $\theta_2$.

Suppose that the following conditions are satisfied

$$\psi(x) \in C(\bar{D})\cap\left[H^{2+\alpha}(\bar{\Omega}_0)\times H^{2+\alpha}(\bar{G}_0)\right],\quad \alpha \in (0,1),$$

$$\Delta\psi \leqslant 0 \ \text{ in } \ \Omega_0\cup G_0,\quad \psi_\rho > 0 \ \text{ in } \ \bar{D}, \tag{1.5}$$

$$\rho = \omega(\theta_1,\theta_2) = \left\{\rho:\ \psi(x) = 1\right\} \in H^{2+\alpha}(\Pi),$$

$$\sum_{k=1}^{3}\left[\psi_{x_k}\right]\cos(n,x_k) \geqslant 0 \ \text{ on } \ \gamma_0,$$

$$\varphi(x,t) \in H^{2+\alpha,1+\alpha/2}(\bar{D}_T),\quad \frac{\partial\varphi}{\partial t} \leqslant 0 \quad \forall\ (x,t) \in \partial B_{R_2}\times(0,T), \tag{1.6}$$

$$\varphi(x,t) \geqslant 1 + d,\quad d > 0,\quad a(\varphi)\varphi_t - \rho^{-2}A\varphi \geqslant 0 \quad \forall\ (x,t) \in \partial B_{R_2}\times(0,T).$$

## 2. CONSTRUCTION OF THE APPROXIMATE PROBLEMS. PROPERTIES OF THE APPROXIMATE SOLUTIONS.

For any $\varepsilon > 0$ define a function $\chi_\varepsilon(x) \in C^\infty(R^1)$ which satisfies:

$$\chi_\varepsilon(x) = 1 \quad \forall\ x \leqslant 1, \quad \chi_\varepsilon(x) = 0 \quad \forall\ x \geqslant 1 + \varepsilon,$$

$$\chi'_\varepsilon(x) \leqslant 0.$$

Let

$$a_\varepsilon(x) = a_1\chi_\varepsilon(x) + a_2\left[1 - \chi_\varepsilon(x)\right].$$

We cut up the cylinder $D_T$ by planes $t = kh$, $k = 1,2, \ldots, N$, $hN = T$, where $T$ is a positive number. We approximate the function $u(x,t)$ by the functions $\{V_{k,\varepsilon,h}(x)\}$, which are the solutions of the following problem

$$\Delta v_{k,\varepsilon,h} - \frac{1}{h}\int_{v_{k-1},\varepsilon,h}^{v_{k,\varepsilon,h}} a_\varepsilon(\tau)\, d\tau = -\frac{\lambda}{h}\left[\chi_\varepsilon(v_{k,\varepsilon,h}) - \chi_\varepsilon(v_{k-1,\varepsilon,h})\right] \tag{2.1}$$

$$\forall\ x \in D,$$

$$v_{k,\varepsilon,h} = 0 \quad \forall\ x \in \partial B_{R_1}, \quad v_{k,\varepsilon,h} = \varphi(x,kh) = \varphi_k(x) \quad \forall\ x \in \partial B_{R_2}, \tag{2.2}$$

$$v_{0,\varepsilon,h} = \psi(x) \quad \forall\ x \in D, \quad k = 1, 2, \ldots, N. \tag{2.3}$$

Along with the approximate solutions $\{v_{k,\varepsilon,h}(x)\}$ we also consider the functions $\{u_{k,\varepsilon,h}(x)\}$ as the approximations. We define these functions as follows:

$$\Delta u_{k,\varepsilon,h} - \frac{1}{h}\int_{u_{k-1},\varepsilon,h}^{u_{k,\varepsilon,h}} a_\varepsilon(\tau)\, d\tau =$$

$$= -\frac{\lambda}{h}\left[\chi_\varepsilon(u_{k,\varepsilon,h}) - \chi_\varepsilon(\psi)\right] + \frac{1}{h}\int^{F_{k-1,\varepsilon,h}} a_\varepsilon(\tau)\, d\tau \tag{2.4}$$

$$\forall\ x \in D,$$

$$\Delta F_{k,\varepsilon,h} - \frac{1}{h}\int_0^{F_{k,\varepsilon,h}} a_\varepsilon(\tau)\, d\tau = -\frac{\lambda}{h}\left[\chi_\varepsilon(\bar{u}_{k,\varepsilon,h}) - \chi_\varepsilon(\Phi)\right]$$

$$\forall \;\; x \in B_{R_2}, \tag{2.5}$$

where $\bar{u}_{k,\varepsilon,h}(x)$ is equal to $u_{k,\varepsilon,h}(x)$ in $D$ and equal to zero in $B_{R_1}$,

$$u_{k,\varepsilon,h} = 0 \;\; \forall\; x \in \partial B_{R_1}, \quad u_{k,\varepsilon,h} = \varphi_k(x) \;\; \forall\; x \in \partial B_{R_2}, \tag{2.6}$$

$$u_{0,\varepsilon,h} = \psi(x) \;\; \forall\; x \in D, \tag{2.7}$$

$$F_{k,\varepsilon,h}(x) = 0 \;\; \text{on} \;\; \partial B_{R_2}. \tag{2.8}$$

**THEOREM 2.1.** Suppose that the conditions (1.5), (1.6) are satisfied. Then $\forall\; \varepsilon > 0$, $\forall\; h > 0$ the problem (2.1)-(2.3) is solvable

$$v_{k,\varepsilon,h}(x) \in H^{2+\alpha}(\bar{D}), \quad v_{k-1,\varepsilon,h}(x) - v_{k,\varepsilon,h}(x) \geqslant 0 \tag{2.9}$$

$$\forall\; x \in \bar{D}, \quad \forall\; k = 1, 2, \ldots, N,$$

and the following estimate holds

$$h\sum_{k=1}^{N} \int_D \left[|\nabla v_{k,\varepsilon,h}|^2 + \frac{\min(a_1,a_2)}{h^2}\left(v_{k,\varepsilon,h} - v_{k-1,\varepsilon,h}\right)^2\right]dx \leqslant C < +\infty, \tag{2.10}$$

where the constant $C$ does not depend upon $\varepsilon$ or $h$.

**THEOREM 2.2.** Suppose that the conditions (1.5) and (1.6) are satisfied and $a_1 \geqslant a_2$. Then $\forall\; \varepsilon > 0$, $\forall\; h > 0$ the problem (2.4)-(2.8) is solvable

$$u_{k,\varepsilon,h}(x) \in H^{2+\alpha}(\bar{D}), \tag{2.11}$$

$$\forall\; x \in \bar{D}, \quad \forall\; k = 1, 2, \ldots, N \quad 0 \leqslant u_{k-1,\varepsilon,h}(x) - u_{k,\varepsilon,h}(x) \leqslant Ch,$$

$$0 \leqslant F_{k,\varepsilon,h}(x) - F_{k-1,\varepsilon,h}(x) \leqslant Ch, \tag{2.12}$$

where the constant $C$ does not depend upon $x$ or $k$ or $\varepsilon$ or $h$.

The solvability can be proved, for example, by means of the variation method. The maximum principle implies the estimations (2.9), (2.11), (2.12), the estimation (2.10) can be obtained just as it is usually done if Rothe's method is applied.

**THEOREM 2.3.** Let $u_{\varepsilon,h}(x,t)$ be piecewise-linear approximations constructed on $\{u_{k,\varepsilon,h}(x)\}$, $\rho = \omega_{\varepsilon,h,\delta}(\theta,t)$ be an equation of the level surface $u_{\varepsilon,h}(x,t) = 1 + c_0 h^{\delta}$, $c_0 > 0$, $\delta \in (0,1/2)$ be some constants,

$$G_{\varepsilon,h,\delta} = \left\{(x,t) \in D_T\colon u_{\varepsilon,h}(x,t) > 1 + c_0 h^{\delta}\right\},$$

$G_{\varepsilon,h,\delta}(\tau)$ be the section of the domain $G_{\varepsilon,h,\delta}$ by the plane $t = \tau$, the conditions (1.5),(1.6) be satisfied, $a_1 \geqslant a_2$. Then

$$\|u_{\varepsilon,h}(x,t)\|_{H^{2+\alpha}(G_{\varepsilon,h,\delta(t)})} \leqslant C_1 < +\infty \tag{2.13}$$

$$\forall\ (x,t) \in G_{\varepsilon,h,\delta} \quad \frac{\partial u_{\varepsilon,h}}{\partial \rho} \geqslant C_2 > 0, \quad -C_3 \leqslant \frac{\partial u_{\varepsilon,h}}{\partial t} \leqslant 0, \tag{2.14}$$

where $C_1$, $C_2$, $C_3$ are some constants which do not depend upon $\delta$ or $\varepsilon$ or $h$; $C_1$ does not depend upon $t$.

## 3. THE EXISTENCE OF THE CLASSIC SOLUTION.

**THEOREM 3.1.** Let $\lim\limits_{\varepsilon,h\to 0} u_{\varepsilon,h}(x,t) = u(x,t)$ be a weak limit of the sequence of functions $\{u_{\varepsilon,h}(x,t)\}$ in $L_p(D_T)$, $p \geqslant 2$, $\lim\limits_{\varepsilon,h\to 0} v_{\varepsilon,h}(x,t) = v(x,t)$ - be a weak limit of the sequence functions $\{(v_{\varepsilon,h}(x,t)\}$ in $\overset{0}{W}{}_2^{1,1}(D_T)$. Then $v(x,t) = u(x,t)$ and for any $\eta(x,t) \in W_2^{1,1}(D_T)$, $\eta(x,t) = 0$,

$$\int\limits_{D_T} \left[\nabla u \nabla \eta + a(u) u_t \eta - \lambda \chi(u) \eta_t\right] dx\, dt = 0,$$

where $\chi(u)$ is the characteristic function of $\Omega_T$.

The proof follows from the uniqueness of the bounded solution of the Stefan problem.

**THEOREM 3.2.** Suppose that the conditions (1.5), (1.6) are satisfied, $a_1 \geqslant a_2$. Then there exists the unique solution of the problem (1.1)-(1.4)

$$u(x,t) \in C(\bar{D}_T) \cap \left[H^{2+\alpha,\,1+\alpha/2}(\bar{\Omega}_T)\times H^{2+\alpha,\,1+\alpha/2}(\bar{G}_T)\right],$$

$$u_\rho(x,t) > 0, \quad u_t(x,t) \leqslant 0 \quad \text{in} \quad \bar{D}_T,$$

free surface $\gamma_T$ is defined by the equation

$$\rho = \omega(\theta,t) \in H^{2+\alpha,\,1+\alpha/2}(\Pi\times(0,T)).$$

The proof of this theorem follows from the uniform estimates obtained in the theorem 2.3 and from the previous theorem.

**REFERENCES.**

[1] A.Friedman. Variational Principles and Free Boundary Problems. John Wiley, New York, (1982).

[2] I.I.Daniljuk. On the Stefan problem, Uspehi Mat. Nauk, (1985), vol.40, N 5, pp.133-185.

[3] A.M.Meirmanov. The Stefan problem. Novosibirsk, Nauka, (1986).

[4] B.V.Basaliy. The Stefan problem, Dokl.Akad.Nauk Ukrain.SSR, Ser.A, (1986), N 11, pp.3-7.

[5] E.V.Radkevich, A.K.Melikulov. Boundary value problems with free boundary, Tashkent, FAN, (1988).

[6] C.Baiocchi. Sur une probleme a frontiere libre traduisant le filtrage de liquides a traverse des milieus poreux. C.R.Acad. Sci.Paris, Ser.A, (1971), v.273, pp.1215-1217.

[7] A.Friedman, D.Kinderlehrer. A one phase Stefan problem - Indiana Univ.Math.J., (1975), v.25, N 11, pp.1005-1035.

[8] D.Kinderlehler, L.Nirenberg. The smoothness of the free boundary in the one phase Stefan problem - Commun. Pure and Appl.Math., (1978), v.31, N 3, pp.257-283.

[9] M.A.Borodin. Theorem of the existence of the solutions of the one phase quasitime-independent Stefan problem, Dokl.Akad. Nauk Ukrain. SSR, Ser.A, (1976), N 7, pp.582-585.

[10] M.A.Borodin. On the solvability of the two-phase time-dependent Stefan problem, Soviet Math.Dokl., vol.25, !1982), N2.

International Series of Numerical Mathematics, Vol. 106, © 1992 Birkhäuser Verlag Basel 

# ASYMPTOTIC THEORY OF PROPAGATION OF NONSTATIONARY SURFACE AND INTERNAL WAVES OVER UNEVEN BOTTOM.

*S.Yu.Dobrokhotov, P.N.Zhevandrov*
*Institute for Problems in Mechanics*
*Moscow 117526, RUSSIA*

*A.A.Korobkin, I.V.Sturova*
*Lavrentyev Institute of Hydrodynamics*
*Novosibirsk 630090, RUSSIA*

The paper is concerned with the consideration of some problems of wave propagation on the surface of fluid with gradually varying depth. The asymptotics have been obtained for the following problems: wave diffraction by a bottom obstacle, evolution of an initial disturbance for one- and two-layer fluids, for gravity - capillary waves and for fluid layer on an elastic base. All the mentioned asymptotics take into account the focusing effects. Their construction is reduced to solution of certain Hamiltonian systems. The final result is obtained with the help of Maslov's canonical operator. For the simplest cases, this result is in agreement with that obtained by the standard ray method [1].

For the Cauchy-Poisson problem, the asymptotics is obtained with the help of the Pearcey functions. The calculation algorithm consists of two stages: first, all the rays coming to the observation point are found, then, second, the free surface elevation is calculated by means of explicit formulas. As compared to the ray method, the algorithm doesn't require additional data. A new method is proposed for calculating the Pearcey functions. The calculations were carried out for the case of nonstationary wave focusing on a single bottom hill.

Key words: inviscid incompressible fluid, linear theory, surface and internal waves, *3-D* problem, uneven bottom.

Some theoretical and numerical results in asymptotic theory of wave propagation in dispersive weakly inhomogeneous media are presented. First, propagation of unsteady surface waves over uneven bottom is considered.The equations describing the waves on the surface of a liquid layer $-B(x)< y < 0$ in the linear approximation and nondimensional variables have the form:

$$\Phi_{yy} + \varepsilon^2\nabla^2\Phi = 0 \quad (-B<y<0), \quad \Phi_y + \varepsilon^2\nabla B\nabla\Phi = 0 \quad (y=-B) \tag{1}$$

$$\varepsilon^2\Phi_{tt} + \Phi_y = 0 \quad (y=0), \quad \Phi = \varphi_1, \ \varepsilon\Phi_t = \varphi_2 \quad (y=0,\ t=0).$$

Nondimensional variables are introduced by the formulae

$$t = t'\sqrt{gH}, \qquad x = x'/L, \qquad y = y'/H, \qquad B(x) = B'(Lx)/H.$$

Here $\Phi(x,y,t)$ is the velocity potential, $x=(x_1,x_2)$, $H$ is the characteristic linear scale in the vertical direction, $L$ is the linear scale in the horizontal direction. Let the parameter $\varepsilon=H/L$ be small. In this case the bottom relief changes gradually.

The problem (1) is equivalent to the Cauchy problem for the operator equation

$$\varepsilon^2\varphi_{tt} + K\varphi = 0 \quad (t>0); \quad \varphi = \varphi_1, \quad \varepsilon\varphi_t = \varphi_2 \quad (t = 0), \tag{2}$$

where $\varphi(x,t) = \Phi(x,0,t)$, $K$ is the 'normal derivative' operator, $K\varphi(x,t) = \Phi_y(x,0,t)$. When $\varepsilon$ is small, the operator $K$ can be expanded into an $\varepsilon$ - power series. The coefficients of this expansion are $\varepsilon$ - pseudodifferential operators. This procedure was proposed by S.Yu.Dobrokhotov [2]. Asymptotic methods developed by V.P.Maslov for p.d. equations can be used for the Cauchy problem (2). Two helpful theorems have been proved in [3,4].

**THEOREM 1.** The solution of (2) exists and is unique in appropriate function spaces.

**THEOREM 2.** Maslov's method leads to the uniform asymptotics of (2).

Some other problems were considered in the same manner.

**1. Two-layer liquid.**

In this case, $\Phi$ in the problem (2) is a vector-function, $K$ is a matrix operator. Wave field asymptotics of surface and internal waves have been constructed [5]. A suitable choice of unknown functions considerably simplifies the final result. The value of the velocity potential on the free surface and the difference between the potentials on the interface (with coefficients equal to the densities of the liquids) are taken.

## 2. Fluid layer on an elastic base.

This problem is connected with tsunami generation by an earthquake. Additional complications are introduced by the presence of a continuous spectrum of elastic waves. Correctness of the initial boundary value problem has been proved, wave asymptotics have been constructed [6].

## 3. The motion of a submerged sphere over an uneven bottom.

It is proved that a sphere can be replaced by a dipole when the sphere radius is small. Wave field asymptotics have been constructed [7].

The construction of the approximate solution of the problem (1) is considered in detail. Initially, the form of the free surface is given by the equation $y = \eta_0(x)$, the function $\eta_0(x)$ rapidly decays in $|x|$. It is necessary to determine the form of the free surface $y = \eta(x,t)$ for $t > 0$ and $\varepsilon \ll 1$.

During the initial stage of motion $(\varepsilon^{-1}t = O(1))$, the depth can be considered to be uniform. The solution of the problem with uniform depth is well-known. For $t = O(1)$, the influence of the bottom topography cannot be neglected, and the amplitude and the length of the wave are assumed to change gradually when the latter passes over the bottom inhomogeneities; however, the wave preserves its structure. This assumption is the basis of the so-called ray method according to which the velocity potential is sought in the form

$$\Phi(x,y,t) = A(x,y,t)e^{\frac{i}{\varepsilon}S(x,t)} + c.c. + O(\varepsilon), \quad t = O(1).$$

The phase function $S(x,t)$ satisfies the equation

$$S_t^2 = |\nabla S|\, th[\,|\nabla S|B(x)]$$

whose characteristic system has the form

$$\frac{dx}{dt} = -\frac{\partial H}{\partial p}, \quad \frac{dp}{dt} = \frac{\partial H}{\partial x}, \quad \frac{dS}{dt} = S_t - q\,\frac{\partial H}{\partial q}. \tag{3}$$

Here $H(x,q) = \sqrt{g\,th(qB(x))}$ is the Hamiltonian, $q = \sqrt{p_1^2+p_2^2}$, $p = (p_1,p_2)$, $p_1 = \partial S/\partial x_1$, $p_2 = \partial S/\partial x_2$. The initial data are obtained by matching of the ray approximation with the solution for uniform depth as $t \to 0$, and they have the form

$$x = 0,\ p_1 = \xi_1,\ \ p_2 = \xi_2,\ \ S = 0,\ \ S_t = H(0,|\xi|),\ \ |\xi| = \sqrt{\xi_1^2+\xi_2^2}.$$

System (3) defines the mapping $\xi_1,\xi_2 \to x_1(\xi_1,\xi_2,t)$, $x_2(\xi_1,\xi_2,t)$. A connected region on the plane $x_1,x_2$, containing the origin where this mapping is one-to-one is denoted by $\Omega(t)$. Then in $\Omega(t)$, the free surface elevation is described by the formula

$$\eta(x,t) = \frac{1}{2\pi}\left|\eta_0^F(\xi)\right|\cdot\left|\frac{\partial x}{\partial \xi}\right|^{-1/2}\cdot\cos\left[\frac{1}{\varepsilon}S(\xi,t) + \theta(\xi)\right] + O(\varepsilon),$$

where $\eta_0^F(\xi) = |\eta_0^F(\xi)|exp[iQ(\xi)]$, $\xi = (\xi_1,\xi_2) = \xi(x,t)$. The function $\xi(x,t)$ is found via the solution of the boundary value problem for the first two equations of (3) under the following conditions: $x = 0$ for $t = 0$, $x = \tilde{x}$ for $t = \tilde{t}$, in this case $\xi(\tilde{x},\tilde{t}) = p(\tilde{x},\tilde{t},t)$ when $t = 0$. This is the so-called 'aim' problem (AP). In $\Omega(t)$, the AP has a unique solution.

For the remaining part of the plane (denote it by $D(t)$), the AP solution is non-unique, hence, the ray approximation fails. The determinant $J_1 = |\partial(x_1,p_2)/\partial(\xi_1,\xi_2)|$ is supposed to be bounded away from zero everywhere in $D(t)$. Then, according to Maslov's method, the approximate solution has the form

$$\eta(x,t,\varepsilon) = \frac{e^{i\delta}}{8\pi^2}\sqrt{\frac{2\pi}{\varepsilon}}\int_{-\infty}^{\infty}\chi(\Sigma_2^t)\frac{\eta_0^F(\xi)}{\sqrt{|J_1|}}\,e^{\frac{i}{\varepsilon}[\hat{S}(x_1,p_2,t)-x_2p_2]}dp_2+c.c.+\ldots$$

The phase function has several saddle points which can coalesce. These points are found from the AP. Maslov's asymptotics needs further simplification and reduction to some standard expressions. These 'standard' asymptotics are classified by catastrophe theory. When the phase function has three saddle points, the asymptotics of the integral can be obtained with the help of the Pearcey functions. This is the case when the obstacle is a single smooth hill without flattening. Then in $D(t)$

$$\eta(x,t,\varepsilon) = \varepsilon^{-\frac{1}{4}}\left[b_0P_0(\gamma,\beta)+\varepsilon^{\frac{1}{4}}b_1P_1(\gamma,\beta)+\varepsilon^{\frac{1}{2}}b_2P_2(\gamma,\beta)\right]e^{\frac{i}{\varepsilon}a_0}+c.c.+O(\varepsilon^2).$$

Here $\gamma = a_1\varepsilon^{-\frac{1}{2}}$, $\beta = a_2\varepsilon^{-\frac{3}{4}}$, $a_i$, $b_i$ are known functions of $x$, $t$,

$i = 0,\ 1,\ 2,\quad P_n(\gamma,\beta) = \int\limits_{-\infty}^{\infty}\sigma^n exp\left[i(\frac{1}{4}\sigma^4 - \beta\sigma^2+\gamma\sigma)\right]d\sigma$ are the Pearcey functions. To use this formula for calculating the free surface elevation at the point $(\tilde{x}_1,\tilde{x}_2)$ at the moment $\tilde{t}$ $(\tilde{x} \in D(\tilde{t}))$, it is necessary i) to find three solutions of the AP, ii) to calculate $a_i,b_i (i=0,1,2)$ by the explicit formulae, iii) to calculate the values of the Pearcey functions. The main body of calculations is associated with the first stage. To solve the AP, the parameter continuation method has been used. A parameter $\alpha$ is introduced in (3) so that the value of $\alpha = 0$ corresponds to the uniform depth when solution of the AP can be constructed analytically. Moving by steps in $\alpha$, the solulion for $\alpha=1$ (for real topography) can be determined.

To calculate the Pearcey functions, the deformable contour method is proposed. The new contour is chosen so that along it the real part of the phase function keeps a constant value, and the imaginary one is positive. Then the integrand rapidly decreases with respect to the distance from the saddle points, and it is sufficient to carry out the integration over small parts of the contour.

Numerical calculations have been performed for the axisymmetric initial elevation

$$\eta_0(x) = a\ exp[-d|x|^2] \qquad (d = 3)$$

and the bottom profile of the form $(x=x_1,y=x_2)$

$$B(x) = 1-b\ exp\{-[c(x-x_o)^2+ y^2]\},\ (x_o = 7,\ c = 0{,}2,\ b = 0{,}3.)$$

The ray and wave front pucture is shown in Fig.1, the isolines of the free surface elevation $\eta/a$ are shown in Fig.2 for $t = 15$ [8]. The presence of the obstacle leads to the concentration of the rays in the vicinity of the $x$-axis, however, for $t = 15$ the intersection of rays doesn't occur. The caustic form for the given form of the obstacle at $t = 25$ is shown in Fig.3. The zone where three rays intersect is very narrow, and for $x = 24$ its width is equal to unity. The free surface elevation for $x = 24$, $t = 25$ is illustrated in Fig.4. Calculations were made with the step $0.05$. The dashed line shows the asymptotic solution for the even bottom.

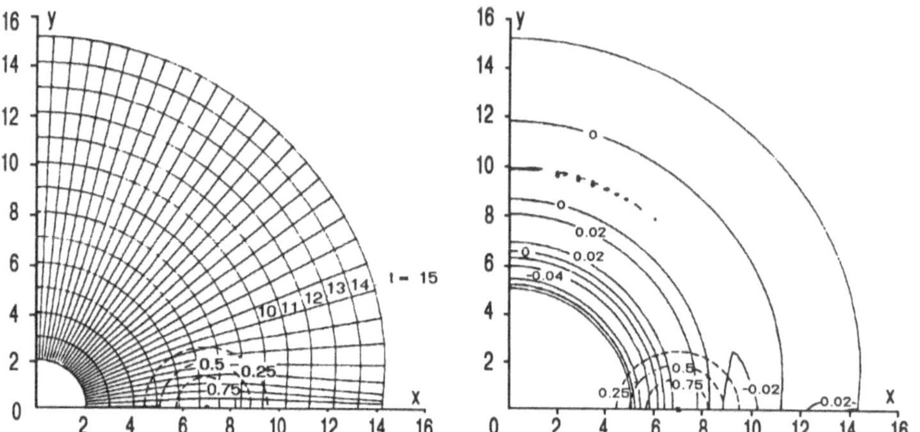


Figure 1. The rays at $|\xi| = 0.05$ and wave fronts for time moments $t = 2,\dots,15$. The dashed curves correspond to the isolines of function $(1+h)/b$.

Figure 2. The isolines of the free surface elevation $\eta/a$ at $t = 15$.

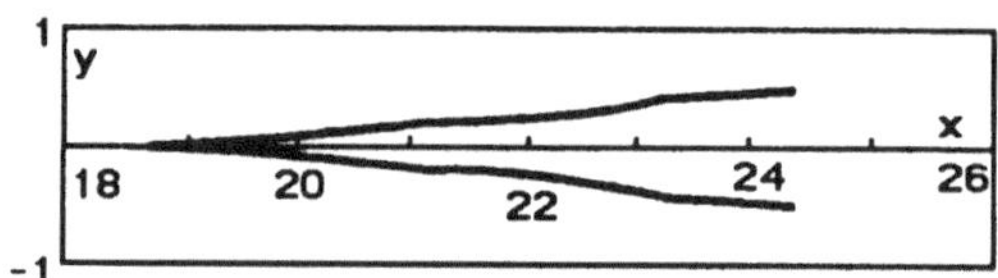


Figure 3. The caustic form at $t = 25$.

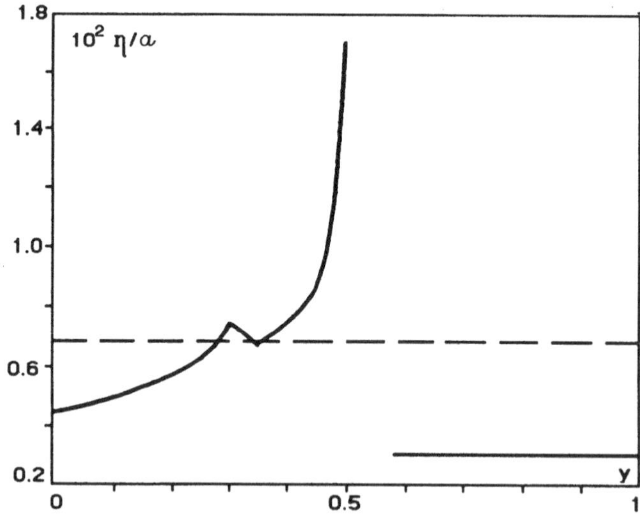


Figure 4. The free surface elevation for $x = 24$, $t = 25$. The dashed line shows the asymptotic solution for the even bottom.

In the vicinity of caustics an abrupt increase in the elevation occurs. Inside this narrow zone the step should be reduced to describe the fine structure of the solution.

It should be noted that this numerical algorithm does not need any additional information for calculations of the free surface elevation between the caustics in comparision with the ray method.

## REFERENCES.

[1] Shen M.C., Keller J.B. Uniform ray theory of surface, internal and acoustic wave propagation in a rotating ocean or atmosphere. SIAM J. Appl. Math, 1975, vol. 28, N 4, pp.857-875.

[2] Dobrokhotov S.Yu. Maslov's methods in the linearized theory of gravity waves on the liquid suface. Dokl. Akad. Nayk, 1983, vol. 269 N 1, pp.76-80 (in Russian).

[3] Dobrokhotov S.Yu., Zhevandrov P.N. Nonstandard characteristics and Maslov operator method in linear problems of unsteady waters waves. Funktsional. Anal. i Prilozhen, 1985, vol. 19, N 4, pp. 43-54, Englisn transl. in Functional Anal. Applic., 1985, vol. 19.

[4] Garipov R.M. On the linear theory of gravity waves: the theorem of existence and uniqueness. Arch. Rat. Mech. Anal, 1967, vol. 24, N 5, pp. 352-362.

[5] Zhevandrov P.N., Isakov R.V. Cauchy-Poisson problem for a two-layered liquid of variable-depth. Mat. zametki, 1990, vol. 47, N 6, pp. 31-44, Englisn transl. in Math Notes, 1990, vol. 47.

[6] Dobrokhotov S.Yu., Tolstova O.L. Application of conservation laws to asymptotic problems for equations with an operator - valued symbol. Mat. zametki, 1990, vol. 47, N 5, pp.148-151(in Russian).

[7] Dobrokhotov S.Yu., Zhevandrov P.N. Surface waves on a liquid of variable depth generated by a moving submerged source. Oscillations and waves in fluids, Gorky, 1988, pp. 32-41 (in Russian).

[8] Korobkin A.A., Sturova I.V. Three-dimensional Cauchy-Poisson problem for a basin with gradually varying depth. Dinamika sploshnoi sredy, 1990, vyp.97, pp.37-47 (in Russian).

International Series of Numerical Mathematics, Vol. 106, © 1992 Birkhäuser Verlag Basel 

# MULTIPARAMETRIC PROBLEMS OF TWO-DIMENSIONAL FREE-BOUNDARY SEEPAGE

*V.N.Emikh*
*Lavrentyev Institute of Hydrodynamics*
*Novosibirsk 630090, RUSSIA.*

The exact solutions were obtained by conformal mappings for several miscellaneous boudary value problems describing two-dimensional steady seepage flows with initially unknown free boundaries. The analytical relations are utilized then to perfome a detailed theoretical analysis of the structure and behaviour of seepage procces under consideration. A special attention is paid to study the proper free boudary seepage problems. In such a way the multivaluedness of their physical significance in the same analytical representation of the solution is established.

Key words: drained soil, conformal mapping, elliptic integral, mapping parameters.

## INTRODUCTION.

A mathematical modelling of two-dimensional steady seepage flows with initially unknown boundary sections reduces to solving miscellaneous boundary value problems of the analytical function theory. In some cases the conformal mapping methods can be used to obtain the effective representation of these problems solutions. The main difficulty on this way consists in the presence of the unknown parameters of conformal mapping in obtained analytical relations. If the flow hydrodynamical characteristics are to be determined in the so-called direct statement, then all the above-mentioned parameters must be precomputed on the basis of the system of equations resulting after assignment of set of governing physical parameters of flow. Realization of such approach as a rule is very laborious.

An inverse order of computations - with partially or completely given mapping parameters - appears to be much simple. In this case, however, some or, respectively, all physical parameters are resulted in calculation process.

Both described ways starting from exact complex potential -

complex coordinate relations are illustrated in this paper by the two particular problems of ground water motion though porous media.

## 1. THE SEEPAGE IN DRAINED SOIL LAYER WITH IMPERVIOUS BED.

Let us consider two-dimensional (in vertical plane) unconfined steady seepage from equidistant flooded stripwise plots on the soil surface, equal in their width, towards equal-discharge drain (i.e.point sinks) which are layed half way between the flooded plots. It is assumed further that the soil layer in which the flow occurs, is underlaid by impervious ground. This scheme can be applied to investigation of problem of salinized soils washing.

Due to periodicity of flow, one can study it within a half of drain space presented in Fig.1. The representation of solution for this case was obtained [2] by conformal mapping of the complex potential region $\omega = \varphi + i\psi$ ( $\varphi$ is the filtration velocity potential, $\psi$ is the flow function) and of the two-sheet velocity hodograph region $1/w$ ( $w = w_x - iw_y$ is the complex velocity of filtration, $z = x + iy$ is the complex coordinate of the filtration region points) to the half- plane $Im\ \zeta \geqslant 0$ (Fig.2) and can be represented as

$$z(\zeta) = z(\zeta_o) + \frac{\omega(\zeta)-\omega(\zeta_o)}{w(\zeta_o)} + M\int_{\zeta_o}^{\zeta} \frac{P(u)[\omega(u)-\omega(\zeta)]}{\sqrt{u^3(g-u)(d-u)^3}}\,du \qquad (1)$$

$$\omega(\zeta) = \frac{2Q}{\pi}\,arsh\sqrt{\frac{(1+b)(g-\zeta)}{(1-g)(b+\zeta)}} + iQ \qquad (0 \leqslant \zeta \leqslant g\ ) \qquad (2)$$

$$\frac{1}{w(\zeta)} = \frac{dz}{d\omega} = \frac{M}{2}\int_{\zeta}^{g} \frac{P(u)du}{\sqrt{u^3(g-u)(d-u)^3}} + i = -M\,\frac{(f+\zeta)\sqrt{g-\zeta}}{\sqrt{\zeta(d-\zeta)}} -$$

$$-\frac{2}{\pi}K\,Z(\alpha,k) + i \qquad (\ 0 \leqslant \zeta \leqslant g\ ) \qquad (3)$$

In these equations the following notations are used:

$$P(u)= u^3 - (2d+f+\alpha E)u^2 + (d+2f+\alpha\Phi)gu - dfg\ , \qquad \alpha = \frac{2\sqrt{d}}{\pi M}\ ,$$

$$\Phi = \int_0^{\pi/2} \frac{\cos^2\vartheta \, d\vartheta}{\sqrt{1-k^2\sin^2\vartheta}}, \quad k=\sqrt{\frac{g}{d}}, \quad \alpha = \arcsin\sqrt{\frac{1-\zeta/g}{1-\zeta/d}}, \quad M > 0$$

Further $2Q$ denote seepage discharge of drain per unit length of it, $K$ and $E$ are complete elliptic integrals of first and second order, respectively, $Z$ - Jacobian *Zeta Function* [1], $b$, $d$, $g$, $f$ - unknown parameters of conformal mapping (s.Fig.2), $\zeta_0$-some fixed value of $\zeta$.The scaled values $z$ and $\omega$ are used which are associated with physical values $z_{ph}$ and $\omega_{ph}$ as $z = z_{ph}/T$, $\omega = \omega_{ph}/\text{æ}T$ while $T$ is the soil layer thickness, æ is the filtration coefficient.

The described exact solution is tedious for engineering computations. More convenient relationships are provided by the approximate flow model which can be obtained from the above-stated model with replacement of unknown depression curve $AG$ by the fixed boundary section $AOG$. A comparison between two models was accomplished by means of calculations [5]. For the exact statement they are conducted in reverse order - with given mapping parameters $b$, $d$ and $g$.

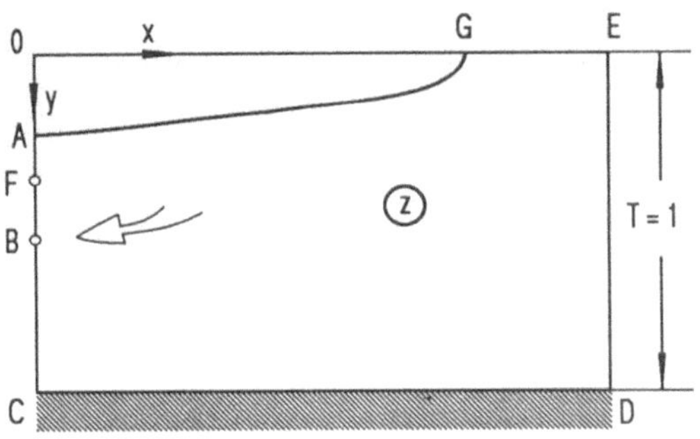

Fig. 1

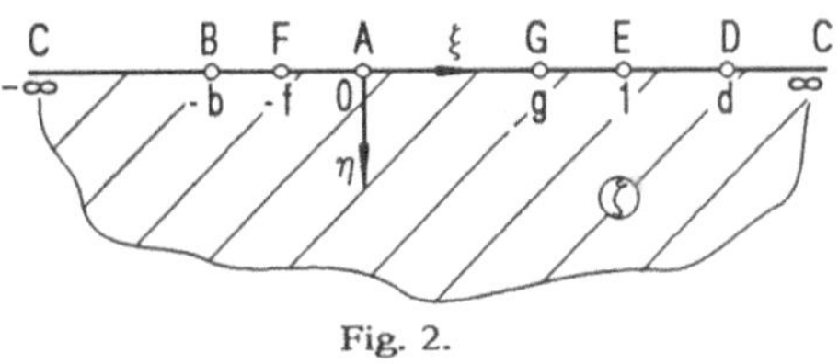

Fig. 2.

One can treat the relationship (1) for $\zeta \in (0,g)$ together with equations (2) and (3) as a complex- parametrical free surface equation and calculate from it the $x,y$-coordinates of points of this initially unknown boundary section.

Let us assume now that the drained water flow in soil is underlaid by resting salt waters and by that the so-called fresh waters fringe is formed.Then, in addition to the free surface, the second unknown boundary appears, i.e. interface between fresh and salt waters.The analytical solution presented above by formulae

(1) - (3), can be extended to this flow model [4]. Investigation of it in direct physical statement is also very complex and at the same time exceedingly interesting and important from hydrodynamical point of view.

## 2. HIGH-PERMEABLE ARTESIAN AQUIFER UNDER SOIL LAYER.

In many cases immediately below drained soil layer, the coarse sand or gravel is deposited containing ground water under pressure. The latter predetermines a constant hydraulic head value along soil bed whereas the seepage flow in the soil layer is subdivided into several difference directed currents. One possible variant of such branching is showed schematically on Fig.3.

For the situation in question, the solution representation of the corresponding boundary value problem was obtained [6] by conformal mapping of the $\omega$-function region and of the region of Zukovsky function $\theta = \omega + iz$ to the half-plane $Im\ \zeta \geqslant 0$ (Fig.4) and can be represented in terms of the scaled values $z$ and $\omega$ as

$$z = i\frac{\Omega-\theta}{1-\varepsilon}, \quad \omega = \frac{\Omega-\varepsilon\theta}{1-\varepsilon}$$

$$\Omega = \frac{(1+b)\sqrt{d-g}}{b+r} M \int_{\zeta}^{g} \frac{(r-u)du}{(b+u)\sqrt{(d-u)(1-u)(g-u)}} + iQ_s =$$

$$= M\left[\frac{1-g}{b+g}\Pi\left(\arcsin\sqrt{\frac{g-\zeta}{1-\zeta}}, \frac{1+b}{b+g}, k'\right) + \right.$$

$$\left. + \frac{r-1}{b+r}F\left(\arcsin\sqrt{\frac{g-\zeta}{1-\zeta}}, k'\right)\right] + iQ_s \quad (-b \leqslant \zeta \leqslant g)$$

$$\left.\begin{aligned}
&\theta = i\frac{(1+b)\sqrt{d}}{b+r} M_o \int_0^{\zeta} \frac{(f+u)du}{(b+u)\sqrt{(d-u)(1-u)u}} + iQ_s = \\
&= M_o\left[\frac{1+b}{b-f} F(\arcsin\sqrt{\zeta}, k_o) - \frac{1+b}{b}\Pi(\arcsin\sqrt{\zeta}, -\frac{1}{b}, k_o\right] + \\
&+iQ_s \quad (0 \leqslant \zeta \leqslant 1) \\
&M = \frac{2Q}{\pi}\sqrt{\frac{(b+g)(b+d)}{(I+b)(d-g)}},\ k' = \sqrt{1-k^2},\ k = \sqrt{\frac{1-g}{d-g}}, \\
&M_o, k_o = (M, k)_{g=0}
\end{aligned}\right\} \quad (4)$$

Here $Q$ and $Q_s$ are seepage discharges of flows to drain $B$ and from canal $GE$, respectively, within the selected flow region (s.Fig.3), $\varepsilon$ is the coefficient of infiltration on the free surface (or one of evaporation from it), $F$ and $\Pi$ are commonly used (s.[1]) symbols of incomplete elliptic integrals of first and, respectively, third kind in Legendre's notation.

With given drain space $2L$, width $2(L-l)$ of canal, depth $\beta$ of point-sink $B$ presenting a drain, its discharge $Q$, the thickness $T$ of soil layer ($T = 1$ after scaling) and hydraulic head on its bed, one can obtain a system of equations with respect to five unknown mapping parameters $b,d,f,g,r$. They were found by means of special computer program. Further such computational algorithm incorporates the calculation of the drain discharge $Q$ by giving hydraulic head value $H_d$ at some point $B_1$ on the drain surface.

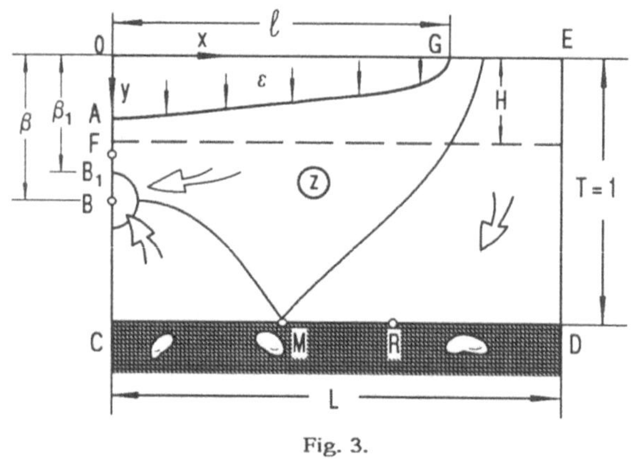

Fig. 3.

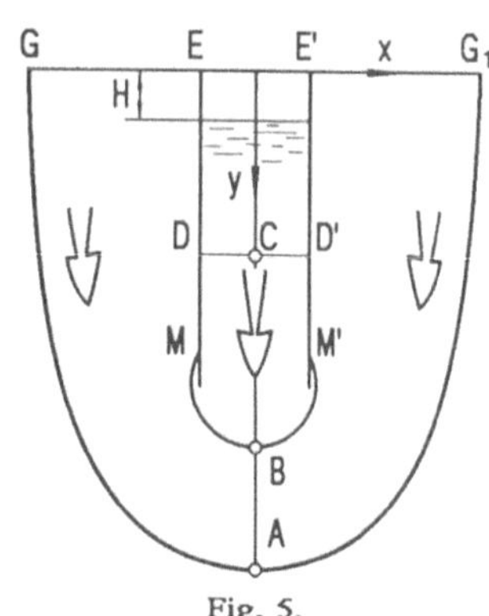

Fig. 5.

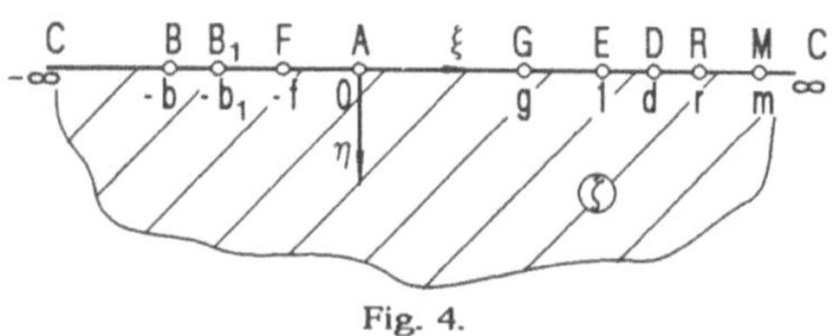

Fig. 4.

Thus for the problem at issue the direct computational scheme was realized. Some flow characteristics other then $Q$, such as coordinates of set of the free surface points, location of the boundary point which separates oppositely flowing seepage currents and discharge values of each partial flows are calculated.

In the course of analysis of mapping parameters finding problem, some limitations are detected on its values. These limitations are concerned with certain restrictions on the governing physical parameters of model, particularly on drainage and infiltration intensity. Otherwise speaking, only under such restrictions the given seepage flow is possible.

Beyond the scope of above-mentioned limitations corresponding to critical regimes of flows, the relations (4) describe just the other hydrodynamical models. One of the latter was treated on the basis of calculations in the paper [6]. It is shown schematically here in Fig.5. One can interpret it as a seepage inflow to drain-sink $B$ from the above-located trench which corresponds to the high-permeable aquifer in the starting model (s.Fig.3), and partly from soil surface flooded on certain width.

Immediately from this scheme but without the trench, one more useful scheme can be obtained, i.e. downcast ground water flow which is partly or completely catched by the vacuum drain. This flow is studied in the paper [7].

A set of similar flows transformations for one specific free-boundary problem is described in the paper [3].

All the above models and some other ones of two-dimensional free boundary seepage flows are described in detail in the monograph [8] which is now in the press.

## REFERENCES.

[1] Byrd P.F., Friedman M.D. Handbook of elliptic integrals for engineers and scientists.- Berlin, Heidelberg, New-York: Springer-Verlag, 1971, 358 p.

[2] Emikh V.N. The solution of problem of the plane steady seepage in drainage washings of soils with impervious bed// Dokl AN USSR, 1975, v.220, N 6, pp. 1289 - 1292 (in Russian).

[3] Emikh V.N. On several hydrodynamical models of drainage// PMM, 1979, v.50, N 1, pp. 168 - 173 (in Russian).

[4] Emikh V.N. Hydrodynamical model of drainage in fresh waters fringe above salt ones// Dokl. AN USSR, 1980, v.252, N 4, pp. 825 - 829 (in Russian).

[5] Emikh V.N. The comparison of approximate and exact models of seepage in drainage washings of soils with impervious bed// Izv. AN SSSR, MJG, 1982, N 3, pp. 168 - 173 (in Russian).

[6]. Emikh V.N. The analysis of two-dimensional steady seepage in a soil layer with high-permeable bed// PMM, 1982, v.46, N 5, pp. 857 - 868 (in Russian).

[7] Emikh V.N. Hydrodynamical model of vacuum intercepting drainage in the downcast ground water flow// PMM,1987, v.50, N.1, pp. 168 - 172 (in Russian).

[8] Emikh V.N. Hydrodynamics of seepage flows with drainage (monograph in Russian; assigned for publication in 1992 y.).

International Series of Numerical Mathematics, Vol. 106, © 1992 Birkhäuser Verlag Basel 

# NONISOTHERMAL TWO-PHASE FILTRATION IN POROUS MEDIA

*R.E. Ewing*
*University of Wyoming, USA*
*P.O. Box 3036, Laramie, Wyoming 82071-3036*

*V.N. Monakhov*
*Lavrentyev Institute of Hydrodynamics,*
*Novosibirsk 630090, RUSSIA*

The paper deals with a mathematical model (MLT model) proposed in [1], [2], which takes into account the effect of heat processes on fluid motion character via the change of viscosity and capillary properties of different inhomogeneous fluid components depending on their temperature and that of porous medium.

The peculiarity of the MLT model is that all the components of its equation, except the Darcy and Laplace laws, are a consequence of the conservation laws in continuum mechanics. In particular, the motion of contact boundaries of inhomogeneous fluid with its immovable components (the Stefan type problem) are described within the framework of this model.

The aim of this paper is to determine the dependence of smoothness of the solutions to the initial boundary value problems for the MLT model (MLT problems) on the coefficients of equations and boundary conditions. The results obtained are applied then, as in [3], [4], to prove convergence of the iterative method for solving the MLT problem and to find velocity of this convergence.

## 1. STATEMENT OF MLT PROBLEM.

A mathematical model of nonisothermal filtration comprises the equation of the equilibrium temperature $\theta$ and the Musket-Leverette model which consists of the Darcy laws connecting the phase velocities $\vec{v}_k$ and $\nabla p_k$, the Laplace law

$$p_0 - p_2 = p_c(x,\theta,s)$$

($p_c$ is the capillary pressure, $(p_0,p_2)$ are the phase pressures) and the phase continuity laws.

Introducing the average pressure

$$p = p_0 - \int_0^s b_2(\theta,s)\, p_{cs}\, ds$$

and making the corresponding transformations of the Musket-Leverette equations result in the following MLT model [1-3]:

$$\begin{cases} m_k u_{kt} + div\, \vec{v}_k = 0, \quad k = 1,\, 2,\, 3, \\ -\vec{v}_k = A_k\, \nabla u_k + B_k \nabla\theta - b_k \vec{v} + \vec{f}_k \end{cases} \tag{1}$$

Here $u_1 = \theta(x,t)$ is the equilibrium temperature of liquid and porous medium; $u_2(x,t) = (s_2 - s_2^0)(1 - s_0^0 - s_2^0)^{-1}$ and $s_2(x,t)$ denote the dynamic and total saturation of a wetting phase; $s_k^0 = const \in (0,1)$, $s_0^0 + s_2^0 < 1$ stand for residual phase saturation; $u_3(x,t) = p$ is the average mixture pressure; $\vec{v}_2$ and $\vec{v}_3 = \vec{v}$ are the filtration velocities of the phase and mixture; $m_2=m_2(x)$ is the efficient porosity, $m_1 = 1$, $m_3 = b_3 = B_1 = \vec{f}_1 = 0$, $b_1=u_1=\theta$, $b_2 = b_2(u_1,u_2)$. The properties of tensors $A_k$ and $B_k$, vectors $\vec{f}_k$, $k = 1, 2, 3$, functions $m_2(x)$, and $b_2(u_1,u_2)$ will be described below.

Instead of $u_2$, we sometimes use the function $u_2^* = m_2 u_2$, to which equation (1) ($k = 2$) with the coefficients $m_2=1$, $A_2^*=m_2A_2$, $B_2$, $-b_2$ and $\vec{f}_2^* = \vec{f}_2 + A_2 u_2 \nabla(m_2^{-1})$ that possess the properties of the initial equation coefficients, corresponds.

Let us take that $Q = \Omega\times[0,T]$, $\Omega \subset R^3$ is the restricted region, $\partial\Omega = \bigcup_1^3 \Gamma^k$, $\Sigma^k = \Gamma^k\times[0,T]$, $\Gamma^1$ and $\Gamma^2$ simulate the regions of pumping, take-off and contact with homogeneous immovable liquid, $\Gamma^3$ corresponds to the contact with impermeable exterior medium.

The initial boundary value problem for $\vec{u}= (u_1,u_2,u_3)$ has the form

$$\begin{cases} m_i u_i = m_i u_i^0, \quad (x,t) \in \Sigma^0 \cup \Sigma^1; \quad \vec{v}_i \vec{n} = \beta_i \psi^0, \quad (x,t) \in \Sigma^2, \\ A_1 \nabla\theta\vec{n} = \beta_0(\theta^0 - \theta), \quad \vec{v}_k \vec{n} = 0, \quad (x,t) \in \Sigma^3, \end{cases} \tag{2}$$

where $i = 1, 2, 3$; $k = 1, 2$; $\beta_1 = -b_1$, $\beta_2 = b_2$, $\beta_3 = 1$; $\Sigma^0 = \{t = 0, x \in \Omega\}$. If $\Gamma^1 = \emptyset$, the conservation law of mixture mass in $\Omega$ leads to conditions

$$\int_\Omega u_3(x,t)\,dx = \int_\Omega \psi^0(x,t)\,dx = 0, \quad t \in [0,T]. \tag{3}$$

The problem (1), (2) is assumed to be called the MLT problem.

## 2. REGULAR MLT POBLEM.

Here the following notations of the norms

$$\|u\|_{q,\Omega} = \|u\|_{L_q(\Omega)}, \quad \|u\|_{q,r,Q} = \|u\|_{L_{q,r}(Q)},$$

$$\|u\|_{V^2(Q)} = \|u\|_{2,\infty,Q} + \|\nabla_x u\|_{2,Q}$$

and spaces $L_q(\Omega)$, $L_{q,r}(Q)$, $V_2(Q)$, $W_q^1(\Omega)$, $W_q^{1,0}(Q)$ are used which were adopted in [3]. Suppose that

$$|u|_\Omega^{(l)} = \|u\|_{C^l(\Omega)}, \quad \alpha = (l - [l]) \in [0,1),$$

$$|u|_Q^{(l,l/2)} = \|u\|_{C^{l,l/2}(Q)}.$$

Introduce the set $\Phi(x,u_1,u_2) = (m_2,\Phi_1,\Phi_2,\Phi_3)$ of the coefficients in system (1); $\Phi_k(x,u_1,u_2) = (A_k,B_k,-b_k,\vec{f}_k)$, $k = 1, 2, 3$ is the set of the coefficients of elliptic operator $(-div\,\vec{v}_k)$; and $\varphi^0(x,t,u_1,u_2) = (\vec{u}^0,\psi^0)$, $(x,t) \in \bar{Q}$ is the set of values of the initial and boundary data, which satisfy conditions [1], [3]:

$$(i)\ |\Phi|_G^{(0)} \leqslant M_0, \quad M_0^{-1} \leqslant \left[m_2,\beta_0,(\bar{A}_k\xi,\xi)\right] \leqslant M_0^{-1}, \quad |\xi| = 1,$$

$$G = \Omega\times(\theta_*,\theta^*)\times(0,1); \quad \bar{A}_k = A_k, \quad k = 1, 3, \quad \bar{A}_2 = a^{-1}(u_2)A_2,$$

$$0 < a(u_2) \leqslant M_0, \quad u_2 \in (0,1); \ (a,A_2,B_2.\vec{f}_2)\Big|_{u_2=0,1} = 0,$$

$$b_2(0) = 0;$$

$(ii)$ $\varphi^0 = (\vec{u}^0,\psi^0)$ for $(x,t) \in \bar{Q}$ possesses the properties

$$0 < \delta_0 \leqslant u_2^0(x,t) \leqslant 1 - \delta_0; \quad 0 < \theta_* \leqslant u_1^0(x,t) \leqslant \theta^* < \infty;$$

$$\left(\|u^0_{1t},u^0_{2t}\|_{1,Q},\|\varphi^0,\varphi^0_x\|_{2,Q},\|\psi^0\|_{2,\Sigma^2}\right) \leqslant M_1.$$

For validity of an analogue of the classic principle of maximum for saturation $u_2(x,t)$

$$0 < \delta \leqslant u_2(x,t) \leqslant 1 - \delta,\ \delta \in (0,1),\quad (x,t) \in \bar{Q}, \tag{4}$$

which provides uniform parabolicity of equation (1) at $k = 2$

$$|ln\ (A_2\xi,\xi)| \leqslant M_2,\quad (x,u_1,u_2) \in \bar{G} \tag{5}$$

it is required, by an analogy with [3], to satisfy conditions

$$\vec{J}_2\vec{n}\ \Big|_\Sigma = 0,\quad \frac{\partial}{\partial\theta}(b_2,\vec{F}_2) = div_x\vec{F}_2 = 0 \quad \forall\ u_2 \notin (\delta,1-\delta) \tag{5*}$$

where $\Sigma = \Sigma^1 \cup \Sigma^2$, $\vec{F}_2 = B_2\nabla\theta + \vec{J}_2$, $\delta > 0$ is the small constant. In a physical sense ($5^*$) means that in filtration of mostly one liquid ($u_2 \leqslant \delta$, $(1-u_2) \leqslant \delta$) the equilibrium temperature $u_1 = \theta$ should be close to that of moving liquid and porous medium saturated by this liquid must be homogeneous.

*Definition.* The vector $\vec{u} = (u_1,u_2,u_3)$ is called the regular solution of the MLT problem if and only if almost everywhere in $\bar{Q}$ $u_1(x,t) \in [\theta_*,\theta^*]$, $u_2(x,t) \in [\delta,1-\delta]$, thus (5) is fulfilled; $(u_1,u_2) \in V_2(Q)$, $u_3 \in W_2^{1,0}(Q)$ and $(u_1,u_2,u_3)$ satisfy the standard integral identities, and on $\Sigma^1$, $\Sigma^2$ expression (2) is satisfied [1], [3].

In [1], the existence of weak solution (without (5)) of the MLT problem has been proved.

## 3. ESTIMATION OF SMOOTHNESS OF REGULAR SOLUTIONS.

**THEOREM 1.** Let us fulfil the assumptions *(i)*, *(ii)*, (5) and

$$|\Phi|_G^{(l)} \leqslant M_3,\ [l] \geqslant 2,\ \alpha = l-[l] > 0;\ |B_3|_G^{(0)} \leqslant \varepsilon, \tag{6}$$

where $\varepsilon > 0$ is the small number. Then, there exists such $\alpha_0 > 0$ that

$$|\vec{u}|^{(l_0,l_0/2)} \leqslant N(Q'),\ l_0 = [l] + \alpha_0,\ \bar{Q}' \subset Q, \tag{7}$$

in this case $Q' = Q$ if additionally we have

$$(iii)\ \partial\Omega \in H_*^{l_1} \quad [3],\ |\varphi^0|^{(l_1, l_1/2)} \leqslant M_4, \quad l_1 = l + 1$$

and the consistency conditions are valid up to the order $[l/2]$.

**Proof.** In conditions $(iii)$, a standard continuation of $\vec{u}$ into a more wide domain $\Omega^* \supset \Omega$ is used. The estimate (7) is valid for $Q' = Q$. Let us restrict ourselves to the case $Q' \subset Q$.

In eqs. (1) for $u_3(x,t)$ the term $B_3\nabla\theta$ is substituted by $B_3\nabla\theta_h$ ($\theta_h$ is the Steklov's averaging $\theta(x,t) = u_1$), the values of $h > 0$ and $N > 0$ are fixed, $\varepsilon > 0$ in (6) is chosen so that

$$\varepsilon\left[\|\theta_h\|_Q^{(2,0)} + \|\theta_{hxxx}\|_{2,Q}\right] \leqslant N. \tag{8}$$

The solution of the problem obtained is, as previously, denoted by $\vec{u} = (u_1, u_2, u_3)$. By virtue of limitedness of $\vec{f}_0 = B_3\nabla\theta_h + \vec{f}_3$, from eq.(1) ($k = 3$) we find $u_3 \in C^{\alpha_0}(\Omega') \cap W^1_{q_0}(\Omega')$, $\alpha_0 > 0$, $q_0 > 2$. Using now equation (1), $k = 1$, we conclude that $u_1 \in C^{\alpha_0}(Q')$ (theorem 8.1 [3]).

Thus, we have

$$|u_k|_{Q'}^{(\alpha_0)} \leqslant N(Q'), \quad \alpha_0 > 0, \quad k = 1, 3. \tag{9}$$

Represent (1) as

$$-m_k u_{kt} + div(A_k \nabla u_k) = h_k, \quad k = 1, 2, 3, \tag{10}$$

where $h_k = \sum_{i,j} q^k_{ij} u_{ix} u_{jx} - \gamma_k \Delta\theta \quad (\gamma_1 = 0)$ is the smooth finite function with a carrier $\Omega_\rho$, $|\Omega_\rho| = \rho$. With the allowance for (9) and consideration of the integral identities corresponding to (10), with applying the multiplicative inequalities [3] to productions $(u_{ix}, u_{jx})$ entering $h_k$, we find

$$J \leqslant C \sum_{k=1}^{3} \left[\varepsilon_k \delta^{-1} \rho^{2\alpha} + \delta\right] \|u_{kxx}\xi^2\|^2_{2,Q} + c_0(\rho),$$
$$J = \sum_1^3 J_k(u_k), \quad J_k(u_k) = \|u_{kxx}\xi^2\|^2_{2,Q} + m_k \|u_k \xi^2\|^2_{V(Q)}, \tag{11}$$
$$\|f\|^2_{V(Q)} = \|f_x\|^2_{2,Q} + \|f_t\|^2_{2,Q}, \ \varepsilon_2 = 0, \quad \varepsilon_1 = \varepsilon_3 = 1.$$

Choosing properly $\rho > 0$ and $\delta > 0$, and covering $Q'$ by a finite number of the regions $Q_\rho$, we find from (11)

$$\|\vec{u}_{xx}\|^2_{2,Q'} + \sum_{j=1}^{2} \|u_j\|^2_{V(Q')} \leqslant N(Q'). \tag{12}$$

From (12) it follows that (9) is valid when $k = 2$. Taking now equation (1), $k = 3$, where $A_3 \in C^\alpha(Q')$ and $|\vec{f}_0| \leqslant M$ $(\vec{f}_0 = B_3\nabla\theta_h + \vec{f}_3)$, we'll obtain $u_{3x} \in L_{q,\infty}(Q')$ $\forall\ q \in (1,\infty)$ (theorem 4.2), from where, using (1), we subsequently find $(u_1,u_2) \in L_q(Q')$ for $k = 1$, $k = 2$ ( theorem 5.3 [3]), i.e.

$$\|\vec{u}_x\|_{q,Q'} \leqslant N(q,Q') \quad \forall\ q \in (1,\infty). \tag{13}$$

Let us differentiate equations (10) for the functions $(u_1,u_2^*,u_3) = \vec{u}$, $u_2^* = m_2u_2$ with the variable $x_i$, $i = 1,\ 2,\ 3$, taking that $f^i = f_{x_i}$ (the sign "*" is omitted). Then the values of $u_k^i$ satisfy the same equations (10) $(m_1 = m_2 = 1)$ with the right-hand sides

$$h_{ix} = div\ \vec{G}_{ik} + F_i(\gamma_k),\quad \vec{G}_{i1} = (\theta\vec{f}_0)^i + \lambda^i\nabla\theta,\quad \vec{G}_{i3} = \vec{f}_0^i,$$

$$\vec{G}_{i2} = (b_2\vec{f}_0 + B_2\nabla\theta + \vec{f}_2)^i,\ F_i(\gamma_k) = \left[div(\gamma_k B_3\nabla u_3)\right]^i;\ \gamma_1 = \theta,$$

$$\gamma_2 = b_2,\ \gamma_3 = 0.$$

For the functionals $J_k(u_k^i)$ in (11), obtain

$$\begin{aligned} &J_k(u_{kx}) \leqslant c_k\|h_k\xi^2\|^2_{2,Q} + c_k^0(Q'), \\ &\|h_k\xi^2\| = \sup_i \|h_{ik}\xi^2\|. \end{aligned} \tag{14}$$

Consider the function entering $h_k$:

$$R_i(\gamma) = \left[div\ \gamma\vec{f}_0\right]^i,\quad \gamma = u_1,\ b_2,\ 1.$$

Let us assume that $div\ \vec{f} \equiv f_x$, $g_{x_i} \equiv g_x$ then

$$R_i(\gamma) = (\gamma f_2)_{xx} + \gamma\left[B_{3xx}\theta_{hx} + 2B_{3x}\theta_{hxx} + B_3\theta_{hxxx}\right].$$

By virtue of estimates (12), (13), we have $(\gamma f_2)_{xx} \in L_2(Q')$ and, in addition, $(B_3\theta_{hxxx}) \in L_\infty(Q')$ according to (8). With this allowed for, for $R_i(\gamma)$ and, hence, for $(h_k\xi^2)$, we have

$$\|h_k\xi^2\|^2_{2,Q} \leqslant \delta c_1\|\vec{u}_{xxx}\xi^2\|^2_{2,Q} + c_0(Q',\delta).$$

Summing up the obtained estimates with respect to "$k$" and choosing small $\delta > 0$ we obtain

$$\|\vec{u}_{xxx}\|^2_{2,Q} + \sum_{j=1}^{2}\|u_j\|^2_{V(Q')} \leqslant N(Q'). \tag{15}$$

From (15), it follows that $\vec{u}_x \in C^{\alpha_0}(\Omega')$, $\alpha_0 > 0$. Then, turning subsequently to eqs. (1) for $u_1$, $u_3$ and after that for $u_2$, we obtain estimates (7), when $l_0 = 2 + \alpha_0$, $\alpha_0 > 0$, which allow the limiting transition over the averaging parameter $h$. The further increase in smoothness $\vec{u}$ $([l] > 2)$ is established in a standard way by differentiation of eqs.(1). The theorem has been proved.

## 4. THE APPROXIMATION MLT(AMLT) PROBLEM.

$$\begin{cases} m_k u^{i+1}_{kt} = div(\Phi^i_k\nabla_k u^{i+1}), \quad k = 1,\ 2; \\ -\nabla\cdot\vec{v}^{i+1} = div\left[A_3\nabla u^{i+1}_3 + B^i_3\nabla\theta^i + \vec{f}^i_3\right] = 0, \end{cases} \tag{16}$$

the values of $\vec{u}^{i+1}$ satisfying conditions (2) $(\Sigma^3 = \emptyset)$ on $\partial\Omega\times[0,T]$. Here $\Phi^i_k = \Phi_k(x,u^i_1,u^i_2)$ and along with the set $\Phi_k = (A_k,B_k,-b_k,\vec{f}_k)$ of the coefficients of elliptic operator $(-div\ \vec{v}_k)$ for $u_k$ the expression $\nabla_k u = \left[\nabla u_k,\nabla\theta,\vec{v},1\right]$ is introduced. As a result, we have $\vec{v}_k = -\Phi_k\nabla_k u$. In the AMLT problem proposed, the linear equations are solved subsequently: first, for $u_3$, then for $u_1$ and, finally, for $u_2$. The analogous iteration method is applied to prove theorem 1. Therefore, we formulate the final result without mentioning proof of theorem 1.

**THEOREM 2.** For solutions $\vec{u}^{i+1}$ the AMLT problem (16) and (2), the statements of theorem 1 are valid.

## 5. ESTIMATION OF VELOCITY OF APPROXIMATION CONVERGENCE.

**THEOREM 3.** Let $B_3 = 0$ and conditions of theorem 1 be fulfilled with $[l] = 2$, in this case $\partial\Omega = \Gamma^1$ or $\partial\Omega = \Gamma^2$.

Then with $i \to \infty$, the functions $\vec{u}^{i+1}$ converge to the classical solution $\vec{u} \in C^{2+\alpha_0, 1+\alpha_0/2}(Q)$, $\alpha_0 > 0$, of the nonlinear MLT problem (1), (2) with

$$\| u_{3i+1}, \nabla u_{3i+1} \|_{2,\infty,Q} + \| u_{1i+1}, u_{2i+1} \|_{V_2(Q)} \leqslant \varepsilon, \tag{17}$$

$$\| \vec{u}_{i+1} \|_{\infty,Q} \leqslant \varepsilon^{\beta}, \quad \beta \in (0,1), \tag{18}$$

where $\vec{u}_{i+1} = \vec{u}^{i+1} - \vec{u}$, $\varepsilon = c\left\{(cT)^i/i!\right\}^{1/2}$ and the constant $c$ depends only on data of the problem.

**Proof.** In notations of section 4 there are

$$m_k \frac{\partial}{\partial t} u_{ki+1} = div\left[\Phi_k \nabla_k u_{i+1} + \vec{F}_k^{\,i}\right], \quad k = 1,\ 2,\ 3, \tag{19}$$

where $\vec{F}_k^i = \Phi_{ki} \nabla_k u^{i+1}$, $\Phi_{ki} = \Phi_k(x, u_1^i, u_2^i) - \Phi_k(x, u_1, u_2)$.

By virtue of (6), we find that

$$|\Phi_{ki}| \leqslant M_3\left(|u_{1i}| + |u_{2i}|\right),$$

from where, in particular for $\vec{F}_3^i$, we obtain

$$\| \vec{F}_3^i \|_{2,\Omega} \leqslant c_0\left(\| u_{1i} \|_{2,\Omega} + \| u_{2i} \|_{2,\Omega}\right) \equiv c_0 y_i(t).$$

Multiplying (19), $k = 3$, by $u_{3i+1}$ and integrating over $\Omega$ result we obtain the estimates

$$\| u_{3i+1}, \nabla u_{3i+1}, \vec{v}_{3i+1} \|_{2,\Omega} \leqslant c y_i(t). \tag{20}$$

Considering now $\vec{F}_1^i$ and $(B_2 \nabla \theta_{i+1} + \vec{F}_2^i)$ as the prescribed vectors, in wiew of (20), we find

$$\| u_{ki+1} \|_*^2 \leqslant c \int_0^t \left(\| u_{ki+1} \|_{2,\Omega}^2 + y_i(\tau) + l_k \| \nabla u_{1i+1} \|_{2,\Omega}^2\right) d\tau,$$

$$\| f \|_*^2 = \| f \|_{2,\Omega}^2 + \int_0^t \| f \|_{2,\Omega}^2 d\tau; \quad k = 1,\ 2; \quad l_1 = 0, \quad l_2 = 1.$$

Substitution of the estimate for $\int_0^t \|\nabla u_{1\,i+1}\|^2_{2,\Omega} d\tau$ from the obtained inequality with $k = 1$ into that with $k = 2$ gives

$$\|u_{k\,i+1}\|^2_* \leqslant c \int_0^t \left(y_{i+1}(\tau) + y_i(\tau)\right) d\tau, \quad k = 1,\ 2. \tag{21}$$

Summation of (21), $k = 1$ and (21), $k = 2$ and neglection of integrals in the left-hand side of equality lead to inequality

$$y_{i+1}(t) \leqslant c_0 \int_0^t \left(y_{i+1}(\tau) + y_i(\tau)\right) d\tau, \quad y_{i+1}(0) = 0,$$

from which, according to the Gronwall inequality [3], we find

$$y_{i+1}(t) \leqslant c \int_0^t y_i(\tau)\, d\tau, \quad (c = c_0 e^{c_0 T}). \tag{22}$$

The method of mathematical induction is used to check up that for $z_i = \sup\limits_{0 \leqslant t \leqslant T} y_i(\tau) = \|u_{1i}\|^2_{2,\infty,Q} + \|u_{2i}\|^2_{2,\infty,Q}$ from (22) the following estimate follows, i.e.

$$z_i \leqslant \varepsilon(i) z_0, \quad \varepsilon(i) = \frac{(cT)^i}{i!},$$

substitution of which into (20), (21) gives (17). With the help of the interpolation inequality

$$\sup_Q |u| \leqslant c(\beta) \left(\|u\|_{V_2(Q)}\right)^{\beta} \left(\|u\|_{C^{\alpha}(\Omega)}\right)^{1-\beta}, \quad \beta \in (0,1),$$

we obtain the estimate (18). The theorem has been proved.

Conisder now (16) as a family of linear systems of equations, in which the set $\Phi^i$ of coefficients depends on the parameter $i$ and, according to theorem 3, the values of $\Phi^i$ converge uniformly as $i \to \infty$. Let us divide the region $\Omega$, $\Omega = \bigcup\limits_1^N \Omega_i$, $|\Omega_i| \leqslant h_1$ and interval $[0,T]$ into parts with the step $h_0$. Let a certain finite-dimensional approximation of the mentioned family correspond to this dividing. Denote a corresponding normed space by

$X = X(\Omega)$ and its finite-dimensional analogue by $X_h = X(\Omega_h)$, $h = (h_0, h_1)$, in this case

$$\|u\|_{X_h} \leqslant \|u\|_X \leqslant |u|_\Omega^{(l)}, \quad [l] \geqslant 2.$$

Then a useful statement, which analogue was used in [4], holds.

**THEOREM 4.** Let the algebraic system of equations, which approximates (16), (2), correspond to this set of problems. There exists the solution of this system, in this case

$$\|\vec{u}^{i+1} - \vec{u}_h\|_{X_h} \leqslant M(h); \quad M(h) \to 0 \quad \text{as} \quad |h| \to 0,$$

where $M$ is independent of $i$.

Then we find such $l \geqslant 2$ (in condition (6)) that $\vec{u}_h$ converges to solution of the initial nonlinear problem (1), (2) with the following estimate

$$\|\vec{u} - \vec{u}_h\|_X \leqslant c\varepsilon^\gamma + M(h), \quad \gamma \in (0,1]. \tag{23}$$

**REFERENCES.**


[1] Bocharov O.B., Monakhov V.N. Boundary-value problems of nonisothermal two-phase filtration in porous media. Dinamika Sploshnoi Sredy, vyp.86 (1988) (in Russian).

[2] Bocharov O.B., Monakhov V.N. Nonisothermal filtration of immiscible fluids with variable resiidual saturation. Dinamika Sploshnoi Sredy, vyp. 88 (1988) (in Russian).

[3] Antontsev S.N., Kazhikhov A.V., Monakhov V.N. Boundary value problems in mechanics of nonhomogeneous fluids. Studies in Mathematics and its Applications, vol.22, North-Holland, 1990.

[4] Antontsev S.N., Papin A.A. Approximate methods of solving two-phase filtration problems. Dokl. Akad. Nauk SSSR, v.247, N 3 (1979) (in Russian).

International Series of Numerical Mathematics, Vol. 106, © 1992 Birkhäuser Verlag Basel 

# EXPLICIT SOLUTION OF TIME-DEPENDENT FREE BOUNDARY PROBLEMS

*Y.E.Hohlov*
*Steklov Mathematical Institute*
*Moscow 117966, RUSSIA*

## 1.INTRODUCTION.

Recently considerable progress has been made on the interfacial dynamics in one-phase Stefan-like problems (e.g. solidification, diffusion-limited aggregation, viscous fingering, dendric formation, etc). The problem under consideration consists in solving the Laplace equation on a moving boundary the local velocity of which is determined by the normal component of gradient of potential field. In the theory of FBP's numerous results about weak solvability were obtained due to the methods of variational inequalities, while the classical solvability in several cases is an intriguing problem. At the same time there is a very few number of the explicit analytic solutions of FBP's.

In the two-dimensional case (see reviews by Bensimon & Kadanoff & Liang & Shraiman & Tang, 1986; Hohlov, 1990; Howison, 1991) the interface dynamics can be described by an evolutionary family of Riemann mapping functions of the unit disk onto regions with the moving boundary.

This paper is concerned with a systematic account of the methods and results of geometric complex analysis in such kind of FBP's and some applications.

In the first part we describe the mathematical model in terms of nonlinear boundary value problem for the Riemann mapping function. Different configurations have been considered, namely the interior problem in a bounded simply connected region, the exterior problem in an unbounded region which is the exterior of a bounded simply connected domain and "channel" problem which describes flows in a channel with the impervious walls. In the second part we discuss the classification of solutions to the model. We consider also some geometrical results about behaviour of solutions.

The study of explicit solutions gives the direction of development in general cases and excellent tests for new mathematical and numerical methods.

## 1. MATHEMATICAL MODEL.

We consider the interfacial dynamics in the following dimensionless one-phase model with the uniform external field. We suppose that the state of "fluid" is described by means of the potential velocity field in time-dependent family of simply connected "phase" regions $D_+(t) \subseteq C$ with the nonempty component $D_-(t) = \overline{C}\backslash D_+(t)$ and a moving interface

$$\partial D(t) = \left\{ (x,y,t) \in C\times I \;\middle|\; \Gamma(x,y,t) = 0 \right\}$$

whose time history needs to be determined as a part of the solution.

Under the set of a priori assumptions we have the velocity field

$$V = grad\ \phi \tag{1}$$

given by the potential function $\phi = \phi(z,t)$ as a solution of continuity equation, that is the Poisson equation

$$-\Delta\phi = \sum Q_j(t)\ \delta(z - z_j), \quad z \in D(t), \tag{2}$$

where the right-hand side terms $Q_j(t)\ \delta(z - z_j)$ are used as a model for a source or sink with the rate $Q_j(t)$ at the point $z_j \in D(t)$.

We assume that the equations are solved together with the constant-surface-tension *dynamic* boundary condition

$$\phi(x,y,t) = const \quad \text{onto } \Gamma(t) \tag{3}$$

and the *kinematic* boundary condition

$$V_n = -\frac{\partial\Gamma}{\partial t}\,|\nabla\Gamma| \qquad \text{onto } \Gamma(t), \tag{4}$$

where $V_n$ is the normal component of velocity of fluid.

We complete the specification for the problem by assuming the initial fluid region

$$D(0) = D_0. \tag{5}$$

The problem can be reformulated in terms of a complex potential $\chi(z,t)$ such that

$$\chi(z,t) = \phi(z,t) + i\phi^*(z,t), \tag{6}$$

where $\phi^*(z,t)$ is the stream function and in terms of auxiliary univalent mapping of the fluid region onto a fixed canonical domain, the unit disk $U = \left\{ z \in C \;\middle|\; |z| < 1 \right\}$. Let us denote such a mapping $z: U \to D_t$ by

$$z = f(\zeta,t), \tag{7}$$

which is normalized by conditions $f(\zeta_0,t) = z_0$, $f'(\zeta_0,t) > 0$, where $\zeta_0 = z_0 = 0$ for the interior problem, $\zeta_0 = 0$, $z_0 = \infty$ for the exterior problem.

**PROBLEM 1.** Find the one-parameter family of mappings (analytic disks) $\Phi = (z,w): U\times R_+ \to \overline{C}^2$ of the unit disk $U$ such that functions $z = f(\zeta,t)$ and $w = g(\zeta,t)$ belong to the class $C^1(\overline{U}\times I)$ and for any $t \in I = (-\alpha,\beta)$ satisfy the next system of nonlinear boundary value problems:

$$\partial_{\bar\zeta} f = \partial_{\bar\zeta} g = 0, \quad \zeta \in U; \tag{8}$$

$$f(\zeta,0) = f_0(\zeta), \; \zeta \in U, \tag{9}$$

$$f(\zeta_j(t),t) = z_j \quad \text{for all} \quad t \in I, \tag{10}$$

$$g_{sing}(\zeta,t) = \sum Q_j(t)(\zeta - \zeta_j(t))^{-1}, \quad \zeta \to \zeta_j(t); \tag{11}$$

$$\mathfrak{I}\left[\zeta g(\zeta,t) + \zeta g_{sing}(\zeta,t)\right] = 0, \quad \zeta \in \partial U, \tag{12}$$

$$\mathfrak{R}\left[\zeta \overline{\frac{\partial f}{\partial \zeta}} \frac{\partial f}{\partial \zeta}\right] = \mathfrak{R}\left[\zeta g(\zeta,t) + \zeta g_{sing}\right], \quad \zeta \in \partial U, \tag{13}$$

where the strength $Q_j(t)$, the position of sinks (sources) $z_j \in \overline{C}$ and the initial domain $D_0 = f(U,0)$ are given.

## 2. STARLIKE SOLUTIONS IN THE DISK.

Let us introduce the starlike domain $D \subseteq C$ with respect to the coordinate origin as the domain such that any segment which connects a point on $\partial D$ with the origin belongs to $D$. The conformal mappings of the unit disk onto the starlike domain are called the *starlike functions*. It is well known from the theory of univalent functions that the evolutionary family $f_t = f(\zeta,t)$ consists of starlike functions in $U$ iff

$$f(\zeta,t) = \theta(t) f_0(\zeta), \tag{14}$$

where $f_0\colon U \to D_0$ is a starlike function.

The next proposition gives the description of rational starlike solutions of the Problem 1.

**THEOREM 1.** The Problem 1 has a unique rational starlike solution

$$f(\zeta,t) = \frac{1}{r_*(D_0)} \sqrt{r_*^2(D_0) + 2\int_0^t Q(\tau)\, d\tau}\; f_0(\zeta) \tag{15}$$

in each case:

a) in the class $S(U) \cap C^1(\overline{U})$ with $P\, div_{\overline{U}}\, f = 0$ this solution is

$$f_0(\zeta) = \zeta; \tag{16}$$

b) in the class of rational functions from $S(U)$ with $P\, div_{\overline{U}}\, f = 1\ (-1)$ this solution is

$$f_0(\zeta) = \frac{\zeta(\zeta + 3)}{\sqrt{3}\,(\zeta + 1)}. \tag{17}$$

There is no other rational starlike solution. The blow-up time $T_*$ is equal to $T_* = \infty$ for the case $Q(t) > 0$. This may be obtained from the relation $\int_0^{T_*} Q(\tau)\, d\tau = -r_*^2 D_0/2$ for $Q(t) < 0$, where

$$r_*(D) = inf \left\{ \Re[\zeta g'(\zeta)\, \overline{g(\zeta)}],\ |\zeta| = 1 \right\}.$$

The uniqueness of solutions of the Problem 1 is supposed in the sense that any other solution is obtained from the solutions (2.1.2)-(2.1.3) due to the action of the rotation and mirror-symmetry operators $R_\theta$ and $Z$, respectively (see 1.6).

## 3. STARLIKE SOLUTIONS IN THE EXTERIOR OF THE DISK

The analogous result is true in the case of unbounded domains which includes the neighbourhood of infinity.

**THEOREM 2.** The 1-Phase FBP 1 has a unique rational starlike solution

$$f(\zeta,t) = \sqrt{r_*^2 + 2\int_0^t Q(\tau)\,d\tau}\,\left[\zeta + \frac{k}{\zeta}\right] \tag{18}$$

in the class $S(U^*)$ with $P\,div_{\overline{U}^*}f = 1\ (\infty)$. There is no other rational starlike solution. Here the value $k \in [0,1]$.

Note that there are no "generalized" solutions of the external problem because the univalent mapping $f\colon U^* \to D(t)$ cannot have singularities on the boundary $\partial U$ at the same time when the interior point $\zeta = \infty$ tends to infinity.

## 4. TRAVELLING-WAVE SOLUTIONS IN THE STRIP.

The case of canonical domains differing from $U$, e.g., a horizontal strip in the case of Hele-Shaw flows in channels with parallel walls $\{y = \pm h\}$ is considered. The equivalent formulation of the Problem 1 for the case of the strip $T = \{z \in \overline{C}\ \big|\ |\Re z| < h\}$ is given for the upper unit half disk $U^+$, because it is more convenient for the illustration of analytic continuation. Note that in all the cases we use the principal branches of $\ln\zeta$ and $\zeta^\alpha$, $\alpha \in R$, such that $\ln 1 = 0$, $1^\alpha = 1$.

First, the aim is to obtain quasistationary solutions of the problem under the assumption

$$P(\zeta,t) = Q(t)\ln(\zeta/i), \quad Q = -2pV, \tag{19}$$

so that the mapping function $f_t = f(\zeta,t)$ in this case satisfies

$$\Re[\partial_t f(\zeta,t)\,\overline{\zeta\partial_\zeta f(\zeta,t)}] = -V, \ \zeta \in \gamma^+ \subset \partial U^+, \tag{20}$$

and

$$\Im f(\zeta,t) = \begin{cases} h & \text{on} \quad \Re z > 0, \quad \Im\zeta = 0, \\ -h & \text{on} \quad \Re z < 0, \quad \Im\zeta = 0; \end{cases} \tag{21}$$

with the initial condition

$$f(\zeta,0) = f_0(\zeta), \tag{22}$$

where $f_0 : U^+ \to D_0 \subseteq T$ is given.

Travelling-wave solutions in the channel $T$ are solutions of the type

$$f(\zeta,t) = Vt - \frac{2h}{\pi} \ln\left(\frac{\zeta}{i}\right) + \Phi(\zeta). \tag{23}$$

We start to solve the problem in classical assumptions that solutions belong to the class $\mathcal{O}(U^+) \cap C^1(\overline{U^+})$. In this case we have the classical Poincare problem in $U$ for the function $\Phi$.

The index of the Poincare problem is equal to $0$, so that the problem has no solutions in the class $\mathcal{O}_R(U) \cap C^1(\overline{U})$, but for the case of $h = \pi/2$ there is the trivial solution $\Phi(\zeta) = a$, $a \in R$.

More convenient classes of solutions of the problem are the Hardy spaces $\mathcal{H}^p(U)$. In this case, boundary values of functions from $\mathcal{H}^p(U)$ exist almost everywhere on $\partial U$.

**THEOREM 3.** The free boundary problem has the unique solution in the class $h(0,a_1,\ldots,a_m)$

$$f(\zeta,t) = Vt - \left(\frac{2h}{\pi} - \lambda\right) \log(\zeta/i) +$$

$$+ \log\left(\prod_{k=1}^{n} (\zeta^{\alpha_k} - a_k)^{\lambda_k}(1 - a_k\zeta^{\alpha_k})^{\lambda_k}\right) + a_0, \tag{24}$$

where $\lambda = \Sigma\lambda_k$, $0 < \alpha_k < 2$.

As a corollary of the Theorem 5, we obtain partial solutions of the FBP for the case without singularities .

**PROPOSITION 1.** The free boundary problem has the unique solution in the class $h(0,-1,1)$

$$f(\zeta,t) = Vt - \lambda \log(\zeta/i) + \lambda_1 \log(1 - \zeta^{\alpha_1}) + \lambda_2 \log(1 + \zeta^{\alpha_2}), \quad (25)$$

where $\lambda = \lambda_1 + \lambda_2$, $0 < \alpha_k < 2$.

## 5. A CLASSIFICATION OF SOLUTIONS IN BOUNDED REGIONS.

In the case of constant pressure conditions, the solution of the initial-value problem with suction for a general initial domain develops a singularity in the moving boundary in finite time, say $t = T_*$; the velocity of the cusp point is infinite. None the less, there are some cases in which the initial interface moves in such a way that cusps are not formed.

*A. Solutions removing all the fluid from a finite region*

This case is straightforward because the problem is time -reversible.

**THEOREM 4.** Let $D_0$ be a bounded domain and suppose that for some $T_*$ the area of the domain $D(T_*)$ is equal to zero:

$$|D(T_*)| = 0; \quad (26)$$

then the initial domain is a disk centered on the sink. This is the only such case.

The solution for the initial domain $D_0 = \left\{ z \in C \;\middle|\; |z| < R_0 \right\}$ is given by the mapping

$$f(\zeta,t) = \zeta \sqrt{R_0^2 + Qt/\pi} \quad (27)$$

and there is a natural stopping time for the model. This is the time $T_* = |D_0|/Q$ when the limiting domain $D(T_*) = f(U,T_*)$ is the point $z = 0$ and the linear measure of the residual region of fluid is equal to zero.

From the complex analytic point of view we have the case in which a continuous family $f_t = f(\zeta,t)$ of conformal mappings converges to $f_* = f(\zeta,T_*) = 0$. This is the "degenerate" case in the

Caratheodory theorem of kernel convergence of sequences of domains and univalent mappings (see, for example, Colusin, 1966 or Pommerenke, 1975).

### *B. Solutions which develop singularities in the moving boundary in finite time*

This was in fact the first type of nontrivial solution for Hele-Shaw flows in bounded regions to be constructed by complex analytic methods (see the review in Hohlov, 1990). The ill-posed character of the model which suction means that the solution develops a singularity, frequently a cusp, in the moving boundary in finite time; the velocity of the cusp point is infinite. The minimum distance from the moving boundary to the sink may or may not vanish at the blow-up time.

**THEOREM 5.** In the case of polynomial mappings of degree $N \geqslant 2$ as an initial mapping function, cusp formation is guaranteed before the moving boundary reaches the sink.

### *C. Solutions in which the moving boundary reaches the sink leaving residual fluid in a finite region*

A time-reversal argument, involving reversal of the sequence of domains created by injection at a source initially situated at the boundary, suggests that this situation is possible and indeed there is a variety of exact solutions for this case (see examples in Hohlov & Howison, 1991).

There is in fact a general result for solutions in which the moving boundary reaches the sink:

**THEOREM 6.** The limiting domain $D(T_*)$ of the continuous family $\{D(t)\}$, $t \in [0,T_*]$ of solutions of the Hele-Shaw flow moving boundary problem (1)-(3) is degenerate to the point $D(T_*) = \{0\}$ when the moving boundary $\partial D(T_*)$ reaches the sink.

The degenerate limiting function means that the model (1)-(6) is incorrect for Hele-Shaw flows in finite regions with a sink at the moving boundary. The pressure $p$ at the sink must have a singularity different from a logarithmic one.

Physically the situation is quite different from case A because the linear measure of the residual region of fluid is not equal to zero as it was there.

## 6. SOME GEOMETRIC PROPERTIES.

We now move on to describe some geometric properties of classical solutions of one-phase FBPs.

The family of initial domains $\{D_0\}$ which includes the coordinate origin is parametrized by univalent holomorphic mappings $f_0\colon U \to D_0$ such that $f_0(0) = 0$, $f'(0) > 0$. The class of the mappings is denoted by $S(U)$.

**THEOREM 7.** Let $D_0 = f_0(U)$ be an initial bounded domain which is parameterized by $f_0 \in S(U)$. If $D(t) = f_t(U)$ is a classical solution of the Problem 1 onto the interval (0,T), then for all $t \in (0,T)$ the next properties are true:

1) the interface $\Gamma(t)$ is an analytic Jordan curve;

2) for initial subdomain $\tilde{D}_0 \subset D_0$ the inclusions

$$a)\ \tilde{D}(t) \subset D(t) \quad for \quad q(t) > 0; \tag{28}$$

$$b)\ D(t_1) \subset D(t_2) \quad for \quad q(t) > 0, \tag{29}$$

$$c)\ D(t_2) \subset D(t_1) \quad for \quad q(t) < 0; \tag{30}$$

are satisfied for all $0 < t_1 < t_2 < T$.

We now give some estimates for the growth of the fluid region in the injection (well-posed) case.

**THEOREM 8.** Let $D_0$ be a bounded domain such that a classical solution of the Hele-Shaw problem with injection from a source of strength $Q$ at the origin and initial domain $D_0$ exist for any $t \in [0,T]$. Then the distance from the moving boundary $\partial D(t)$ to the source is bounded below by

$$dist(D(t),0) \geqslant \frac{1}{4}\sqrt{R^2(D_0) + Qt/\pi}\ , \tag{31}$$

where $R(D_0)$ is the conformal radius of the initial domain with respect to the origin.

In the injection case (shrinking bubble), we can derive an estimate on the dimensions of the bubble similar to that given above for finite regions. In fact, we have

**THEOREM 9.** Let a solution of the bubble problem with $Q > 0$ and initial domain $B_0$ exist for any $t \in [0,T]$. Then the linear measure of the shrinking bubble at time t is estimated by the following bound:

$$diam\ B(t) \leqslant 4\sqrt{C^2(B_0) - Qt/\pi},$$

where $C(B_0)$ is the capacity of the initial bubble.

REFERENCES

[1] L.de Branges. CTA MATH., 154 (1985), pp.137-152.

[2] Galin L.A. Dokl. Akad. Nauk SSSR, 47 (1945), pp.246-249.

[3] Golusin G.M. Geometric theory of functions of a complex variable, 2nd ed., "Nauka", Moscow, 1966 (Russian), English transl. AMS, 1969.

[4] Hohlov Yu. E. Time-dependent free boundary problems: Explicit solutions I, Prepr. MIAN, No.14 (1990a), 52 pp.

[5] Hohlov Yu.E. Soviet Math. Dokl., 315 (1990b), pp.80-83.

[6] Hohlov Yu.E., Howison S.D. Quart. J. Mech., Mech. Appl. Math., 1991 (to appear).

[7] Howison S.D. SIAM J. Appl. Math., 45 (1986a), pp.20-26.

[8] Howison S.D. Proc. Roy. Soc. Edinb., A102 (1986b), pp.141-148.

[9] Howison S.D., Ockendon J.R., Lacey A.A. Quart. J.Mech. Appl. Math. , 38 (1985), pp.343-360.

[10] Howison S.D., Lacey A.A., Ockendon J.R. Quart. J. Mech., Appl. Math., 41 (1988), pp.183-193.

[11] Kufarev P.P., Dokl. Akad. Nauk SSSR, 60 (1948), pp.1333-1334.

[12] Lacey A.A., Howison S.D., Ockendon J.R., Wilmott P.P. Quart. J. Mech. Appl. Math., 1990 (to appear).

[13] Polubarinova-Kochina P.Ya. Dokl. Akad. Nauk SSSR, 47 (1945), pp.254-257.

[14] Pommerenke Chr. Univalent Functions, Vanderhoeck and Ruprecht, Gottingen, 1975.

[15] Saffman P.G., Taylor G.I., Proc. Roy. soc., A245 (1958), pp.312-329.

[16] Taylor G.I., Saffman P.G., Quart. J. Mech. Appl. Math., 12 (1959), N 3, pp.265-279.

International Series of Numerical Mathematics, Vol. 106, © 1992 Birkhäuser Verlag Basel 

# NONEQUILIBRIUM PHASE TRANSITIONS IN FROZEN GROUNDS

*I.A.Kaliev*
*Lavrentyev Institute of Hydrodynamics*
*Novosibirsk 630090, RUSSIA*

At first, we discuss what is meant by equilibrium phase transition and nonequilibrium one. For equilibrium phase transition the concentration of an unfrozen water $w$ coincides with Heaviside's function $w(x,t) = H(u(x,t))$, here $u(x,t)$ is the temperature,

$$H(u) = \begin{cases} 1, & u > 0, \\ \in [0,1], & u = 0, \\ 0, & u < 0. \end{cases}$$

$u = 0$ is the melting temperature. This means that under change of temperature $u$ the concentration of water $w$ takes instantaneously a new value corresponding to the new temperature. Meanwhile in real processes some finite time is required for achieving equilibrium between ice and an unfrozen water in the ground.

One mathematical model of water freezing and melting with phase relaxation is considered. In this model, a real concentration of the water $w$ is connected with an equilibrium concentration $H(u)$ by the equation

$$\frac{\partial w}{\partial t} - \lambda\Delta w = \tau^{-1}\Big[H(u) - w\Big], \ (x,t) \in Q_T = Q\times(0,T), \ Q \subset \mathbb{R}^n, \ 0<T<\infty, \tag{1}$$

where $\tau$, $\lambda$ are the positive constants, $\tau$ is the relaxation time, $\lambda$ is the coefficient of water migration.

The temperature $u$ satisfies the equation

$$\frac{\partial u}{\partial t} - k\Delta u = -\,l\tau^{-1}\Big[H(u) - w\Big], \ (x,t) \in Q_T. \tag{2}$$

Here $k$, $l$ are the positive constants, $k$ is the heat conduction, $l$ is the latent heat of the phase water-ice transition.

It is required to find functions $u(x,t)$, $w(x,t)$ satisfying equations (1), (2) and following initial and boundary conditions

$$u\Big|_{t=0} = u_0(x), \quad w\Big|_{t=0} = w_0(x), \; x \in Q, \tag{3}$$

$$u\Big|_{S_T} = u_s(x,t), \quad w\Big|_{S_T} = w_s(x,t), \; (x,t) \in S_T = \partial Q \times (0,T). \tag{4}$$

For short, we shall call problem (1) - (4) the problem A. The problem A was investigated in [1], where theorems of existence and uniqueness from spaces $u, w \in W_q^{2,1}(Q_T)$, $1 < q < \infty$ were proved.

Some other problems will be considered in this paper except the problem A.

**THE PROBLEM B.** It is required to find functions $u(x,t)$, $w(x,t)$ defined in the domain $Q_T$ satisfying equations

$$\frac{\partial u}{\partial t} = k\Delta u - l\,\frac{\partial w}{\partial t}, \quad \frac{\partial w}{\partial t} = \frac{1}{\tau}\Big[H(u) - w\Big],$$

initial conditions (3) and the first boundary condition in (4). The problem B was investigated in [2], where theorems of existence and uniqueness from classes $u \in W_q^{2,1}(Q_T)$, $w, w_t \in L_\infty(Q_T)$ were proved.

**THE PROBLEM D.** Let us denote

$$u + lH(u) = U, \quad u + \frac{\lambda l}{k}\,H(u) = \bar{u}.$$

$U$, $\bar{u}$ are connected by the relations

$$U = \begin{cases} \bar{u}, & \bar{u} \leqslant 0, \\ \bar{u}\,\dfrac{k}{\lambda}, & 0 \leqslant \bar{u} \leqslant \dfrac{\lambda l}{k}, \\ \bar{u} + \dfrac{l(k-\lambda)}{k}, & \bar{u} \geqslant \dfrac{\lambda l}{k}. \end{cases}$$

The problem D is formulated by the following way. It is required to find the function $\bar{u}(x,t)$ in $Q_T$ from the equation

$$\frac{\partial}{\partial t}\left[U(\bar{u})\right] = k\Delta\bar{u}$$

and conditions

$$\bar{u}\Big|_{t=0} = \bar{u}_0(x), \quad x \in Q, \quad \bar{u}\Big|_{S_T} = \bar{u}_s(x,t), \quad (x,t) \in S_T.$$

Theorems of existence and uniqueness of the solution of the problem D from the class

$$\bar{u} \in W_2^{2,1}(Q_T) \cap L_\infty(Q_T) \cap L_\infty(0,T;W_2^1(Q))$$

are proved in [1].

**THE PROBLEM C.** (Stefan problem). Assume that $U = u + lH(u)$. It is required to find the functions $U$ and $u$ from the equation

$$\frac{\partial}{\partial t}\left[U(u)\right] = k\Delta u$$

and conditions

$$U\Big|_{t=0} = U_0(x), \quad x \in Q, \quad u\Big|_{S_T} = u_s(x,t), \quad (x,t) \in S_T.$$

The existence and uniqueness of a weak solution of the problem C are proved, for example, in [3]:

$$u \in W_2^{1,1}(Q_T) \cap L_\infty(Q_T) \cap L_\infty(0,T;W_2^1(Q)), \; U \in L_\infty(Q_T).$$

Problems A, B, D, C are connected as follows. Solutions of the problem A converge to the solution of the problem B (in a certain sense) under $\lambda \to 0$ [1]. Solutions of the problem B converge to the solution of the problem C under $\tau \to 0$ [2]. Solutions of the problem A converge to the solution of the problem D under $\tau \to 0$ [1]. And, at last, solutions of the problem D converge to the solution of the problem C under $\lambda \to 0$ [1].

New properties of solutions of the problems A, B, D will be formulated below.

**THEOREM** 1 (the comparison theorem). Let $\left\{u_i(x,t), w_i(x,t)\right\}$, $i = 1, 2$ be two solutions of the problem A with initial and boundary data $\left\{u_{io}, w_{io}, u_{si}, w_{si}\right\}$ and

$$u_{01}(x) \geqslant u_{02}(x), \quad w_{01}(x) \geqslant w_{02}(x), \quad x \in Q;$$

$$u_{s1}(x,t) \geqslant u_{s2}(x,t), \quad w_{s1}(x,t) \geqslant w_{s2}(x,t), \quad (x,t) \in S_T;$$

Then

$$u_1(x,t) \geqslant u_2(x,t), \quad w_1(x,t) \geqslant w_2(x,t), \quad (x,t) \in Q_T.$$

At first, the comparison theoremS are proved for regularized problems with $H(u)$ replaced by the smooth function $H_n(u)$. The result of theorem 1 is obtained by the limit procedure $(n \to \infty)$.

The same theorems are true for problems B, D. The comparison theorem for Stefan problem has been well-known earlier [4].

The stationary problem corresponding to A is denoted by $A_s$. It is required to find the functions $\tilde{u}(x)$, $\tilde{w}(x)$ from

$$k\,\Delta\tilde{u} = \frac{l}{\tau}\left[H(\tilde{u}) - \tilde{w}\right], \quad \lambda\Delta\tilde{w} = -\frac{1}{\tau}\left[H(\tilde{u}) - \tilde{w}\right] \quad in \quad Q, \tag{5}$$

$$\tilde{u}\,\Big|_S = \tilde{u}_s(x), \; \tilde{w}\,\Big|_S = \tilde{w}_s(x), \; x \in S = \partial Q. \tag{6}$$

A pair of functions $\left\{\tilde{u}, \tilde{w}\right\}$ such as $(i)$ $\tilde{u}$, $\tilde{w} \in W_q^2(Q)$, $1 < q < \infty$; $(ii)$ equations (5) fulfilled almost everywhere in $Q$; $(iii)$ boundary conditions (6) accepted as traces of functions from an indicated class, is called a solution of the problem $A_s$.

**THEOREM** 2. Let the bound $S \in O^2$ and boundary data $\tilde{u}_s$, $\tilde{w}_s$ be such that the functions $y$, $z \in W_q^2(Q)$ satisfying conditions

$$y\,\Big|_S = \tilde{u}_s(x), \quad z\,\Big|_S = \tilde{w}_s(x); \; 0 \leqslant \tilde{w}_s \leqslant 1, \; x \in S = \partial Q, \quad \text{exist.}$$

Then the problem $A_s$ has the unique solution, and estimates

$$0 \leqslant \tilde{w}(x) \leqslant 1, \quad x \in Q,$$

$$\|\tilde{u}\|^{(2)}_{q,Q} \leqslant C_1 \left[ 1 + \|y\|^{(2)}_{q,Q} \right], \quad \|\tilde{w}\|^{(2)}_{q,Q} \leqslant C_2 \left[ 1 + \|z\|^{(2)}_{q,Q} \right],$$

are true, where $C_1$, $C_2$ are positive constants. If additionally the condition $\|\tilde{u}_S\|_{\infty,S} = M < \infty$ is fulfilled, then $\|\tilde{u}\|_{\infty,Q} \leqslant M$.

Theorem 2 is proved similar to theorem 1 in [1].

The stationary problems corresponding to B, D, C are denoted by $B_S$, $D_S$, $C_S$, by the way, problems $B_S$, $C_S$ coincide. Solutions of the problems $B_S$, $D_S$ are defined in the natural way and theorems of existence and uniqueness are proved.

In the problem $B_S$ the uniqueness theorem is proved for the function u (temperature), but not for the internal energy $U = u + lH(u)$.

**THEOREM 3.** Let the conditions, under which the solutions of the problems A, $A_S$ exist, be fulfilled, and

$$\lim_{t\to\infty} u_S(x,t) = \tilde{u}_S(x), \quad \lim_{t\to\infty} w_S(x,t) = \tilde{w}_S(x), \quad x \in S.$$

Then the solution of the problem A converges to the solution of the problem $A_S$ under $t \to \infty$:

$$\lim_{t\to\infty} u(x,t) = \tilde{u}(x), \quad \lim_{t\to\infty} w(x,t) = \tilde{w}(x), \quad x \in Q.$$

Theorem 3 is proved by using the barrier functions with application of theorem 1.

The same theorems are true for problems $\{B, B_S\}$, $\{D, D_S\}$.

Problems $A_S$, $B_S$, $D_S$, $C_S$ are interconnected as problems A, B, D, C. Solutions of the problem $A_S$ converge to the solution of the problem $B_S$ under $\lambda \to 0$. Solutions of the problems $B_S$ and $C_S$ coincide. Solutions of the problem $A_S$ converge to the solution of the problem $D_S$ under $\tau \to 0$. Solutions of the problem $D_S$ converge to the solution of the problem $C_S$ under $\lambda \to 0$.

Let us formulate in detail the theorem of convergence of solutions of the problem $A_S$ to the solution of the problem $B_S$ under $\lambda \to 0$. Solutions and boundary data of the problem $A_S$

are denoted with the index $\lambda$: $\tilde{u}_\lambda$, $\tilde{w}_\lambda$, $\tilde{u}_{s\lambda}$, $\tilde{w}_{s\lambda}$. Functions $y$, $z$ defined in theorem 2 will also have the index $\lambda$: $y_\lambda$, $z_\lambda$. Solution and boundary data of the problem $B_s$ are denoted as $\tilde{u}$, $\tilde{u}_s$:

$$\Delta\tilde{u} = 0, \quad \tilde{u}\Big|_S = \tilde{u}_s.$$

Let the function $\tilde{u}_s$ be such that the function $y \in W_q^2(Q)$ which satisfies the condition $y\Big|_S = \tilde{u}_s$ exists.

**THEOREM 4.** Let conditions of theorem 2 be fulfilled and

$\|y_\lambda - y\|_{q,Q}^{(2)} \to 0$ under $\lambda \to 0$.

Then the solution of the problem $A_s$ converges to the solution of the problem $B_s$ in the following sense:

$\tilde{u}_\lambda \to \tilde{u}$ almost everywhere in $Q$, weak in $W_q^2(Q)$,

$\tilde{w}_\lambda \to H(\tilde{u})$ $*$-weak in $L_\infty(Q)$.

Theorem 4 is proved similar to the theorem 2 in [1].

Passage to the limit formulated in this paper may be illustrated by the following scheme:

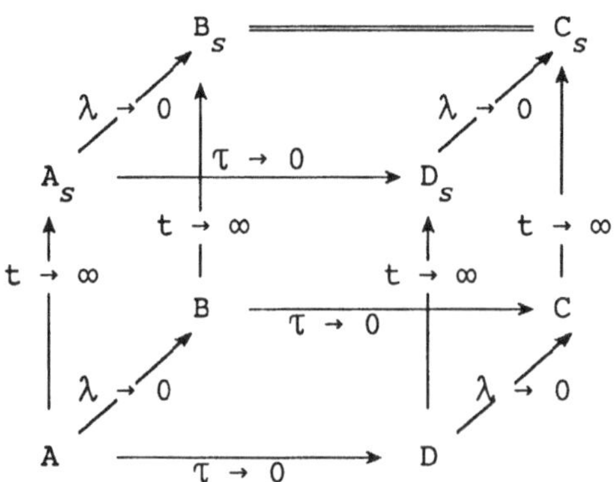

## REFERENCES.

[1] I.A.Kaliev, S.T.Mukhambetzhanov, E.N.Razinkov. Correctness of one mathematical model of nonequilibrium phase transition of water in porous medium //Dinamika Sploshnoi Sredy, Sb. nauch. tr./ Akad. Nauk SSSR, Sibirsk. Otd., Institut Gidrodinamiki, 1989, vyp. 93,94, pp.46-59 (in Russian).

[2] I.A.Kaliev, E.N.Razinkov. About Stefan problem with phase relaxation //Dinamika Sploshnoi Sredy, Sb. nauch. tr./ Akad. Nauk SSSR, Sibirsk. Otd., Institut Gidrodinamiki, 1989, vyp. 91, pp.21-36 (in Russian).

[3] A.M.Meirmanov. The Stefan Problem./ Novosibirsk: "Nauka", 1986 (in Russian).

[4] A.Friedman. The Stefan problem in several space variables. - Trans. Amer. Math. Soc., 1968, v.132, p.51-78, Correction, 1969, v.142, p.557.

International Series of Numerical Mathematics, Vol. 106, © 1992 Birkhäuser Verlag Basel 

# SYSTEM OF VARIATIONAL INEQUALITIES ARISING IN NONLINEAR DIFFUSION WITH PHASE CHANGE

*N.Kenmochi*
*Department of Mathematics, Faculty of Education, Chiba University, Yayoi-cho 1-33, Chiba 260, Japan*

*M.Niezgodka*
*Institute fo Applied Mathematics and Mechanics, Warsaw University, Banacha 2, 00-913 Warsaw,Poland*

*partially supported by Systems Research Institute of the Polish Academy of Sciences*

We discuss a new type of phase-field models applicable to processes of nonlinear diffusion with phase change. The models are set up as systems consisting of a quasilinear parabolic partial differential equation and a parabolic variational inequality, governing the temperature distribution $u(t,x)$ and an order parameter $w(t,x)$. We give some results on the global existence, uniqueness and asymptotic behavior of the solution $\{u,w\}$ as $t \to \infty$.

## 1.INTRODUCTION.

We study a class of nonlinear parabolic systems including variational inequalities that in particular arise from the modeling of solid-liquid phase transitions. Such physical processes are mathematically described by coupled systems in terms of the (Kelvin) temperature $u = u(t,x)$ and the order parameter $w = w(t,x)$ (a state variable that is capable of distinguishing individual phase involved), defined on $Q := (0,+\infty)\times\Omega$, with a bounded domain $\Omega \subset R^3$ occupied by the system. In the case when the order parameter $w$ is nonconserved, a phase-field model was proposed by Fix [3] and further developed by Caginalp [1] in the form

$$u_t + lw_t - \Delta u = f(t,x) \quad \text{in} \quad Q, \tag{1.1}$$

$$\alpha\xi^2 w_t - \Delta w + g(w) = u \quad \text{in} \quad Q, \tag{1.2}$$

where the melting temperature is supposed to be $0$, the coefficients $l$, $\alpha$ and $\xi$ represent the latent heat, relaxation time and interactive length, respectively, $f$ is the rate of a given heat source, and the function $g(w)$ is $N$-shaped so that the primitive $G(w)$ of $g(w)$, with $G(0) = 0$, has a double-well form. The system (1.1)-(1.2) with appropriate boundary and initial conditions was theoretically studied by Elliott & Zheng [2]. Recently, Penrose & Five [8] proposed a more general phased field model, essentially reducing to the form

$$\rho(u)_t + \lambda(w)_t - \Delta u = f(t,x) \quad \text{in} \quad Q, \tag{1.3}$$

$$w_t - k\Delta w + g(w) = \lambda'(w)u \quad \text{in} \quad Q; \tag{1.4}$$

where $\rho\colon R \to R$ is a strictly increasing function, $\lambda\colon R \to R$ is a smooth function, $\lambda'$ denotes the derivative of $\lambda$ and $k$ is a positive constant. For this model, under suitable boundary and initial conditions, Zheng [10] and Spekels & Zheng [9} proved a result on the global existence and uniqueness of solutions and discussed some aspects of their asymptotic behavior as $t \to \infty$.

In this paper we propose to extend the Penrose-Fife model of non-smooth components in the order parameter part. We shall consider a system consisting of the equation

$$\begin{aligned} &\rho(w)_t + \lambda(w)_t - \Delta u = f(t,x) \quad \text{in} \quad Q, \\ &\frac{\partial u}{\partial n} + h_0 u = h(t,x) \quad \text{on} \quad \Sigma = (0,+\infty)\times\partial\Omega, \\ &u(0,\cdot) = u_0 \quad \text{in} \quad \Omega, \end{aligned} \tag{1.5}$$

and variational inequality

$$\begin{aligned} &w(t,\cdot) \in K = \left\{z \in H^1(\Omega);\ \sigma_* \leqslant z \leqslant \sigma^* \text{ a.e. on } \Omega\right\} \quad \text{for} \quad t > 0, \\ &\int_\Omega \left[w_t(t,x) + g(w(t,x)) - \lambda'(w(t,x))u(t,x)\right]\left[w(t,x) - z(x)\right] dx + \\ &+ \int_\Omega k\nabla w(t,x)\cdot\nabla(w(t,x) - z(x)\Big] dx \leqslant 0 \\ &\qquad \text{for all } z \in K \text{ and a.e. } t \geqslant 0, \\ &w(0,\cdot) = w_0 \quad \text{in} \quad \Omega, \end{aligned} \tag{1.6}$$

where $h_0 > 0$ is a given constant, $h$ is a given function on $\Sigma$, $u_0$, $w_0$ are given initial data and $\sigma_*$, $\sigma^*$ are given constant with $\sigma_* < 0 < \sigma^*$. In the system (1.5)-(1.6), $u$ represents the temperature shifted so that $0$ is the melting point and $w$ is a non-conserved order parameter scaled so that $w$ assumes values in the segment $[\sigma_*,\sigma^*]$; the regions with $w = \sigma^*$ and $w = \sigma_*$ are then to be interpreted as pure liquid and pure solid phases, respectively, while their component, with $\sigma_* < w < \sigma^*$, as the mushy phase. As was in detail discussed in [1,8], equations (1.2) and their more general form (1.4), subject to initial and homogeneous Neumann boundary conditions, refer to the Landau-Ginzburg free energy functional $F_0(w;u)$, which is $u$ - dependent and given by

$$F_0(w;u) = \int_\Omega \left\{ \frac{k}{2} |\nabla w|^2 + G(w) - \lambda(w)u \right\} dx, \quad w \in H^1(\Omega).$$

Since we assume physically well-founded hypothesis that the order parameter $w$ never exceeds the segment $[\sigma_*,\sigma^*]$, a natural form of the free energy $F(w;u)$ is given by

$$F(w;u) = \begin{cases} F_0(w;u), & \text{if } w \in K, \\ +\infty, & \text{otherwise}, \end{cases} \tag{1.7}$$

i.e.,

$$F(w;u) = F_0(w;u) + I_K(w),$$

where $I_K(\cdot)$ is the indicator functions of the convex set $K$, and the variational inequality (1.6) can thus be derived as a kinetic relation of the Landau-Ginzburg type.

Actually, in the equilibrium case (provided that $u = u(x)$), $w$ is a local minimizer of $F(w;u)$ with respect to $w$, hence it is a solution of the elliptic variational inequality

$$w \in K,$$

$$\int_\Omega k\nabla w\cdot\nabla(w - z)\, dx + \int_\Omega \left[g(w) - \lambda'(w)u\right](w - z)\, dx \leqslant 0 \quad \text{for all } z \in K.$$

Therefore, in the non-equilibrium case, the kinetic relation of Landau-Ginzburg type would be of the form (1.6).

In this paper, we give some results on global existence, uniqueness and asymptotic stability of solutions to the phase field model (1.5)-(1.6). We consider the problem with $\sigma_*$ and $\sigma^*$ being functions of $(t,x)$. Throughout the paper, we for simplicity assume that $k \equiv 1$.

## 2. MAIN RESULTS.

We shall use the following notation:

$$H = L^2(\Omega),\ (v,z) = \int_\Omega vz\,dx \quad \text{for} \quad v,\ z \in H;$$

$$V = H^1(\Omega),\ a(v,z) = \int_\Omega \nabla v\cdot\nabla z\,dx \quad \text{for} \quad v,\ z \in V;$$

$$\Gamma = \partial\Omega,\ (v,z)_\Gamma = \int_\Gamma vz\,d\Gamma \quad \text{for} \quad v,\ z \in L^2(\Gamma).$$

We denote by $(P)$ the system (1.5)-(1.6) with $K$ in the form of a time-dependent convex constraint set

$$K(t) = \left\{z \in H^1(\Omega);\ \sigma_*(t,\cdot) \leqslant z \leqslant \sigma^*(t,\cdot) \quad a.e.\ on\ \Omega\right\}.$$

This problem $(P)$ is discussed under the following hypotheses $(H1) \sim (H5)$:

(H1) $\rho: R \to R$ is a Lipschitz continuous increasing function with Lipschitz continuous inverse $\rho^{-1}: R \to R$.

(H2) $\lambda: R \to R$ and $g: R \to R$ are Lipschitz continuous functions, with the first time derivative $\lambda'$ of $\lambda$ continuous on $R$.

(H3) $n_0$ is a positive number; ; $h: R_+ = [0,+\infty) \to L^2(\Gamma)$ such that $h' \in L^1(R_+;L^2(\Gamma))$ and

$$h - h_\infty \in L^2(R_+;L^2(\Gamma)) \quad for\ some \quad h_\infty \in L^2(\Gamma);$$

$$f - f_\infty \in L^2(R_+;H) \quad for\ some \quad f_\infty \in H.$$

(H4) $\sigma_*$, $\sigma^* \in L^\infty(R_+\times\Omega)\cap L^\infty(R_+;V)\in W^{1,2}_{loc}(R_+;V)$ such that

$\sigma^* - \sigma_* \geqslant \sigma_0$ a.e. on $R_+\times\Omega$ for a constant $\sigma_0 > 0$ an

$\sigma'_*$, $\sigma^{*\prime} \in L^2(R_+,V)\cap L^1(R_+,V)$ ; we put

$$\sigma_\infty := \lim_{t\to\infty} \sigma_*(t) \quad \text{and} \quad \sigma^\infty = \lim_{t\to\infty} \sigma^*(t) \quad \text{in } V,$$

(H5) $u_0 \in V$ and $w_0 \in K(0)$.

We say that a couple of functions $\{u,w\}$ is a solution of $(P)$ on the time-interval $[0,T]$, $0 < T < +\infty$ if $u: [0,T] \to H$ and $w: [0,T] \to V$ satisfy:

(S1) $u,\ w \in W^{1,2}(0,T;H)\cap L^\infty(0,T;V)$;

$$\text{(S2)} \quad \begin{cases} \left(\rho(u)'(t)+\lambda(w)'t,z\right)+a(u(t),z)+(hh\text{-}0\text{-}h(t),z)_\Gamma=(f(t),z) \\ \text{for any } z \in V \text{ and a.e. } t \in [0,T] \\ u(0) = u_0, \end{cases}$$

$$\text{(S3)} \quad \begin{cases} w(t) \in K(t) \quad \text{for any} \quad t \in [0,T] \\ \left(w'(t)+g(w(t))-\lambda'(w(t))u(t),w(t)-z\right)+a(w(t),w(t)-z)\leqslant 0, \\ \quad \text{for any } z \in K(t) \text{ and a.e. } t \in [0,T] \\ w(0) = w_0. \end{cases}$$

We also say that a couple of functions $u: R_+ \to V$ and $w: R_+ \to V$ is a solution of $(P)$ on $R_+$, if $\{u,w\}$ is a solution of $(P)$ on $[0,T]$ for every finite $T > 0$.

Our main results are stated in the following theorems. The first theorem is concerned with the global existence and uniqueness of the solution to $(P)$.

**THEOREM 2.1.** Assume that the hypotheses (H1) ~ (H5) hold. Then there exists a unique solution $\{u,w\}$ of (P) on $R_+$.

Next, we consider the steady-state problem $(P)^\infty$ corresponding to $(P)$:

$u_\infty \in V$,

$$a(u_\infty,z) + (h_0u_\infty - h_\infty,z)_\Gamma = (f_\infty,z) \quad \text{for any} \quad z \in V; \tag{2.1}$$

$$w_\infty \in K_\infty = \left\{ z \in V;\ \sigma_\infty \leqslant z \leqslant \sigma^\infty \quad a.e. \quad on\ \Omega \right\},$$
$$a(w_\infty, w_\infty - z) + \left[ g(w_\infty) - \lambda'(w_\infty) u_\infty, \vec{w}_\infty - z \right] \leqslant 0 \quad for\ any\ z \in K_\infty. \tag{2.2}$$

Then, since $h_0$ is a positive constant, problem (2.1) has a unique solution $G(v) - \lambda(v)u_\infty$. Moreover, as the functional

$$v \to \frac{1}{2} |\nabla v|_H^2 + \int_\Omega \left\{ G(v) - \lambda(v) u_\infty \right\} dx$$

is coercive on $K_\infty$, problem (2.2) has at least one solution $w_\infty$, which in general is nonunique because of the double-well form of $u_\infty$. The second result is concerned with the asymptotic stability of solutions $\{u,w\}$ of $(P)$ as $t \to +\infty$.

**THEOREM 2.2.** Assume that $(H1) \sim (H5)$ hold, and let $\{u,w\}$ be the solution of $(P)$ on $R_+$. Then

*(i)* $u(t) \to u_\infty$ in $H$ and weakly in $V$ as $t \to \infty$, where $u_\infty$ is the unique solution of (2.1).

*(ii)* Let $\omega(w)$ be the limit set of $w(t)$ in $H$ as $t \to +\infty$, i.e.

$$\omega(w) = \left\{ z \in H;\ w(t_n) \to z \text{ in } H \ (as\ n \to \infty) \text{ for some } t_n \text{ with } t_n \uparrow +\infty \right\}.$$

Then, $\omega(w) \neq 0$ and $\{u_\infty, w_\infty\}$, with any $w_\infty \in \omega(w)$, is a solution of $(P)^\infty$.

With the same notations as in (ii) of Theorem 2.2, it is clear that $w$ is asymptotically convergent in $H$ as $t \to +\infty$ and only if $\omega(w)$ is a singleton. Besides, to provide that $\omega(w)$ be a singleton, it is sufficient for (2.2) to have a unique solution $w_\infty$. This property essentially depends on the solution $u_\infty$ of (2.1) which can be controlled by the asymptotic data $f_\infty$ and $h_\infty$.

**THEOREM 2.3.** In addition to $(H1) \sim (H5)$, assume that $\sigma_\infty$ and $\sigma^\infty$ are constant on $\Omega$, $\lambda' > 0$ on $[\sigma_\infty, \sigma^\infty]$ and $f_\infty \in L^2(\Omega)$. Let $\{u,w\}$ be the solution of $(P)$ on $R_+$. Then there are constants $M^\infty > 0$ and $M_\infty < 0$ such that

*(i)* if $h_\infty \geqslant M^\infty$ a.e. on $\Gamma$, then the constant function $w_\infty \equiv \sigma^\infty$ is the unique solution of (2.2), and $w(t) \to \sigma^\infty$

in $H$ and weakly in $V$ as $t \to +\infty$.

*(ii)* if $h_\infty \leqslant M_\infty$ a.e. on $\Gamma$, then the constant function $w_\infty \equiv \sigma_\infty$ is the unique solution of (2.4), and $w(t) \to \sigma_\infty$ in $H$ and weakly in $V$ as $t \to \infty$.

The assertions of Theorem 2.3 are phenomenologically justified; in particular, there are solid-liquid phase transition processes where, by keeping the temperature sufficiently high or low, the system approaches one of pure phases as $t \to +\infty$.

## 3. OUTLINE OF THE PROOFS.

In what follows we apply some abstract results on nonlinear evolution equations involving time-dependent subdifferentials in Hilbert spaces (cf.[4,17]).

For each $t \geqslant 0$, define

$$\varphi^t(z) := \begin{cases} \frac{1}{2}|\nabla z|_H^2 + \frac{h_0}{2}|z|^2_{L^2(\Gamma)} - \left(h(t),z\right)_\Gamma, & \text{if } z \in V, \\ +\infty, & \text{otherwise;} \end{cases}$$

$$\psi^t(z) := \begin{cases} \frac{1}{2}|\nabla z|^2, & \text{if } z \in K(t), \\ +\infty, & \text{otherwise.} \end{cases}$$

Then, with subdifferentials $\partial\varphi^t$ and $\partial\psi^t$ in $H$, *(S2)* and *(S3)* can respectively be expressed as

$$\rho(u)'(t) + \lambda(w)'(t) + \partial\varphi^t(u(t)) = f(t) \quad \text{a.e. } t \in [0,T],$$
$$u(0) = u_0, \tag{3.1}$$

$$w'(t) + \partial\psi^t(w(t)) \ni \lambda'(w(t))u(t) - g(w(t)) \quad \text{a.e. } t \in [0,T],$$
$$w(0) = w_0. \tag{3.2}$$

**Proof of Theorem 2.1.** We fix $0 < T < \infty$, and consider the following two problems for each $\tilde{w} \in W^{1,2}(0,T;H)$ and $\tilde{u} \in L^\infty(0,T;V)$:

$$\rho(u)'(t) + \partial\varphi^t(u(t)) = f(t) - \lambda(\tilde{w})'(t) \quad \text{a.e. } t \in [0,T],$$
$$u(0) = u_0, \tag{3.3}$$

$$w'(t) + \partial\psi^t(w(t)) \ni \lambda'(w(t))\tilde{u}(t) - g(w(t)) \quad a.e.\ t \in [0,T],$$

$$w(0) = w_0. \tag{3.4}$$

According to the abstract results of [4] and [19], the above problems have unique solutions $u$ and $w$ in $W^{1,2}(0,T;H)\cap L^\infty(0,T;V)$, respectively. Clearly, the couple $\{u,w\}$ gives a solution of $(P)$ on $[0,T]$, if $w = \tilde{w}$ in (3.3) and $u = \tilde{u}$ in (3.4). We can find such solutions $u$ of (3.3) and $w$ of (3.4) by using Schauder's fixed point theorem for compact operators in $C([0,T];H)$, provided that $T > 0$ is assumed small enough. Thus $(P)$ has a local in time solution. The global existence of a solution can be shown by making use of the usual energy inequality associated to system (3.3) - (3.4). Finally, the uniqueness of a solution can be shown with the use of Gronwall's inequality.

**Proof of Theorem 2.2.** The crucial step is to prove the following global estimates:

$$u,\ w \in L^\infty(R_+;V), \quad u',\ w' \in L^2(R_+;H). \tag{3.5}$$

These estimates are obtained immediately from an inequality derived by computing

$$\Big\{(3.1)\times u(t)\Big\} + \Big\{(3.2)\times w'(t)\Big\} + \delta\ \Big\{(3.1)\times u'(t)\Big\}$$

for a sufficiently small $\delta > 0$. Once global estimate (3.5) is obtained, we see assertions $(i)$ and $(ii)$ by the standard arguments of $\omega$-limit set.

**Proof of Theorem 2.3.** By the standard comparison result for linear parabolic equations, we see that the solution $u_\infty$ of (2.1) diverges uniformly to $+\infty$ on $\Omega$ as $h_\infty \to +\infty$ uniformly on $\Gamma$. Therefore for a certain sufficiently large $M^\infty > 0$, it follows that the solution $u_\infty$ corresponding to $h_\infty$ with $h_\infty \geqslant M^\infty$ on $\Gamma$ satisfies

$$g(r) - \lambda'(r)u_\infty(x) < 0 \quad \text{for all } r \in [\sigma_\infty,\sigma^\infty] \text{ and a.e. } x \in \Omega.$$

The constant function $w_\infty \equiv \sigma^\infty$ is thus a solution of (2.2). Besides, problem (2.2) admits on other solution, whence (i) of Theorem

2.2 implies that the $\omega$-limit set $\omega(w)$ is a singleton, i.e., $w(t) \to \sigma^{\infty}$ in $H$ and weakly in $V$ as $t \to \infty$. Thus the assertion $(i)$ of the theorem has been obtained. Assertion $(ii)$ can be obtained similarly. For detailed proofs we refer to the author's forthcoming papers [5,6].

## REFERENCES.

[1] G.Caginalp. An analysis of a phase field model of a free boundary, Arch. Rat. Mech. Anal. 92 (1986), 205-245.

[2]. C.Eliott and S.Zheng. Global existence and stability of solutions to the phase field equations, in: Free Boundary Problems, K.-H.Hoffmann and J.Sprekels, eds.,Intern.Ser.Numer. Math.Vol.95 Birkhauser Verlag, Basel, (1990) 48-58.

[3]. G.J.Fix. Phase field methods for free boundary problems, in: Free Boundary Problems: Theory and Appplications, A.Fasano and M.Primicerio, eds., Pitman. London, (1983) 580-589.

[4]. N.Kenmochi. Solvability of nonlinear evolution equations with time-dependent contstraints and applications, Bull.Fac.Education, Chiba Univ., 30 (Part II) (1981), 1-87.

[5]. N.Kenmochi and M.Niezgodka. Evolution system of nonlinear variational inequalities arising from phase change problems, preprint.

[6]. N.Kenmochi and M.Niezgodka. Systems of parabolic variational inequalities arising in phase transitions, preprint.

[7]. N.Kenmochi and I.Pawlow. A class of nonlinear elliptic-parabolic equations with time-dependent constraints, Nonlinear Anal.TMA, 10 (1986), 1181-1202.

[8]. O.Penrose and P.C.Fife. Thermodynamically consistent models of phase-field type for the kinetics of phase transitions, Physica D, 43 (1990), 44-62.

[9]. J.Sprekels and S.Zheng. Global smooth solutions to a thermodynamically consistent model of phase-field type in higher space dimensions, preprint, Univ.Essen, (1991).

[10]. S.Zheng. Global existence for a thermodynamically consistent model of phase field type, IMA Preprint Series, No.735, Univ. Minnesota, Minneapolis, (1990).

# CONTACT VISCOELASTOPLASTIC PROBLEM FOR A BEAM

A.M.Khludnev
Lavrentyev Institute of Hydrodynamics
Novosibirsk 630090, Russia

The aim of this paper is to investigate the problem of a contact between a beam and a rigid punch. The constitutive law for the beam describes viscoelasticity and plasticity properties. The problem contains two restrictions of inequality type. The first restriction has a geometrical character and describes a non-penetration condition, the second one is of mechanical nature and corresponds to yield condition. The main result of the paper is a proof of existence theorem. As a rule, the existence theorems in plastic problems provides a solvability of some corollaries obtained from initial exact formulation of the problem. In so doing, boundary conditions partly may be lost ([1-4]). A main peculiarity of the result obtained in this paper is the following. It provides a solvability of the exact formulation of the problem. In particular all boundary conditions will be fulfilled. A similar result for a beam with another constitutive law was established in [5].

Key words: contact problem, variational inequality, regular solution, viscoelastoplasticity.

The formulation of the problem is as follows. Let $I=(0,1)$. It is required to find the functions $v, w, m, n, \xi$, satisfying in $Q = I\times(0,1)$ the following relations

$$w - v\varphi_x \geqslant \varphi, \tag{1}$$

$$(m_{xx} + f)(\bar{w} - w) + (n_x + g)(\bar{v} - v) \leqslant 0 \tag{2}$$

$$\forall\ \bar{v},\bar{w}, \quad \bar{w} - \bar{v}\varphi_x \geqslant \varphi,$$

$$n_t - v_x = 0, \tag{3}$$

$$m_{txx} - w_{xx} = \eta, \tag{4}$$

$$|m| \leqslant k, \quad \eta(\bar{m} - m) \leqslant 0 \quad \forall\, \bar{m}, \ |\bar{m}| \leqslant k, \tag{5}$$

$$n = n^0, \ m = m^0, \ t = 0, \tag{6}$$

$$v = w = m = 0, \ x = 0, \ 1. \tag{7}$$

Here $v$, $w$ are horizontal and normal displacements of beam points; $m$, $n$ are bending moment and tangential force, respectively. The equation $y = \varphi(x)$ describes a punch shape; $f, g \in L^2(I)$ are given exterior forces.

Let us introduce the convex sets

$$K = \left\{ m \in H_0^1(I) \,\Big|\, |m| \leqslant k \text{ on } I \right\}, \ k = const > 0,$$

$$B = \left\{ (v,w) \,\Big|\, v,w \in H_0^1(I) \,, \ w - v\varphi_x \geqslant \varphi \text{ on } I \right\}.$$

We assume that the initial data $m^0$, $n^0$ are chosen satisfying the following equations in $I$

$$m^0_{xx} + f = 0, \quad n^0_x + g = 0 \tag{8}$$

and moreover $(1+\varkappa)m^0 \in K$, $\varkappa = const > 0$ . It is assumed also that $\varphi \in H^2(I)$; $\varphi(0)$, $\varphi(1) \leqslant 0$. In this case the functions $v^0, w^0$ exist such that $(v^0, w^0) \in B$ and $w^0_{xx} \in L^2(I)$. It should be noted from (3), (6) that $n(t) = n^0 + \int_0^t v_x d\tau$. In particular, taking into account (8) one has $n_x = -g + \int_0^t v_{xx} d\tau$ . The main result of this paper may be formulated as follows.

**THEOREM.** Let the above-mentioned conditions be fulfilled. Then the functions $v$, $w$, $m$, $n$, $\xi$ exist such that

$$m, \ m_t, \ v \in L^\infty(0,T;H_0^1(I)), \quad n, \ n_t \in L^\infty(0,T;L^2(I)),$$

$$w \in L^2(0,T;H_0^1(I)), \ \eta \in L^2(0,T;H^{-1}(I)),$$

$$m(t) \in K, \ (v(t),w(t)) \in B \ \text{ a.e. on } (0,T),$$

$$\int_0^T \langle m_{xx}+f,\bar{w}-w\rangle dt + \int_0^T \langle n_x+g,\ \bar{v}-v\rangle dt \leqslant 0$$

$$\forall\ \bar{v},\bar{w} \in L^2(0,T;H_0^1(I)),\ (\bar{v}(t),\bar{w}(t)) \in B,$$

$$n_t - v_x = 0,$$

$$m_{txx} - w_{xx} = \eta,$$

$$\int_0^T \langle \eta,\bar{m}-m\rangle dt \leqslant 0 \quad \forall\ \bar{m} \in L^2(0,T;H_0^1(I)),\ \bar{m}(t) \in K,$$

the boundary conditions (6) being fulfilled.

**Proof.** Let $p(m) = m - \pi m$, where $\pi$ is the projector of the space $L^2(I)$ onto the set $\left\{m \in L^2(I)\ \middle|\ |m| \leqslant k \text{ a.e. on } I\right\}$. Let also $(h_1(v,w),h_2(v,w)) = (v,w) - \pi_0(v,w)$, $\pi_0$ projects the space $[L^2(I)]^2$ on the set of functions from this space satisfying the restriction $w - v\varphi_x \geqslant \varphi$.

We consider the regularized boundary-value problem with positive parameters $\varepsilon$, $\delta$, $\lambda$

$$\varepsilon v_t - \int_0^t v_{xx}\, d\tau + \delta^{-1}h_1(v,w) = 0, \tag{9}$$

$$\varepsilon w_t - \varepsilon w_{xx} - m_{xx} + \delta^{-1}h_2(v,w) = f, \tag{10}$$

$$- m_{txx} + w_{xx} + \lambda^{-1}p(m) = 0, \tag{11}$$

$$v = v^0,\quad w = w^0,\ m = m^0,\quad t = 0, \tag{12}$$

$$v = w = m = 0,\qquad x = 0,1. \tag{13}$$

A general scheme of reasonings is the following. First of all, the solvability of (9)-(13) will be established for fixed parameters. Then the passages to the limit $\varepsilon \to 0$, $\lambda \to 0$, $\delta \to 0$ will be justified. Let us state a priori estimates of the solution of (9)-(13). To do this we multiply (9)-(11) by $v - v^0$, $w - w^0$, $m - m^0$ respectively and integrate over I. The following differential inequality will be obtained

$$\frac{1}{2}\frac{d}{dt}\left\{ \varepsilon\|v(t)\|^2 + \varepsilon\|w(t)\|^2 + \|m_x(t)\|^2 + \|\int_0^t v_x d\tau\|^2 \right\} +$$

$$+ \frac{\varepsilon}{2}\| w_x(t)\|^2 \leqslant \varepsilon\langle v_t(t),v^0\rangle + \varepsilon\langle w_t(t),w^0\rangle + \frac{1}{2}\|m_x(t)\|^2 +$$

$$+ \frac{1}{2}\| \int_0^t v_x d\tau \|^2 + \langle m_{tx}(t),m_x^0 \rangle + c.$$

Here we have used the first equation (8) and the nonnegativity of the terms connected with the operators $(h_1,h_2)$, $p$. Constant $c$ is uniform all over $\varepsilon \leqslant \varepsilon_0$, $\delta$, $\lambda$. The brackets $\langle\cdot,\cdot\rangle$ mean a scalar product in $L^2(I)$ and duality between $H_0^1(I)$ and $H^{-1}(I)$, $\|\cdot\|$ is the norm in $L^2(I)$.

The obtained differential inequality may be integrated so far as the functions $v^0$, $w^0$, $m^0$ do not depend on $t$. As a result, we obtain

$$\max_{0\leqslant t\leqslant T}\left\{ \varepsilon\|v(t)\|^2 + \varepsilon\|w(t)\|^2 + \|m_x(t)\|^2 + \| \int_0^t v_x d\tau \|^2 \right\} \leqslant c, \quad (14)$$

$$\varepsilon\|w_x\|^2_{L^2(Q)} \leqslant c \quad (15)$$

uniformly over $\varepsilon \leqslant \varepsilon_0$, $\delta$, $\lambda$. It follows from (9) as $t = 0$ that $v_t(0) = 0$. Besides, the equations (10), (11) give $w_t(0) = w^0_{xx} \in L^2(I)$, $m_{txx}(0) = w^0_{xx}$ as $t = 0$, i.e. $m_t(0)$ is bounded in $H^2(I)\cap H_0^1(I)$. Let us differentiate the equations (9)-(11) with respect to t and multiply by $v_t$, $w_t$, $m_t$. By view of inequalities

$$\langle(h_{1t},h_{2t}),\ (v_t,w_t)\rangle \geqslant 0\ ,\ \langle p(m)_t,\ m_t\rangle \geqslant 0,$$

valid a.e. on $(0,T)$ (see [6], ch. 3, §6) we derive

$$\frac{1}{2}\frac{d}{dt}\left\{ \varepsilon\|v_t(t)\|^2 + \varepsilon\|w_t(t)\|^2 + \|v_x(t)\|^2 + \|m_{tx}(t)\|^2 \right\} +$$

$$+ \varepsilon\|w_{tx}(t)\|^2 \leqslant 0.$$

After integrating this inequality the following estimates can be found

$$\max_{0\leqslant t\leqslant T}\left\{ \varepsilon\|v_t(t)\|^2 + \varepsilon\|w_t(t)\|^2 + \|v_x(t)\|^2 + \|m_{tx}(t)\|^2 \right\} \leqslant c, \quad (16)$$

$$\varepsilon\|w_{tx}(t)\|^2_{L^2(Q)} \leqslant c. \quad (17)$$

Let us establish one more estimate. Multiplication (9)-(11) by $v - v^o$, $w - w^o$, $m - m^o$, respectively, gives that $\lambda^{-1}\langle p(m), m-m^0\rangle$ are bounded in $L^2(0,T)$ uniformly over $\varepsilon \leqslant \varepsilon^0$, $\delta$, $\lambda$. We introduce the functional $q(m) = \| m - \pi m \|^2/2$. Its derivative can be found by formula $q'(m) = p(m)$. Let us choose $\hat{m} \in L^\infty(Q)$ such that $|\hat{m}(t) + m^0| \leqslant k$ a.e. on $(0,T)$. Then the convexity of $q$ gives

$$\lambda^{-1}\langle p(m),\hat{m}\rangle \leqslant \lambda^{-1}\langle p(m),m-m^0\rangle + \lambda^{-1}\langle q(\hat{m} + m^0) - \lambda^{-1}q(m).$$

Taking into account the above-mentioned boundedness of $\lambda^{-1}\langle p\langle m), m-m^0\rangle$ we obtain

$$\lambda^{-1}p(m) \text{ are bounded in } L^2(0,T;L^1(I)).$$

Consequently the inclusion $L^1(I) \subset H^{-1}(I)$ and (11) provide the following estimate

$$w \text{ are bounded in } L^2(0,T;H^1_0(I)), \quad (18)$$

which is uniform over $\varepsilon \leqslant \varepsilon_0$, $\delta$, $\lambda$. Now we can prove the solvability of the problem (9)-(13) for fixed parameters using Galerkin method. It is of importance that the above-established estimates (14)-(17) could be "reestablished" for Galerkin's solutions. It gives an opportunity to prove the solvability of Galerkin's equations and to justify the passage to the limit over Galerkin's number. The estimate (18) could be obtained in the similar way from (9)-(13) after proving the solvability of (9)-(13). Therefore, the solution of the problem (9)-(13) for fixed parameters exists such that

$$v,\ w,\ v_t,\ w_t \in L^\infty(0,T;L^2(I)),$$

$$w,\ w_t \in L^2(0,T;H^1_0(I)), \quad v,\ m,\ m_t \in L^\infty(0,T;H^1_0(I)).$$

Let us justify the passages to the limit. At every step of reasonings the solution will be supplied by an appropriate symbol without mentioning the dependence on the rest parameters. First of all, we consider the case $\varepsilon \to 0$. The solution of (9)-(13) will be denoted by $v^\varepsilon$, $w^\varepsilon$, $m^\varepsilon$. We may assume that a subsequence possesses the following properties as $\varepsilon \to 0$

$$
\begin{aligned}
&\varepsilon v^\varepsilon,\ \varepsilon v_t^\varepsilon,\ \varepsilon w^\varepsilon,\ \varepsilon w_t^\varepsilon \to 0 && \text{*-weakly in } L^\infty(0,T;L^2(I)),\\
&\varepsilon w^\varepsilon \to 0 && \text{weakly in } L^2(0,T;H_0^1(I)),\\
&v^\varepsilon,\ m^\varepsilon,\ m_t^\varepsilon \to v^\lambda,\ m^\lambda,\ m_t^\lambda && \text{*-weakly in } L^\infty(0,T;H_0^1(I)),\\
&w^\varepsilon \to w^\lambda && \text{weakly in } L^2(0,T;H_0^1(I)),\\
&m^\varepsilon \to m^\lambda && \text{strongly in } L^2(Q),\\
&\int_0^t m_x^\varepsilon\, d\tau \to \int_0^t m_x^\lambda\, d\tau && \text{weakly in } L^2(Q).
\end{aligned}
$$

It follows from (9)-(11) after passage to the limit $\varepsilon \to 0$ that limiting functions $v^\lambda$, $w^\lambda$, $m^\lambda$ satisfy the relations

$$- \int_0^t v_{xx}^\lambda d\tau + \delta^{-1} h_1(v^\lambda, w^\lambda) = 0, \tag{19}$$

$$- m_{xx}^\lambda + \delta^{-1} h_2(v^\lambda, w^\lambda) = f, \tag{20}$$

$$- m_{txx}^\lambda + w_{xx}^\lambda + \lambda^{-1} p(m) = 0. \tag{21}$$

The justification of the convergence in nonlinear terms $h_i(v^\varepsilon, w^\varepsilon)$ could be done using the monotonicity of the operator $(h_1, h_2)$. Let us fulfil the passage to limit $\lambda \to 0$ in (19)-(21). It can be assumed that a subsequence $v^\lambda$, $w^\lambda$, $m^\lambda$ with previous notations possesses the following properties as $\lambda \to 0$

$$
\begin{aligned}
&v^\lambda,\ m^\lambda,\ m_t^\lambda \to v^\delta,\ m^\delta,\ m_t^\delta && \text{*-weakly in } L^\infty(0,T;H_0^1(I)),\\
&w^\lambda \to w^\delta && \text{weakly in } L^2(0,T;H_0^1(I)),\\
&\lambda^{-1} p(m^\lambda) \to \eta^\delta && \text{weakly in } L^2(0,T;H^{-1}(I)),\\
&\int_0^t v_x^\lambda\, d\tau \to \int_0^t v_x^\delta\, d\tau && \text{weakly in } L^2(Q).
\end{aligned}
$$

It can be easy seen from (20) that as $\lambda \to 0$

$$m^\lambda \text{ are bounded in } L^2(0,T;H^2(I)\cap H_0^1(I)).$$

This boundedness is not uniform over $\delta$, in general, of course. Therefore, we may additionally assume that for every fixed $\delta$

$$m^\lambda \to m^\delta \qquad \text{strongly in } L^2(0,T;H_0^1(I)). \tag{22}$$

It follows from (21) that

$$- m^\delta_{txx} + w^\delta_{xx} + \eta^\delta = 0 \ . \tag{23}$$

Besides, (21) and monotonicity provide the validity of the inequality

$$\int_0^T \langle -m^\lambda_{txx} + w^\lambda_{xx} \ , \ \bar{m} - m^\lambda \rangle \, dt \geqslant 0.$$

We can pass to the limit here on the basis of above-written convergence and get

$$\int_0^T \langle \eta^\delta, \ \bar{m} - m^\delta \rangle \, dt \leqslant 0 \quad \forall \ \bar{m} \in L^2(0,T;H_0^1(I)), \ \bar{m}(t) \in K \text{ a.e.} \tag{24}$$

Thus, the limiting functions $v^\delta$, $w^\delta$, $m^\delta$, $\eta^\delta$ satisfy (23), (24) and equations

$$- \int_0^t v^\delta_{xx} \, d\tau \ + \delta^{-1} h_1(v^\delta, w^\delta) = 0 \ , \tag{25}$$

$$- m^\delta_{xx} + \delta^{-1} h_2(v^\delta, w^\delta) = f. \tag{26}$$

The inclusion $m^\delta(t) \in K$ follows immediately from (21).

At last, we demonstrate the opportunity of the passage to limit $\delta \to 0$. Choosing a subsequence with the same notation we assume

$$\begin{aligned} v^\delta, \ m^\delta, \ m^\delta_t \ &\to v, \ m, \ m_t && *\text{-weakly in } L^\infty(0,T;H_0^1(I)), \\ w^\delta &\to w && \text{weakly in } L^2(0,T;H_0^1(I)), \\ \eta &\to \eta && \text{weakly in } L^2(0,T;H^{-1}(I)), \\ \int_0^t v^\delta_x \, d\tau &\to \int_0^t v_x \, d\tau && \text{weakly in } L^2(Q). \end{aligned}$$

Let us multiply (25), (26) by $\bar{v} - v^\delta$, $\bar{w} - w^\delta$, respectively and sum up with (24). It should be noticed that the term $\int_0^T \langle m^\delta_{xx}, w^\delta \rangle dt$ will be absent in the obtained relation. Therefore after passage to the limit we get the equation

$$-m_{txx} + w_{xx} + \eta = 0$$

and the inequality

$$\int_0^T \langle m_{xx} + f,\ \bar{w} - w \rangle dt + \int_0^T \langle \int_0^t v_{xx} d\tau,\ \bar{v} - v \rangle dt +$$

$$+ \int_0^T \langle \eta,\ \bar{m} - m \rangle dt \leqslant 0 \quad \forall\ \bar{v},\ \bar{w},\ \bar{m} \in L^2(0,T;H^1_0(I)),$$

$$(\bar{v}(t),\ w(t)) \in B, \qquad \bar{m}(t) \in K \ \text{ a.e. on } (0,T).$$

We may denote

$$n(t) = n^0 + \int_0^t v_x \, d\tau. \tag{27}$$

The functions $(\bar{v},\bar{w})$ , $\bar{m}$ are chosen independently, therefore, it follows from the previous relations that all equations and inequalities written in Theorem are fulfilled. We have to note also that all passages to limit $\varepsilon \to 0$, $\lambda \to 0$, $\delta \to 0$ keep the initial condition $m = m^0$. The initial condition for $n$ follows from (27). Theorem has been completely proved.

## REFERENCES.


[1] Suquet P.M. Evolution problems for a class of dissipative materials, Quart. of Appl.Math. (1981), v.38, N 4, pp.391-414.

[2] Johnson C. Existence theorems for plasticity problems, J.Math.Pures et Appl., 1976, v.55, N 4, pp.431-444.

[3] Temam R. Problemes mathematiques en plasticite, Gauthier - Villars, Paris, 1983.

[4] Khludnev A.M. On variational inequalities in contact plastic problems. Differential Equations, (1988), v.24, N 9, pp.1622-1628 (in Russian).

[5] Hoffmann K.-H., Khludnev A.M. Contact elastoplastic problem for a beam (to appear).

[6] Lions J.L. Some methods of solving nonlinear boundary-value problems, Moscow, Mir Publisher (1972), 587 p.

# APPLICATION OF A FINITE-ELEMENT METHOD TO TWO-DIMENSIONAL CONTACT PROBLEMS

*S.N.Korobeinikov, V.V.Alyokhin*
*Lavrentyev institute of Hydrodynamics*
*Novosibirsk 630090 USSR*

The paper is concerned with the solution of two-dimensional nonlinear contact problems of solid mechanics by the finite element method. The Lagrange multiplier method so as penalty function one are used. The solutions of some 2/D contact problems are presented.

Key words: finite element method, contact problems.

## 1. INTRODUCTION.

Recently a finite-element method has been widely applied for solving nonlinear problems of solid mechanics [1] including the class of contact problems. Their solution is based on pre-determining the unknown boundary between two (or more) bodies and contact forces of interaction between them.

As the contact problem belongs to the class with restrictions, the Lagrange multiplier [2-7] and penalty function [6-9] methods are the most popular ones used for solving such problems.

Each of the above methods is advantageous or disadvantageous over other ones. In the Lagrange multiplier method, a kinematic contact condition is fulfilled, however, introducing the Lagrange multipliers results in a complicated formulation of equations. As far as the penalty function method is concerned, the number of equations is invariable when introducing contact conditions. However, the kinematic contact condition is difficult to be satisfied exactly in a numerical algorithm, since introducing a high penalty coefficient leads to an ill conditioning of a stiffness matrix, and for a small penalty coefficient, the kinematic contact condition is poorly fulfilled.

In the present paper, the contact problems are solved both with the use of the Lagrange multiplier method and penalty function one. Both of them have been introduced into the PIONER program [10] and may be used in solving nonlinear static problems and dynamic ones by integration of equations of motion with the help of implicit schemes. The solutions of some 2/D contact problems are presented.

## 2. FORMULATION OF CONTACT PROBLEMS.

Let two bodies $B^1$ and $B^2$ be in contact as a result of application of prescribed loads and displacements. This means that both of them have a mutual boundary on which the condition of non-penetration of one body into another is fulfilled,

$$g = ({}^t\mathbf{x}^2 - {}^t\mathbf{x}^1)\cdot\mathbf{n} \geqslant 0 \tag{1}$$

where $g$ is the normal gap between the bodies, ${}^t\mathbf{x}^1$ and ${}^t\mathbf{x}^2$ are the vectors of the positions of $B^1$ and $B^2$ points at the moment $t$, $\mathbf{n}$ is the unit vector of the normal to the contact surface.

In the contact zone, there develop normal contact forces,

$$t_n \equiv \mathbf{t}\cdot\mathbf{n} \leqslant 0, \tag{2}$$

where $\mathbf{t}$ is the vector of contact traction, $\mathbf{n}$ is the outer normal to the body boundary. At the same time, tangential contact tractions $t_t$ may be both positive or negative. These tractions are to obey some friction law. We'll use the Coulomb's friction law here. A static friction coefficient is denoted by $\mu_s$, a dynamic one is denoted by $\mu_d$, $\mu_d \leqslant \mu_s$. Let a tangential contact traction $t_t$ be obtained from the solution of the problem. According to the Coulomb's friction law, it is assumed that there is no relative motion between the contacting particles of the bodies as long as the inequality is fulfilled,

$$|t_t| \leqslant \mu_s\cdot|t_n|. \tag{3}$$

When the latter is not fulfilled, contacting particles come into relative motion. Tangential forces $t_t$ must thereby obey the equality

$$|t_t| = \mu_d\cdot|t_n|. \tag{4}$$

The relative motion comes to an end if

$$|t_t| < \mu_d\cdot|t_n| \tag{5}$$

and we have to follow whether inequality (3) is fulfilled or not.

## 3. GEOMETRICAL CONDITIONS OF CONTACT BETWEEN BODIES.

A numerical solution of the contact problem is obtained by the finite-element method. In so doing, the contacting bodies are covered by finite elements with appropriate nodes. If there is a pair of contacting bodies, one of them will be called a contactor and the other will be called a target. The contact problem is solved so that the nodes of the contactor do not penetrate the boundary of the target. At the same time, the nodes of the target may penetrate the contactor [2].

Before starting the solution of the contact problem, the zones of possible contact between the contactor and target may be distinguished in the form of segments connected by the nodal points.

When solving the contact problem, it is checked at each time step whether the nodal points are in the zone of possible contact or not. If the penetration takes place, the algorithm of solving contact problems holds true, thus the nodal points of the zone of possible contact of the contactor must be on the boundary of the target as a result of an iteration process. The contact forces acting upon the boundary are to be determined.

## 4. SOLVING A CONTACT PROBLEM BY A LAGRANGE MULTIPLIER METHOD.

The incremental equilibrium equations to solve a contact problem are presented in [2]. They may be written as follows:

$$\begin{array}{r} \\ \text{NEQ}\{ \\ \\ \text{NEQC}\{ \end{array} \overset{\qquad\text{NEQ}\qquad\qquad\qquad\text{NEQC}}{\begin{bmatrix} {}^{t+\Delta t}K^{(i-1)} & {}^{t+\Delta t}B^{(i-1)} \\ & \\ {}^{t+\Delta t}B^{(i-1)T} & 0 \end{bmatrix}} \begin{bmatrix} \Delta U^{(i)} \\ \\ \Delta\Lambda^{(i)} \end{bmatrix} = \begin{bmatrix} {}^{t+\Delta t}R_1^{(i-1)} \\ \\ {}^{t+\Delta t}\Delta_c^{(i-1)} \end{bmatrix} \tag{6}$$

where

$${}^{t+\Delta t}R_1^{(i-1)} = {}^{t+\Delta t}R - {}^{t+\Delta t}F^{(i-1)} + {}^{t+\Delta t}R_c^{(i-1)}. \tag{7}$$

Here $\Delta U^{(i)}$ is the displacement increment vector for the iteration $i$; $\Delta\Lambda^{(i)}$ is the contact force increment vector for the iteration $i$; ${}^{t+\Delta t}K^{(i-1)}$ is the tangential stiffness matrix of the bodies $B^1$ and $B^2$ (without taking into account contact forces); ${}^{t+\Delta t}\mathbf{B}^{(i-1)}$ is the tangential contact stiffness matrix obtained by means of taking into account contact forces; ${}^{t+\Delta t}\mathbf{F}^{(i-1)}$ is the internal nodal force vector, equivalent to the element stresses in the bodies $B^1$ and $B^2$; ${}^{t+\Delta t}\mathbf{R}$ is the vector of applied external forces; ${}^{t+\Delta t}\mathbf{R}_c^{(i-1)}$ is the contact force vector; ${}^{t+\Delta t}\Delta_c^{(i-1)}$ is the overlap vector; NEQ is the total number of degrees of freedom of the bodies $B^1$ and $B^2$; NEQC is the total number of contact equations = $2\times$(a total number of nodes of the contactor being in stick contact)+(a total number of nodes of the contactor being in slide contact); the left superscript denotes the moment of time with respect to which a given value is considered; the right superscript in brackets denotes the number of iteration; index "$T$" denotes the transposition operation.

System (6) in [2] is solved directly with the use of $\mathrm{LDL}^T$ factorization [1] of the matrix in the left-hand side. However, there arise some difficulties: *(i)* first, zeros on the main diagonal may lead to a not positive definite matrix, *(ii)* second, since the value of NEQC is variable on each iteration, both the size and profile of the matrix change in the left-hand side of system (6).

System (6) may be re-written as

$$
\begin{aligned}
&{}^{t+\Delta t}\mathbf{K}^{(i-1)}\cdot\Delta\mathbf{U}^{(i)} + {}^{t+\Delta t}\mathbf{B}^{(i-1)}\cdot\Delta\Lambda^{(i)} = {}^{t+\Delta t}\mathbf{R}_1^{(i-1)},\\
&{}^{t+\Delta t}\mathbf{B}^{(i-1)T}\cdot\Delta\mathbf{U}^{(i)} = {}^{t+\Delta t}\Delta_c^{(i)}.
\end{aligned}
\tag{8}
$$

After further transformations of equations (8), analogous to those made in [7], we obtain

$$\mathbf{B}^T\cdot\mathbf{K}^{-1}\cdot\mathbf{B}\cdot\Delta\Lambda = \mathbf{B}^T\cdot\mathbf{K}^{-1}\cdot\mathbf{R}_1 - \Delta_c, \tag{9}$$

$$\Delta\mathbf{U} = \mathbf{K}^{-1}\cdot(\mathbf{R}_1 - \mathbf{B}\cdot\Delta\Lambda). \tag{10}$$

Here the indices of matrices and vectors, which are the same as in system (8) have been omitted.

Since in the explicit form, the matrix $\mathbf{B}^T\cdot\mathbf{K}^{-1}\cdot\mathbf{B}$ in the left-hand side of (9) is not calculated, system (9) is solved by the conjugate gradient method [7]. After the solution of (9) has been obtained, the calculated value of $\Delta\Lambda$ is substituted in the right-hand side of system (10) and, as a result, the solution for $\Delta U$ is obtained.

At each iteration for the moment of time $t + \Delta t$, we successively solve the system of equations (9) and (10) as long as the solution converges. Following [7], this procedure of the solution of the contact problem will be called the one-level iteration procedure. A two-level iteration method [7] may also be used. Its essence is that to satisfy an equilibrium state, material and geometrial nonlinearities are considered in an outer cycle, while nonlinearity introduced by solving contact problems is considered in an internal iteration cycle.

In the latter case, the following system of equations is solved:

$$B^T \cdot K^{-1} \cdot B \cdot \Delta\Lambda = - \Delta_c, \tag{11}$$

$$\Delta U = - K^{-1} \cdot B \cdot \Delta\Lambda. \tag{12}$$

The internal cycle iterations are continuing so long as the convergence condition of the iteration process at overlaps is fulfilled. In so doing, only the elements of the matrix B are recalculated. In calculating the vector of contact forces ${}^{t+\Delta t}R_c^{(i-1)}$ we pass to the external iteration cycle. Since, when solving a contact problem in the internal cycle the vector $\Delta\Lambda$ becomes small, in the external cycle the following equation is solved according to (6):

$${}^{t+\Delta t}K^{(i-1)} \cdot \Delta U^{(i)} = {}^{t+\Delta t}R - {}^{t+\Delta t}F^{(i-1)} + {}^{t+\Delta t}R_c^{(i-1)} \tag{13}$$

## 5. SOLVING A CONTACT PROBLEM BY A PENALTY FUNCTION METHOD.

Let us formulate the solution of the contact problem by the penalty function method for the case of an absolute sliding ($\mu_s = \mu_d = 0$). The incremental equilibrium equations taking into account contact forces are in the form [6]

$$\left[ {}^{t+\Delta t}K^{(i-1)} + {}^{t+\Delta t}K_c^{(i-1)} \right] \cdot \Delta U^{(i)} = {}^{t+\Delta t}R_1. \tag{14}$$

Here $^{t+\Delta t}\mathrm{K}_c^{(i-1)}$ is the tangential contact stiffness matrix (dimension of NEQ × NEQ) which is formed by a direct summing of local contact stiffness matrices $^{t+\Delta t}\mathrm{K}_{c,e}^{(i-1)}$, formed after the nodes of the contactor have come into contact with the segments of the target. For a contact element, the matrix $^{t+\Delta t}\mathrm{K}_{c,e}^{(i-1)}$ is in the form

$$^{t+\Delta t}\mathrm{K}_{c,e}^{(i-1)} = \omega \cdot \mathrm{B}_e \cdot \mathrm{B}_e^T, \tag{15}$$

where $\mathrm{B}_e$ is the local vector whose form is as in solving the problem by the Lagrange multiplier method without taking into account friction, $\omega$ is the penalty parameter. The local vector of the contact forces takes the form

$$^{t+\Delta t}\mathrm{R}_{c,e}^{(i-1)} = -\,\omega \cdot {}^{t+\Delta t}g_e^{(i-1)} \cdot \mathrm{B}_e, \tag{16}$$

where $^{t+\Delta t}g_e^{(i-1)}$ is the normal component of the overlap vector. From (16) it follows that the value of a contact force is directly proportional to the penalty value and the normal penetration depth. Thus, the use of the penalty function method is equivalent to the introduction of fictitious springs acting along the normals towards the segments of the target with the Young's modulus equal to the penalty value $\omega$.

From (16) it follows that the overlap $^{t+\Delta t}g_e^{(i-1)}$ decreases with increasing the penalty $\omega$. In theoretical investigations [9] of the solution convergence of the penalty function method algorithm to the solution of the basic contact problem, the penalty parameter tends to infinity. However, in the numerical solution, high value of $\omega$ may result in a ill conditioning of the stiffness matrix, and as $\omega$ decreases the overlap in the numerical solution will increase . Therefore, the parameter $\omega$ is to be used with care.

Since the local contact stiffness matrix is determined only when the contactor node comes into contact with the target segment, the profile of a global tangential stiffness matrix may vary at each iteration (without changing the dimension). This circumstance makes the solution of the contact problem somewhat more complicated as compared to ordinary problems without taking into account contact interactions. In other respects, the solution algorithm for the contact problem by the penalty function method is the same as in solving nonlinear problems [1].

## 6. SOLVING DYNAMIC CONTACT PROBLEMS.

The above approaches to solve contact problems are valid for quasi-static ones. When solving dynamic contact problems by the implicit integration scheme, the modification of this algorithm from statics to dynamics is such as in the case of solving nonlinear problems without the contact conditions.The allowance for dynamic terms results in that the effective tangential stiffness matrix ${}^{t}\hat{K}$ is considered instead of the tangential stiffness matrix ${}^{t}K$ and the effective vector of external loads ${}^{t+\Delta t}\hat{R}$ - instead of the vector ${}^{t+\Delta t}R$ [1].

The approach to solving dynamic contact problems is used in [3]. However, in this case the same difficulties arise as in numerical modelling of shock loading. In using the nondissipative schemes of integration of the motion equations, spurious oscillations take place in the numerical solution, which are due to that we are unable to represent accurately enough the contribution of higher modes to the solution. The implicit Newmark scheme with the parameters $\delta = 0.5$, $\alpha = 0.25$ is in particular such a numerical scheme. In [3], to solve dynamic contact problems it is recommended to use the Newmark dissipative scheme with the parameters $\delta = \alpha = 0.5$.

## 7. EXAMPLES OF NUMERICAL CALCULATION.

The above-mentioned algorithms of solution of the contact problems by the Lagrange multiplier method and the penalty function method are introduced into a computer program PIONER [10]. Further, the solutions of some test problems obtained with the help of this program are presented.

### THE ROD IMPACT UPON A RIGID TARGET.

Let us consider an elastic rod in a free flight that impacts upon a rigid wall (Fig.1a). In a finite-element modelling, the rod is supposed to be in the plane stress state. To model one-dimensional deformation, all degrees of freedom in the $y$ axis are assumed to be absent and in the $x$ axis the rod is divided into 20 elements (Fig.1b).

The problem is considered in a geometrically linear statement. The numerical solution of the problem is compared with the analytical one.

At the moment of impact, the compression wave is generated, which propagates from the right-hand end as far as the left-hand end in $t_1 = 0.01$ sec. After reflection from the left-hand end, the rarefaction wave forms that reaches the right-hand end at $t_2 = 0.02$ sec. Then the rod rebounds from the rigid wall and moves with velocity $V_0$ in a free flight in the opposite direction.

The problem is solved numerically using the implicit Newmark scheme. The time step $\Delta t = 0.0005$ sec is chosen for the wave to run one element in one time step. The numerical solution of the problem is presented before the time step $t/\Delta t = 45$ (that is after rebounding the free rod flight is considered for five time steps). To solve this problem by the Lagrange multiplier method, the one-level iteration procedure was used. The problem solution by the penalty function method is obtained with the penalty parameter $\omega = 10^7$ MN/m. Numerical solutions obtained by both methods almost coincide (Fig.1c). For solution, the different parameters of the Newmark scheme were used. It is seen from comparison of calculation results (Fig.1) that the use of the parameters $\delta = 0.7$, $\alpha = 0.36$ gives the less spurious oscillations in the solution than the use of the parameters suggested in [3].

## HERTZ PROBLEM.

Let us consider the problem of indentation of an elastic cylinder into a rigid half-plane. The problem is taken to be two-dimensional, under the plane strain conditions. The cylinder radius $R$ equals 10 m, the general vertical load $F$ is applied to the central cylinder axis (Fig.2a). By virtue of symmetry, a quarter of the cylinder is considered. The finite-element model of the cylinder is close to those ones presented in [2,8]. For the zone of possible contact, the finite elements with a quadratic approximation of geometry and displacements were used. The expected contact zone of the contactor consists of six segments and coincides with a cylinder boundary part so that one segment is pulled over the boundary of one element. The expected contact zone of the target consists of one segment that is on the boundary of a rigid body.

To avoid the middle cylinder line distortion that takes place when applying the concentrated force $F$ into the cylinder center, the action of this force is substituted by the prescribed displacement on the middle cylinder line. The general displacement of the middle cylinder line is brought to the displacement $\delta_0 = 0.2718\ m$. The corresponding theoretical force $F = 3972\ MN$ [8]. In numerical solution, the applied force is calculated by integration of contact pressures along the contact boundary. Since in the analytical solution only the vertical displacement component is considered, in the numerical solution all the degrees of freedom of nodes in the $x$-coordinate are eliminated. This corresponds to the sticking contact. The numerical solutions of the contact problem are obtained by the penalty function method with the penalty parameter $\omega = 10^9\ MN/m$ and by the Lagrange multiplier one. 65 uniform steps were made along the given displacement $\delta_0$. The vertical load $F^*$ obtained by numerical integration of the contact pressure $P$ over the whole actual contact zone corresponds to the displacement $\delta_0$.

A comparison of the numerical and analytical solutions is performed for the contact pressures $P$ dependent on the current contact length $b$. An analytical solution is represented by the quarters of circles drawn in the axes $(b/b_{max}, P/P_{max})$, where $b_{max}$ is the general contact length and $P_{max}$ is the maximum contact pressure corresponding to the applied force $F^*$. The pressure values determined numerically in the middle of segments of the contact zone of the contactor are denoted by points (numerical solutions obtained by the penalty function method and the Lagrange multiplier one almost coincide). As is seen from the plots (Fig.2b), the numerical solutions agree quite well with the analytical one.

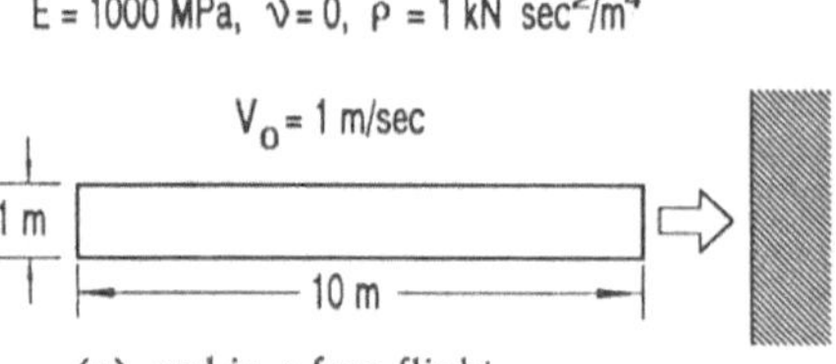


(a) rod in a free flight

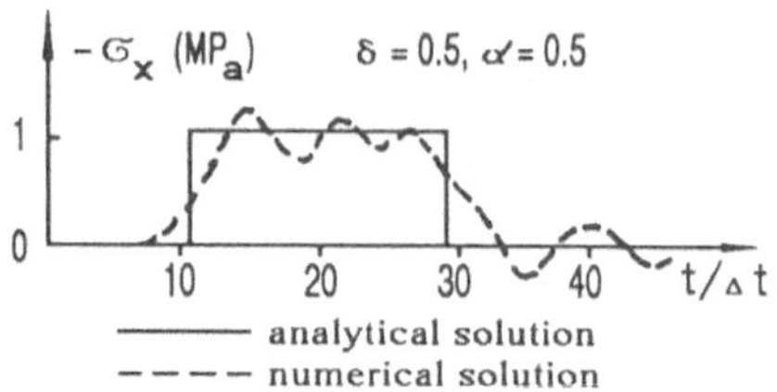


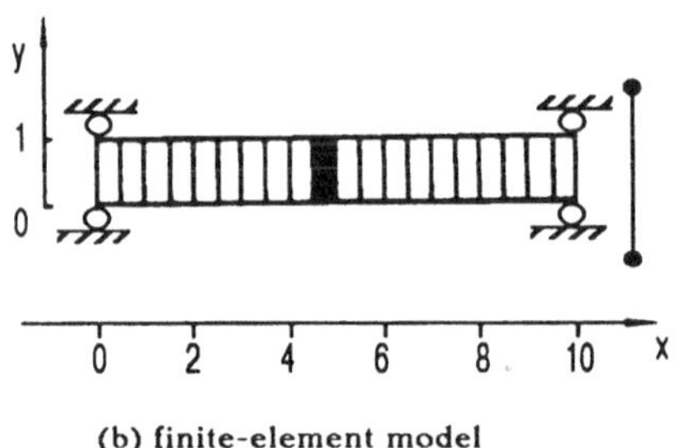


(b) finite-element model

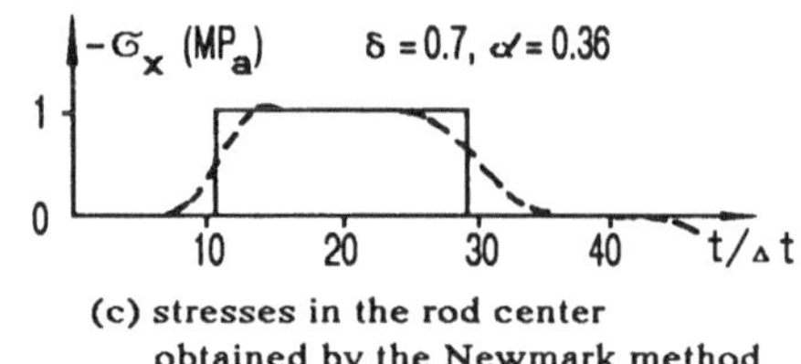


(c) stresses in the rod center obtained by the Newmark method

Fig. 1. Rod impact upon a rigid wall.

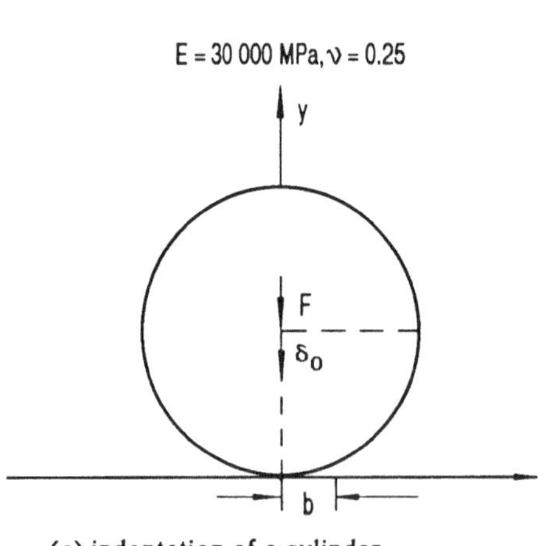


(a) indentation of a cylinder into a rigid half-plane

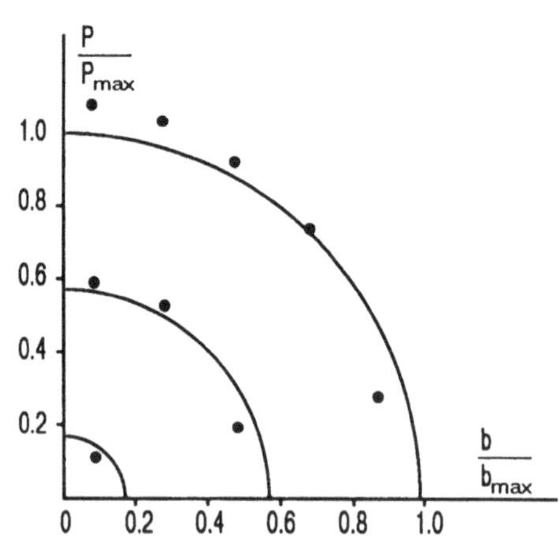


(b) comparison of numerical and analytical solutions

Fig. 2. Hertz problem.

## REFERENCES

[1] K.J.Bathe. Finite element procedures in engineering analysis. Prentice-Hall, Englewood Cliffs, N.J., 1982.

[2] K.J.Bathe, A.B.Chaudhary. A solution method for planar and axisymmetrical contact problem. Int. J. Numer. Meth. Eng., 1985, vol.21, pp. 65-88.

[3] A.B.Chaudhary, K.J.Bathe. A solution method for static and dynamic analysis of three-dimensional contact problems with friction. Computers & Structures, 1986, vol.24, No.6, pp. 855-873.

[4] T.J.R.Hughes et al. A finite-element method for a class of contact-impact problems. Comp. Meth. Appl. Mech. Eng., 1976, vol.8, No.3, pp. 249-276.

[5] N.G.Burago, V.N.Kukudzhanov. Solution of elastoplastic problems by the finite-element method. Institute of Mechanics Problems, USSR Academy of Sciences, preprint No.326, 1988, 63 p.

[6] F.M.Guerra, R.V.Browning. Comparison of two slideline methods using ADINA. Computers & Structures, 1983, vol.17, No.5-6, pp. 819 - 834.

[7] B.Nour-Omid and P.Wriggers. A two-level iteration method for solution of contact problems. Comp. Meth. Appl. Mech. Eng., 1986, vol.54, pp. 131-144.

[8] N.Chandrasekaran, W.E.Haisler, R.E.Goforth. Finite element analysis of Hertz contact problem with friction. Finite Elements in Analysis and Design, 1987, vol.3, pp. 39-56.

[9] N. Kikuchi, J.T.Oden. Contact problems in elastostatic. In: Finite Elements: Special problems in solid mechanics, vol.4 (ed. by J.T.Oden and G.Carey), Prentice-Hall, Englewood Cliffs, N.J., 1984, pp. 158-212.

[10] S.N.Korobeinikov, V.P.Agapov, M.I.Bondarenko, A.N.Soldatkin. The general purpose nonlinear finite element structural analysis program PIONER, Trans. Int. Conf. Numer. Methods and Applications, Sofia, 1989, pp. 228-233.

# COMPUTATIONS OF A GAS BUBBLE MOTION IN LIQUID

*V.A.Korobitsyn*
*Research Institute of Applied Mathematics and Mechanics, Tomsk 634050, RUSSIA*

The rigid body motion in a liquid is determined by complex interaction of bodies and gas cavities. The contact region of bodies and liquid changes essentially due to gas bubbles evolution. The numerical modelling of this interaction meets with difficulties generated by unsteady nonlinear interaction, complex geometric picture of boundaries altering in time, on which the boundary conditions should be satisfied, change of the solution area connection, resource constraint of electronic computer.

The aim of the work is to create an effective computation algorithm of potential flows of incompressible ideal liquid with free surface in terms of homogeneous difference scheme for which the different laws of conservation of mass, impulse and energy are valid and to illustrate algorithm effectiveness on a number of problems. The flow in unbounded fields is modelled on the meshes with a finite number of nodes where computational boundary conditions appear on difference boundaries, modelling the infinity. Boundary conditions on "difference infinity" follow from the energy conservation law. The numerical modelling results of gas bubble evolution in a liquid are presented.

1. Vortex-less axisymmetric flow of heavy liquid occurs in the area constrained with rigid and free boundaries, interfaces. The boundaries on infinity are also possible. Due to axis symmetry of a problem the solution is considered in the area of meridianal section of a liquid $Q(q_1,q_2,t)$, $t > 0$, at the plane $(q_1,q_2)$, $H_3 > 0$, $\Sigma$ - gas-liquid interface, $g$ - gravity vector parallel to $Ox$ - symmetry axis. $H_i$ - Lame functions of curvilinear coordinate system $q_i$ $(i = 1, 2, 3)$, associated with cylindrical coordinate system $(x,r,\psi)$ by nondegenerate transformation.The velocity potential $\varphi(q_i,t)$ in $Q$ satisfies Laplace equation in axisymmetric orthogonal curvilinear coordi-

nate system

$$\left[ \frac{\delta}{\delta q_1} (H_2 H_3 H_1^{-1} \frac{\delta\varphi}{\delta q_1}) + \frac{\delta}{\delta q_2} (H_3 H_1 H_2^{-1} \frac{\delta\varphi}{\delta q_2}) \right] / H_1 H_2 H_3 = 0. \quad (1)$$

At the axis of symmetry $H_3=0$ and rigid boundaries $\frac{\delta\varphi}{\delta n} = 0$. Approaching the infinity the velocity $\mathbf{W} = \Delta\varphi$ and the velocity potential $\varphi$ tend to a zero. At the boundary $\Sigma$ two boundary conditions are performed: kinematic one

$$\frac{dq_i}{dt} = H_i^{-2} \frac{\delta\varphi}{\delta q_i} \quad (2)$$

and dynamical condition of pressure continuity in the vicinity of boundary

$$\frac{\delta\varphi}{\delta t} - 0.5 |\, \Delta \cdot \varphi \,|^2 + p/\rho + gx = f(t) \quad (3)$$

Gas pressure $p$ is considered to change by adiabatic law

$$p(t) = p(0)\ [V(t)/V(0)]^{-\gamma}, \quad (4)$$

where $V(t)$ - the gas cavity volume, $p(0)$, $V(0)$ - pressure and gas volume at initial moment of time, $\gamma$ - adiabatic constant.

2. Let the rigid boundaries of flow domain $Q$ be parallel to coordinate lines. In the domain $Q$ we'll introduce the associated orthogonal non-uniform mesh of nodes correlated with rigid boundaries. Potential and pressure are prescribed at the nodes of this mesh; the physical components of velocities were determined at shift meshes; the $w_i$ - component was determined at the point being half-step of a mesh distant from the node in descending direction of variable $q_i$. Derivatives are approximated by central difference relations. The difference scheme, approximating the equations (1)-(4), has the second approximation order at a mesh with constant spatial steps. The laws of conservation of mass, impulse, energy and vortex as well as a maximum principle hold for this scheme.

The gas-liquid interface will be defined by the points of intersection of a boundary line (surface) with mesh lines. The kinematic condition (2) will be satisfied, if the interface points at each time step are identified with the liquid particles the velocities of which are determined by the potential of the nearest points of a mesh ( form ). The approximation of the conditions at infinity does not disturb the energy and impulse balance. It follows from the kinetic energy conservation that the normal velocity in mesh space $L_2$ is a bounded quantity. It means the stability of a scheme of linear flows and its convergence due to approximation.

To avoid folds of the interfaces the time step must be bounded with the condition

$$\tau < \frac{| R_a - R_b |^2}{( U_b - U_a , R_a - R_b )},$$

where $a$, $b$ are the adjacent liquid particles of interface with velocities $U_a$ , $U_b$ and radius-vectors - $R_a$ , $R_b$ .

The algorithm was tested by Rayleigh problem of collapse of a spherical bubble in incompressible liquid and has shown a good agreement with analytical solution.

Consider the problem of plane interface stability of heavy incompressible liquid being above more light one, i.e. the plane problem of Rayleigh-Taylor instability. The comparison of its numerical solution with Garabedyan solution has shown (Fig. 1) that the peak velocity more readily entered the asymptotic regime with which its growth was linearly dependent on time. The velocity of the upper point of a bubble coincides somewhat less satisfactorily; it partly exceeds the asymptotic value, 0.339, obtained by Garabedyan and shown with horizontal dashed line. As a result the value 0.4 has been obtained. It will be noted that overestimated value of the asymptotic velocity of a bubble was also obtained in numerical computations [1], [2], [3].

3. Let there be given two coaxial round cylindrical cavities with different radii $R_1 > R_2$ with the second cavity (lesser radius) adjoining the hole with radius $R_2$in a plane bottom of the first bottom. The height of the second cavity is $L_2$. (The cylindrical tube with a widening). At the initial time $t = 0$ the second cavity was filled with a gas under pressure $p(0)$ and in the first cavity there is a layer of quiescent heavy liquid with a

thickness - $L_1$ and density - $\rho$. It is considered that $p(0) > \rho g L_1$, where $g$ is the vector length of $\mathbf{g}$ - acceleration due to gravity directed to the second cavity.

Let us take the values $R_2$, $\rho g L_1$, $R_2(gL_1)^{-0.5}$, $R_2(gL_1)^{0.5}$ as scales of length, pressure, time and potential, respectively. The following parameters were selected: $R_1 = 2.95$; $R_2 = 1$; $L_1 = 20$; $L_2 = 4$; $p(0) = 4$; $\gamma = 1.4$ . On the rigid boundaries the condition of non-flowing was approximated.

It follows from the computations that the flow begins from formation of expanding bubble above the section of the second cavity; in the sequential phase of compression there occurs the breaking of the cavity into two parts with appearing the second bubble. The breaking moment is characterized by hydraulic impact acting on rigid boundaries. Further the liquid flows in the second cavity and the lift of the pulsing bubble to a free surface occurs. At the moment of arrival of a liquid at a bottom and filling-in the second cavity the water hammer occurs as well. When filling-in the cavity the expanding gas leaves the section and further breaking occurs again and from here the count comes to an end. Forms of interface of gas bubbles during filling-in of the cavity are shown in fig.2. The form of interface at initial moment $t = 0$ is indicated by a dotted line. The interruptive line with two points indicates the boundary on the phase of a bubble collapse at the moment $t = 5.17$ when liquid in the plane $x = 0$ breaks through to the center of a bubble. The dashed-dotted line corresponds to $t = 15.48$, the liquid has two gas bubbles, and the liquid jet has already reached the bottom of a cylindrical cavity. The upper bubble has turned into two-bonded torus-like one. The moment $t = 16.81$ is denoted by a solid line. The bottom of a cavity has already covered with a liquid, the lower bubble is being displaced into a cavity with a greater radius and the upper bubble is slowly lifting to a free surface.

The plot of the lower bubble dimensionless volume is presented in fig.3(solid line). The volume breakage at the moment $t = 9.22$ is explained by separation of a part of gas mass from the main bubble.

4. Let there be a circular cylindrical cavity with a depth $L_2$, coaxial with a symmetry axis of ellipsoid, in on oblate ellipsoid of revolution with a height $2x_s$. At an initial moment of time, $t = 0$, the cavity is filled with a gas under pressure $p(0) > \rho g L_1$ and the ellipsoid is in a heavy incompressible liquid, $L_1$ is the distance from a free surface of the liquid to the upper part of the ellipsoid.

The problem will be solved in coordinate system $\sigma$ , $\tau$ ,$\varphi$ of the oblate ellipsoid of revolution $x = a\sigma\tau$, $y = r\cos\varphi$, $z = r\sin\varphi$. The cylindrical coordinate is $r = y^2 + z^2 = a^2\ (1+\sigma^2)(1-\tau^2)$. The ellipsoid surface is a coordinate surface $\sigma = \sigma_s$ and the surface of cylindrical cavity is a part of coordinate surface $\tau = \tau_R$ .The bottom of cylindrical cavity is a part of coordinate surface $\sigma=\sigma_b=x_b/a$ , $x_b=x_s-L_2$ . Radius of cavity at the upper part of ellipsoid is expressed as

$$R^2_{max} = a^2(1 + \sigma_s^2)(1 - \tau_R^2),\ \tau_R^2 = 1 - R^2_{max}a^{-2}(1 + \sigma_s^2)^{-1}.$$

The computations of evolution of a gas bubble fully occurring in the curvilinear cylindrical cavity at initial moment were carried out. The initial gas pressure $p = 4$, $L_1 = 11.0$, $L_2 = 4$, $a = 16$, $x_s = 16$. Under the effect of excessive pressure the gas expands forming a bubble, the edge of which moves over the ellipsoid. After phase expansion of bubble there occurs compression of the bubble. A diagram of the gas cavity volume change (dashed line) has been presented in Fig.3. In Fig. 4 the forms of gas-liquid interface at the time $t = 1.18,\ 1.64,\ 4.11,\ 8.41$ have been shown. As early as at the first step of cavity expansion the effect on Rayleigh - Taylor instability solution $(t = 8.41)$ begins to dominate noticeably. The comparison with computation in rectangular coordinates allows the conclusion that significant instability effects significantly the computations carried out with the use of curvilinear coordinates.

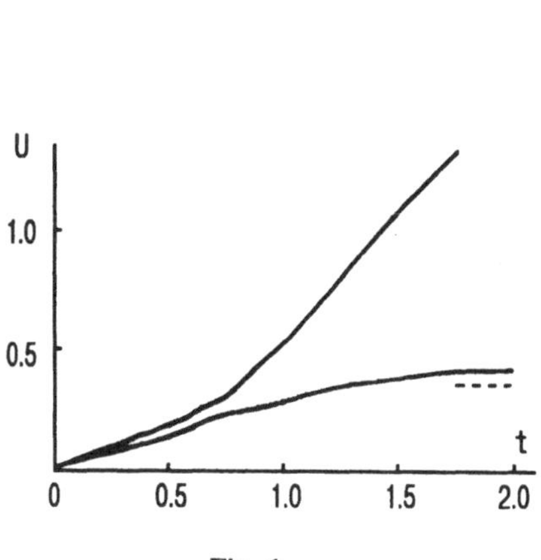


Fig. 1.

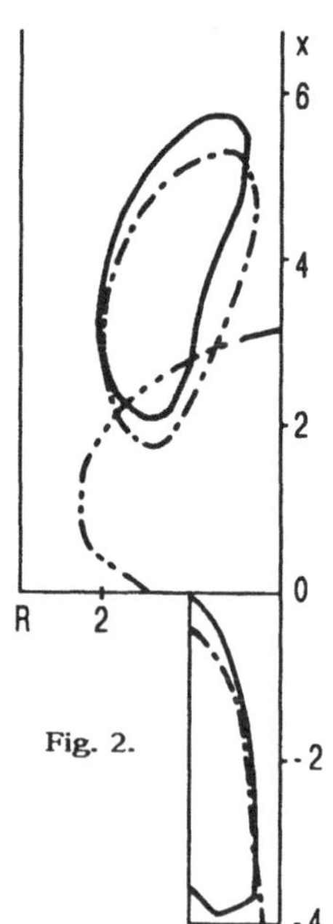


Fig. 2.

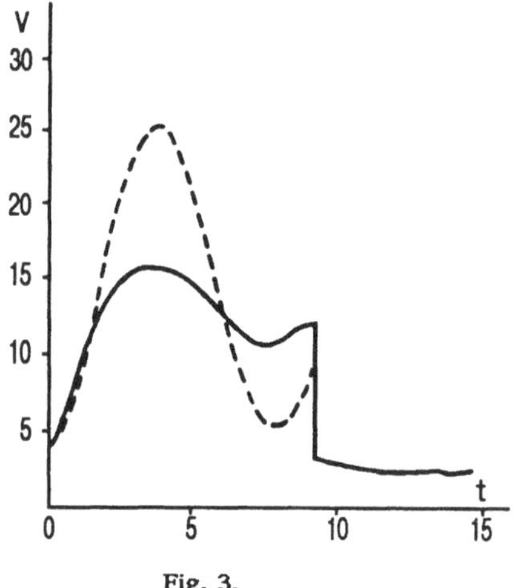


Fig. 3.

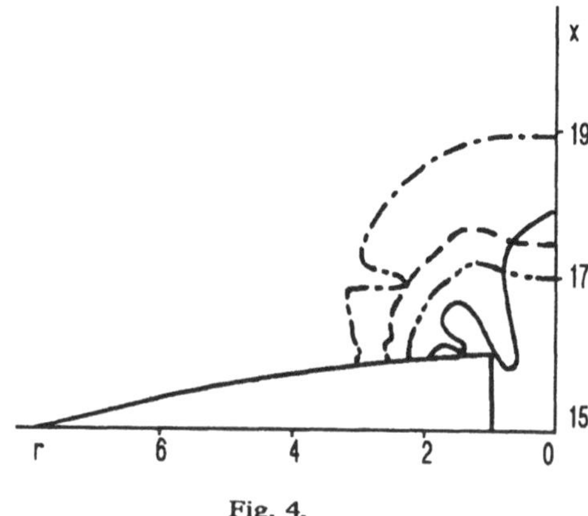


Fig. 4.

## REFERENCES.

[1] Prosperetti, A. & Jacobs, J.W. A numerical method for potential flows with a free surface.//J. of Comput. Phys. 1983. v. 51, N 3, pp. 365-386.

[2] Gasilov, V.A. Goloviznin, V.M. Taran, M.D. at al. On numerical modelling Rayleigh - Taylor instabilityin incompressible liquid. Preprint N 70. M.: IPMatem AN SSSR, 1979. (Institute of Applied mathematics of the USSR Academy of Sciences).

[3] Hirt, C.W. Cook, J.L. Butler, T.D. A Lagrangian method for calculating the dynamics of an incompressible fluid with free surface //J Comput. Phys, 1970, v.5, N 2, pp.103-124.

# WAVES ON THE LIQUIDS-GAS FREE SURFACE IN THE PRESENCE OF THE ACOUSTIC FIELD IN GAS

*I.A.Lukowsky, A.N.Timoha*
*Institute of Mathematics,*
*Kiev 252601, UKRAINE*

We investigate the low-gravity problem on wave motions of the liquid $(Q_2)$ - gas $(Q_1)$ free surface $\Sigma$ $(\xi(x,y,z,t) = 0)$, that is placed in a limited volume $Q = Q_1 \cup Q_2$ $(W(x,y,z) < 0)$. The high-frequency vibrator is contained in gas on

$$S_0 \subset S_1 = \partial Q_1 \backslash \Sigma \subset \partial Q.$$

It creates the acoustic field. This problem is being studied in nonlinear hydrodynamics formulation under hypothesis about idealized and potential motions of mediums. Frequency range of acoustic field is in $1 \div 3$ kHZ with Bond number $0 \div 30$.

## 1. THE FORMULATION OF A PROBLEM.

Let $\varphi_i$ be speed potential of a liquid and a gas respectively, $p_i$ - pressure, $\rho_i$ - density, $\nu$- frequency of influence, $l$ - characteristic dimension of a body, $t_* = 1/\nu$ - characteristic time. Then the problem on the liquid-gas joint motions take the form:

$$\rho_i = \left( \frac{p_i}{p_{0i}} \right)^{1/\gamma_i}; \ \rho_{it} + div\,(\rho_i \nabla\varphi_i) = 0 \quad \text{in } Q_i,$$

$$\rho_i \nabla\left[ \varphi_{it} + 0{,}5(\nabla\varphi_i)^2 + Bo\, x/\nu_*^2 \right] = -\nabla p_i \ \text{ in } Q_i; \quad i = 1,\, 2 \qquad (1)$$

$$\partial\varphi_j/\partial n = 0 \text{ on } S_j; \quad \partial\varphi_j/\partial n = -\xi_t\,/|\nabla\xi| \quad \text{on } \Sigma; \quad j = 1,2,$$

$$-p_2 + \nu_*^{-2}(K_1 + K_2) = -p_1 \frac{\rho_{01}}{\rho_{02}} \quad \text{on } \Sigma;$$

$$\frac{(\nabla W, \nabla\xi)}{|\nabla W|} = \cos\alpha\, |\nabla\xi| \ \text{ on } \partial\Sigma,$$

$$\rho_1\, \partial\varphi_1/\partial n = \frac{\sup V_0}{c\,\mu_0} \frac{V_0}{\sup V_0} \frac{\mu_0}{k} \ \sin t \quad \text{on } S_0,$$

where $\sigma$ is the coefficient of surface tensions, $g$ - the gravity acceleration, $\gamma_i$ - the constants, $\alpha$ - the angle of moistening, $K_i$ - the main curvatures $\Sigma$, $|\mu_0| \sim 1$, $Bo = gl^2 \rho_{02}/\sigma$ is the Bond number, $\nu_*^2 = \nu^2 l^3 \rho_{02}/\sigma$, $k$ is the wave number of the acoustic field in a gas, $V_0(x,y,z)\sin \nu t$ - normal speeds on $S_0$.

Frequency ranges and a power of acoustic influence have such quantity, that $\varepsilon = sup|V_0|/(c\mu_0) \ll 1$ and

$$\rho_{01}/\rho_{02} = \mu_1\varepsilon, \ |\mu_1| \sim 1; \qquad \nu_*^{-2} = \mu\,\mu_1\,\varepsilon^3, \quad \mu \sim 1. \tag{2}$$

## 2. METHOD OF THE STUDY OF THE PROBLEM (1), (2).

We use the perturbation theory and the principle of division of motions. Number $\varepsilon \ll 1$ exists only in the case of inhomogeneous condition of the problem. This number is proportional to the Mach number of acoustic field in gas. The principle of division of motions allows proceeding from the initial hydrodynamics problem to the new approximate nonlinear model, which describes free oscillations in low-frequency range.

We introduce quick-time $t$ and slow-time $\tau = \varepsilon^{3/2}t$. The dependence of the functions $\varphi_i$, $p_i$ and $\xi_i$ is defined in the form $\varphi_i = \varphi_i(x,y,z,t,\tau)$, $p_i = p_i(x,y,z,t,\tau)$, $\xi = \xi(x,y,z,t,\tau)$. We express the formal solution of the problem (1),(2) in the form:

$$\varphi_i = \sum_k \varepsilon^{k/2}\,\varphi_i^{(k/2)}; \ p_i = \sum_k \varepsilon^{k/2}\,p_i^{(k/2)}; \ \xi = \sum_k \varepsilon^{k/2}\,\xi_{(k/2)}. \tag{3}$$

Then the solving of the original initial problem is reduced to the solving of the sequence of problems relative to $\varphi_i^{(k/2)}$, $p_i^{(k/2)}$, $\xi_{(k/2)}$, $i = 1,2$, $k = 0,1,2\ldots$ . Let us introduce also functions

$$\varphi_2(x,y,z,\tau) = \langle \varphi_2 \rangle_t = \varepsilon^{3/2}\,\varphi_2^{(3/2)}(x,y,z,\varepsilon^{3/2}t) + o(\varepsilon^{3/2});$$

$$\xi(x,y,z,\tau) = \langle \xi \rangle_t + o(\varepsilon^2) = \xi_0(x,y,z,\tau),$$

which describe slow free oscillations of the liquid with a free surface $\Sigma(\xi(x,y,z,\tau) = 0)$. The model similar to the corresponding problem on free oscillations of a limited liquid volume in the exact theory of nonlinear waves is obtained after averaging in the sequence of recursion problems with respect to $\varphi_i^{(k/2)}$, $p_i^{(k/2)}$, $\xi_{(k/2)}$, $i = 1,2$, $k = 0,1,2,\ldots$:

$$\Delta\varphi_2 = 0 \text{ in } Q_2; \quad \partial\varphi_2/\partial n = 0 \text{ on } S_2;$$

$$\partial\varphi_2/\partial n = -\xi_\tau/|\nabla\xi| \quad \text{on } \Sigma,$$

$$\varphi_{2\tau} + 0{,}5(\nabla\varphi_2)^2 + \mu\,\mu_1\Big[\,Bo\,x - (K_1 + K_2)\,\Big] +$$

$$+\,0{,}25\,\mu_1\Big[\,k^2(\Phi_1)^2 - (\nabla\Phi_1)^2\,\Big] = const \quad \text{on } \Sigma; \tag{4}$$

$$(\nabla W, \nabla\xi)/|\nabla W| = \cos\alpha\,|\nabla\xi| \text{ on } \partial\Sigma; \qquad \int_{Q_2} dQ = const,$$

$$\Delta\Phi_1 + k^2\Phi_1 = 0 \quad \text{in } Q_1; \qquad \partial\Phi_1/\partial n = 0 \quad \text{on } S_1 \cup \Sigma,$$

$$\partial\Phi_1/\partial n = \mu_0\,V(x,y,z)/k \quad \text{on } S_0, \tag{5}$$

where $V(x,y,z) = V_0(x,y,z)/\ \sup\ |V_0(x,y,z)|$,

$$\Phi_1(x,y,z,\tau) = \varphi_1^{(1)}(x,y,z,\tau,t)/\sin t$$

is the wave function of acoustic field in gas. The balance surface $\Sigma_0$ ($\varphi_2 = 0$; $\Phi_1 = \Phi_1(x,y,z)$; $\xi = \xi_0(x,y,z)$) is found from the problem of capillary-acoustic balance mode (KABM). This problem is a generalization of the well-known capillary problem for our case [1]:

$$\mu\Big[\,Bo\,x - (K_1 + K_2)\,\Big] + 0{,}25\Big[\,k^2(\Phi_1)^2 - (\nabla\Phi_1)^2\,\Big] = const \quad \text{on } \Sigma_0,$$

$$(\nabla W, \nabla\xi_0)/|\nabla W| = \cos\alpha\,|\nabla\xi_0| \text{ on } \partial\Sigma_0; \qquad \int_{Q_2} dQ = const, \tag{6}$$

In (6) $\Phi_1$ satisfies (5) under the condition that $\partial\Phi_1/\partial n = 0$ on $\Sigma_0$. (If $V = 0$, then $\Phi_1 = const$ and (5),(6) is the capillary problem).

**3. PROPER OSCILLATIONS OF LIQUID WITH RESPECT TO KABM. STABILITY OF KABM.** We define the perturbation motions of a free surface $\Sigma$ in the form $x = H_0(y,z) + H(y,z,\tau)$, where equation $x = H_0(y,z)$ determines KABM. The problem on small (linear) oscillations of a liquid with respect to the KABM, that follows from (4),(5), is formulated as regards functions $x = H(y,z,\tau)$ and $\varphi_2 = \varphi_2(x,y,z,\tau))$. We present functions $H$ and $\varphi_2$ in the form $H(y,z,\tau) = exp(i\lambda\tau)\, h(y,z)$, $\varphi_2(x,y,z,\tau) = i\lambda\, exp(i\lambda\tau)\, \varphi(x,y,z)$. Then h and $\varphi$ are found from the following problem on proper (normal) oscillations of a system:

$$\Delta\varphi = 0 \quad \text{in } \langle Q_2\rangle; \qquad \partial\varphi/\partial n = 0 \quad \text{on } \langle S_2\rangle,$$

$$\partial\varphi/\partial n = h \,/ \sqrt{1 + (\nabla H_0)^2} \quad ; \quad -\lambda^2 \varphi + \mu_1\mu\, Ah = 0 \quad \text{on } \Sigma_0, \tag{7}$$

where $\Sigma_0(\xi(x,y,z)=0)$ - RABM, and operator $A$ has the form:

$$Ah = -div\left[\nabla h/(1 + (\nabla H_0)^2)^{1/2} - (\nabla h, \nabla H_0)\nabla H_0/(1 + (\nabla H_0)^{3/2}\right] +$$

$$+(2\mu)^{-1}\left\{ k^2\, \Phi_1\Phi_{1x}\, h - (\nabla\Phi_1, \nabla\Phi_{1x})\, h + k^2\, \Phi_1\, \Phi - (\nabla\Phi_1, \nabla\Phi)\right\} + Boh,$$

$$\int_{\Sigma_0} H\, dydz = 0;$$

$$W_{xx}H - W_zH_z - W_yH_y - \left(W_{zx}H_{0z} + W_{yx}H_{0y}\right)H = \tag{8}$$

$$= \cos\alpha \left( \frac{(\nabla H, \nabla H_0)}{(1 + (\nabla H_0)^2)^{1/2}} |\nabla W| + \sqrt{1 + (\nabla H_0)^2}\, \frac{W_xW_{xx} - W_yW_{yx} - W_zW_{zx}}{|\nabla W|} \right)$$

on $\partial\Sigma_0$,

$\Phi$ being obtained from the solution of a boundary problem:

$$\Delta\Phi + k^2\Phi = 0 \quad \text{in } \langle Q_1\rangle; \qquad \partial\Phi/\partial n = 0 \quad \text{on } \langle S_1\rangle \cup S_0,$$

$$\partial\Phi/\partial n = - \Big( \Phi_{1xx}h - \Phi_{1z}h_z - \Phi_{1y}h_z - \tag{9}$$
$$- \Big[ \Phi_{1xy}H_{0y} + \Phi_{1xz}H_{0z} \Big] h \Big) / (1 + (\nabla H_0)^2)^{1/2} \quad \text{on } \Sigma_0,$$

The proper problem (7)-(9) has parameter $\lambda^2$ in surface condition. We find the proper frequency parameter $\lambda$ of slow normal oscillations of free surface of liquid from this proper problem.

The following theorems are proved in [2]:

**THEOREM 1** (on the spectrum of operator $A$).

Let $H_0 \in C^2(\Sigma_0)$, $\Phi_1 \in C^2(\langle Q_1 \rangle \cup \Sigma_0)$, then

1. The linear self-conjugate operator $A$ has the real point spectrum, that consists only of proper values.
2. The multitude of proper functions $\{h_n\}$ - basis in $L_2(\Sigma_0)$.
3. The multitude of proper values of operator $A$ has the only limit point + ∞. The multitude of negative proper values is finite.

**THEOREM 2.** Let $H_0 \in C^2(\Sigma_0)$, $\Phi_1 \in C^2(\langle Q_1 \rangle \cup \Sigma_0)$, when $\Big[(\lambda_n^2, h_n, \varphi^{(n)})$- solution of the problem (7)-(9)$\Big]$:

1. $\forall n \quad \lambda_n^2 \in \mathbb{R}$, $\{h_n\}$ - basis in $L_2(\Sigma_0)$.
2. The multitude $\{h_n: \lambda_n^2 < 0\}$ is finite.

**THEOREM 3** (the spectral sign of stability).

Let $H_0 \in C^2(\Sigma_0)$, $\Phi_1 \in C^2(\langle Q_1 \rangle \cup \Sigma_0)$. Then RABM $\Sigma_0$ is stable if and only all proper values of the operator $A$ are positive.

## 4. EXAMPLE.

Let

$$Q = \Big\{ W(y,z) < 0,\ h_1 > x,\ -h_2 < x),\quad S_0(x = h_1) \Big\}$$

- right round cylinder, $\alpha = \pi/2$ be the angle of moistening,

$V_0(y,z) = V_0 = const$. We select $l$ as a radius of the cylinder, $\mu_0 = - sin(kh_1)$.

The problem (7)-(9) is solved by Fourier method. Proper values of the problem (7)-(9) are determined by the formula

$$\lambda^2_{pq} = \mu_1 æ_{pq}\ th(æ_{pq}h_2) \left[\mu(Bo + æ^2_{pq}) - 0,5 \begin{cases} cth(\xi h_1)/\xi, & k^2 < æ^2_{pq}, \\ -ctg(\xi h_1)/\xi, & k^2 > æ^2_{pq} \end{cases}\right],$$

$$\xi = \sqrt{|k^2 - æ^2_{pq}|},$$

modes of free surface oscillations take the form

$$h_{pq} = J(æ_{pq}r)\ sin(cos)p\theta,\ J'_p(æ_{pq}) = 0.$$

If there is no acoustic field, modes of oscillations are the same, proper values of a problem (squares of proper frequencies) are calculated ($t_* = \left(\varepsilon^{1/2}\nu\right)^{1/2}$ - characteristic time) by the formula [3]

$$\tilde{\lambda}^2_{pq} = \mu_1\mu\ æ^2_{pq}\ th(æ_{pq}h_2)\ (Bo + æ^2_{pq}).$$

In the work [1] the stability ranges of capillary balance modes ($\tilde{\lambda}^2_{pq} > 0$) and the stability ranges of KABM ($\lambda^2_{pq} > 0$) are shown for this example. We also indicate in [4] the ranges of numbers $k$, when $\lambda^2_{pq} > 0\ \ \forall\ pq$, but $\exists\ \chi\beta$: $\tilde{\lambda}^2_{\chi\beta} < 0$ (stabilization of the liquid-gas free surface).

## 5. THE VARIATIONAL FORMULATION OF A PROBLEM. THE PRINCIPLE OF BATEMAN-LUKE.

Let $Q$ be convex piece-smooth limited domain $Q \subset \mathbb{R}^3$, $\left\{\Sigma(t):\ (\xi(t,x,y,z) = 0)\right\}$ - the set of smooth single-connected surfaces, which divides $Q$ into two subdomains $Q_1(t)$ and $Q_2(t)$. Let also $\varphi_i(t,x,y,z) \in C^1([t_1,t_2] \times Q)$, where $t_1$ and $t_2$ are any fixed numbers, and $\rho_i = \rho_i(p_i):(0,\infty) \to (0,\infty)$ - certain smooth strictly monotonically increasing function,

$$P_i(p_i) = \int_{p_{0i}}^{p_i} \rho_i^{-1} dp, \quad p_i = P_i^{-1}(A_i)$$

- inverse to $P_i(p_i)$:

$$p_i = P_i^{-1}(A_i); \; A_i = -\varphi_{it} - 0,5(\nabla\varphi_i)^2 - \mu\,\mu_1\varepsilon^3 x. \qquad (10)$$

(In the investigated case $\rho_i = (p_i / p_{0i_1})^{1/\gamma_i}$, $\gamma_i \neq 1$ and function $P_i(p_i)$ has the inverse function $P_i^{-1}$, which takes the form

$$p_i = P_i^{-1}(A_i) = \left[(\gamma_i - 1) A_i / (p_{0i}\gamma_i) + 1\right]^{\gamma_i/(\gamma_i - 1)} \quad ).$$

We examine the Bateman-Luke functional

$$B(\xi,\varphi_i) = \int_{t_1}^{t_2} L\, dt \qquad (11)$$

with the Lagrange function

$$L = \mu\,\mu_1\varepsilon^3\left[-|\Sigma| - \cos\alpha\,|S_1|\right] + \int_{Q_2(t)} p_2\, dQ + \qquad (12)$$

$$+ \mu_1\varepsilon\left[\int_{Q_1(t)} p_1\, dQ + \frac{\mu_0}{k}\int_{S_0} \varepsilon\, V(x,y,z)\,\varphi_1 \sin t\, ds\right]$$

where $|\cdot|$ denotes areas of surfaces.

THEOREM 4. The problem of finding of stationary points of functional (10)-(12) with smooth variations $\delta\xi$ and $\delta\varphi_i$:

$$\delta\xi\Big|_{t_1,t_2} = 0; \qquad \delta\varphi_i\Big|_{t_1,t_2} = 0$$

is equivalent to the solving of the problem (1)-(2).

We carry out the averaging in the problem (10)-(12) and obtain the variational principle of KABM stability [2]:

**THEOREM 5.** Let $H_0 \in C^2(\Sigma_0)$, $\Phi_1 \in C^2(<Q_1> \cup \Sigma_0)$, $H_0$ and $\Phi_1$ satisfy the condition (5). Then the problem of finding of strict maximum of functional U by $H_0$, when the condition (5)

$$U(H_0,\Phi_1) \xrightarrow[H_0]{(5)} max$$

(where

$$U(H_0,\Phi_1) = < B > = \varepsilon\ const + \varepsilon^3\Big\{ \mu\, \mu_1 \Big(-|\Sigma_0| - \cos\alpha\ |<S_1>| - \int_{<Q_2>} Box\ dQ \Big] + \mu_1\Big( \int_{<Q_1>} (k^2(\Phi_1)^2 - (\nabla\Phi_1)^2)dQ + \mu_0/k \int_S V(x,y,z)\ \Phi_1 ds\Big]\Big\})$$

is equivalent to the problem of finding of the stable KABM.

The formulation of the theorem 5 allows constructing the algorithm of solving the KABM problem, using direct methods of mathematical physics [1].

**REFERENCES.**


[1] Lukowsky I.A., Timoha A.N. Nonlinear dynamics of liquid-gas free surface in the presence of acoustic field in gas. Stable regime of motion, Kiev, 1988, 39p. (Preprint/ Ac. of Sc. Ukr.SSR. In-t of mathematics: 88.9).

[2] Lukowsky I.A., Timoha A.N. One class of the boundary problems in theory of surface waves// Ukr. mathem. journ, 1991, 43, N 3, pp. 359-364.

[3] The zero-gravity hydromechanic / Edited by A.D.Myshkies, M.:Nauka, 1978, 504p.

[4] Lukowsky I.A., Timoha A.N. On stabilization of liquid-gas freesurface when interacting with acoustic field in gas// News of Ac.of Sc. USSR. Mech. of fluid and gas, 1991, N 3, pp. 80-84.

# SMOOTH BORE IN A TWO-LAYER FLUID

*N.I. Makarenko*
*Lavrentyev Institute of Hydrodynamics*
*Novosibirsk 630090, RUSSIA*

A plane steady two-layer flow is considered. The bifurcation theory is used to prove the existence of a one-parametric set of solutions of Euler equations describing propagation of a smooth bore along the interface between layers. The wave profile is an analytical curve which exponentially tends to two horizontal asymptotes. The motion may be realized for Froude numbers beyond the spectrum of linear waves; the flow degenerates to a piecewise uniform flow with a critical Froude numbers as the wave amplitude vanishes.
Key words: two-layer fluid, steady internal waves, smooth bore.

One of peculiarities of the stratified fluid motion is that localized disturbances in the form of transitions from one stationary state to the other, i.e. continuous bores or surges, can propagate along with solitary waves. An approximate description of these nonlinear internal waves is based on long-wave asymptotic expansions [1,2]. Quite recently, the existence of an exact solution of hydrodynamic Euler equations that describes a smooth bore travelling in the quiescent two-layer fluid has been proved [3]. In this case it turned out that the mentioned wave regime is realized only for a certain density-depth relation of layers. In the case of piecewise uniform flow before the wave the approximate theory predicts a possibility of bore existence without the essential restrictions on density and depth [2]. The present paper gives a favourable answer to this point.

## 1. STATEMENT OF THE PROBLEM.

Let us consider the steady plane irrotational flow of a two-layer fluid with the densities $\rho_1 > \rho_2$ (the subscript "1" regards the lower layer of a more heavy fluid). The fluid is bounded by an even bottom from below and by a rigid cover from above, at a distance $H$ from the bottom. The state before the wave, to which the flow tends asymptotically as $x \to -\infty$, is assumed to be a piecewise constant flow with the velocities $U_j$ and heights $H_j$ in the layers $(H_1 + H_2 = H)$. The basic dimensionless parameters are the Froude numbers

$$F_j = \frac{U_j}{\sqrt{gH_1(\rho_1-\rho_2)/\rho_j}}$$

and ratio $r = H_2/H_1$ of the undisturbed upper-layer depth to that one of the lower layer.

The flow region with the unknown interface is transformed into a fixed region by passing into a plane of the independent variables $(x,\psi)$, where $\psi$ is the streamline function. The scales of values can be chosen such that strips $\Pi_j = \mathbb{R}\times T_j$ with $T_1 = (0,1)$, $T_2 = (1,h)$, where $h = 1 + r$ is the dimensionless total fluid depth, will correspond to the layers. In these variables, the form of streamlines $y = \psi + w(x,\psi)$ is unknown. The function $w$ characterizing the deviation from the uniform flow before the wave should satisfy the system of equations

$$\frac{\partial}{\partial x}\frac{w_x}{1 + w_\psi} - \frac{1}{2}\frac{\partial}{\partial\psi}\frac{1 + w_x^2}{(1 + w_\psi)^2} = 0 \quad \text{in} \quad \Pi = \Pi_1 \cup \Pi_2,$$

$$w = 0 \quad \text{at} \quad \psi = 0 \quad \text{and} \quad \psi = h, \tag{1}$$

$$[w] = 0, \quad \left[F^2\left(w_\psi + \frac{1}{2}(w_\psi^2 - w_x^2)\right)(1 + w_\psi)^{-2}\right] + w = 0 \quad \text{at} \quad \psi = 1,$$

$$w, \nabla w \to 0 \quad \text{as} \quad x \to -\infty.$$

Here it is denoted that $F = F_j$ in $\Pi_j$, and the square brackets are used as the discontinuity jump symbol.

A continuous bore changes the piecewise constant flow as $x \to -\infty$ into an analogous state with $x \to +\infty$ and the same discharges in the layers, however, with the other levels of the streamlines (conjugate flow in the sense [4]). If the dimensionless wave amplitude $a = \lim_{x\to+\infty} w(x,1)$ is introduced, then the distribution of the streamlines behind the wave has the form $w = al(\psi)$, where $l$ is the continuous piecewise linear function, $l(\psi) = \psi$ when $0 \leqslant \psi \leqslant 1$ and $l(\psi) = (h-\psi)/r$ when $1 \leqslant \psi \leqslant h$. The parameters $F_j$ and $a$ aren't independent, there exists their simple connection which follows from the presence of the first integral in problem (1). By virtue of the conservation law for a horizontal momentum flux, any of its solutions should satisfy relation

$$\int_0^h F^2 \frac{w_\psi^2 - w_x^2}{1 + w_\psi} d\psi = w^2(x,1). \tag{2}$$

Within the limit, as $x \to +\infty$, this integral together with the second boundary condition in (1) at $\psi = 1$ results in two independent relations. From these relations it follows that the flow with a non-zero amplitude $a$ is possible only if we take

$$|F_1| = (1 + a)/\sqrt{h}, \quad |F_2| = (r-a)/\sqrt{h} \quad (-1 < a < r). \tag{3}$$

The figure (3) in a plane $(F_1,F_2)$ is the square having the nodes in the points $(\pm\sqrt{h},0)$ and $(0,\pm\sqrt{h})$; a double symmetry of a set of admissible parameter values is due to reversibility of flow direction in each layer. In view of translation invariance in $x$ of problem (1) the condition that fixes the wave profile is required. For example, it is possible to choose the coordinate axes such that shadowed areas $S_1$ and $S_2$ are equal (Fig.1).

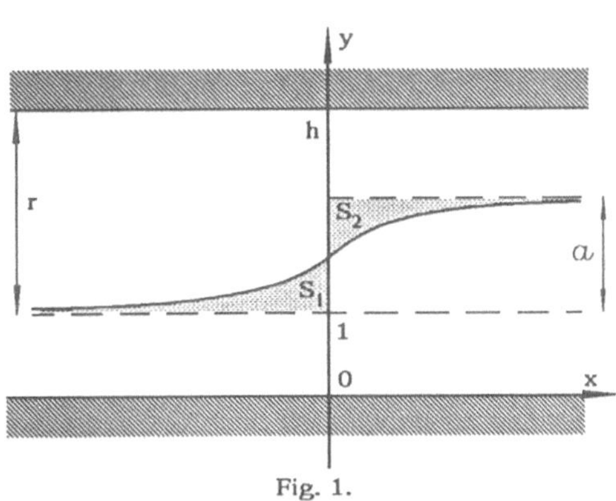

Fig. 1.

Thus, the problem of bore propagation is to find the solution of equations (1) with the Froude numbers (3), that satisfies the above-formulated areas equality condition.

## 2. APPROXIMATE SOLUTION AND AUXILIARY LINEAR PROBLEM.

The exact solution is found in the neighbourhood of its long-wave approximation:

$$w(x,\psi,a) = aw_0(x^*,\psi) + a^2u(x^*,\psi,a), \quad x^* = \varepsilon x.$$

As the small parameter $\varepsilon > 0$, it is convenient to use the decay exponent of $w$ at infinity as $x \to -\infty$. In this case, as is seen from an analysis of the linear problem of small disturbances in a piecewise uniform flow with the Froude numbers (3), the parameters $\varepsilon$ and $a$ are related by the equality

$$(1 + a)^2\varepsilon \,\mathrm{ctg}\, \varepsilon + (r - a)^2\varepsilon \,\mathrm{ctg}\, r\varepsilon = 1 + r.$$

Hence, it follows, in particular, that $\varepsilon = \varepsilon(a)$ has the same smallness order as the value of $a$ has, when $a \to 0$. Development of solution in a power series of $a$ and substitution into (1) give the main term of asymptotics in the form $w_0(x,\psi) = \theta(x)l(\psi)$, where the function $\theta$ satsisfies the nonlinear ordinary differential equation

$$\theta'' = 2\theta^3 - 3\theta^2 + \theta.$$

The bounded nonzero solution of this equation, that vanishes as $x \to -\infty$ and satisfies the areas equality condition, has the form $\theta(x) = (1 + th(x/2))/2$. It is worth of noting that this approximation has the same wave amplitude as the unknown exact solution. This promotes essentially determination of the function $u$ since it must be vanishing as $x \to +\infty$. The boundary value problem for $u(x,\psi,a)$ is written as follows

$$\begin{aligned} &\varepsilon^2u_{xx} + u_{\psi\psi} = f \quad \text{in } \Pi, \\ &u = 0 \quad \text{at } \psi = 0 \text{ and } \psi = h, \\ &[u] = 0, \ [F^2u_\psi] + u = \varphi \quad \text{at } \psi = 1, \\ &u, \nabla u \to 0 \quad \text{as } |x| \to \infty. \end{aligned} \tag{4}$$

The specific form of the right-hand parts

$$f = div(f_1(\nabla w,a),f_2(\nabla w,a)) \quad \text{and} \quad \varphi = \varphi(\nabla w,a)$$

is defined by nonlinear terms in (1).

Solvability of problem (4) is valid as a result of its reducing to the equivalent operator equation with a contraction operator in an appropriate function space. With this aim, an auxiliary linear problem is considered first, in which it is required to find the solution of equations (4) for the given $f$ and $\varphi$. It is obtained in the explicit form with the help of the Fourier transformation in $x$. In the strip $\Pi_j$, $\Pi_j = \mathbb{R}\times T_j$, we have

$$\hat{u}(\xi,\psi) = \int_{T_j} G(\psi,\chi,\varepsilon\xi)\hat{f}(\xi,\chi)\,d\chi -$$

$$- \frac{1}{\Delta(\varepsilon\xi)}\left\{ K(\varphi,1,\varepsilon\xi)\,\hat{\varphi}(\xi) + \int_0^h F^2(\chi)K(\psi,\chi,\varepsilon\xi)\hat{f}(\xi,\chi)\,d\chi\right\}.$$

Here $G(\psi,\chi,z)$ is the Green function of operator $D_\psi^2 - z^2I$ with boundary conditions $G = 0$ at $\psi = 0$, $\psi = 1$ and $\psi = h$. The kernel $K$ is prescribed by the formula

$$K(\psi,\chi,z) = \begin{cases} sh\ zl(\psi)\ sh\ zl(\chi)/sh^2 z, & 0 < \psi < 1, \\ sh\ rzl(\psi)\ sh\ rzl(\chi)/sh^2 rz, & 1 < \psi < h, \end{cases}$$

and $\Delta$ has the form

$$\Delta(z) = F_1^2 z\ cth\ z + F_2^2\ z\ cth\ rz - 1.$$

Zeros of the meromorphic function $\Delta(z)$ in a complex plane $z$ (there is a countable set of them) give the poles of $\hat{u}$. All zeros lie on the coordinate axes, on the real axis the most large number of them being two, depending on $F_1$ and $F_2$. There are the real roots if and only if the inequality

$$F_1^2 + r^{-1}F_2^2 \leqslant 1 \tag{5}$$

is valid. That inequality gives a continuous spectrum of linear problem (4). Ellipse (5) is inscribed into square (3) tangential at four points $F_1 = \pm\, 1/\sqrt{h}$, $F_2 = \pm r/\sqrt{h}$ corresponding to the amplitude value $a = 0$ (Fig. 2).

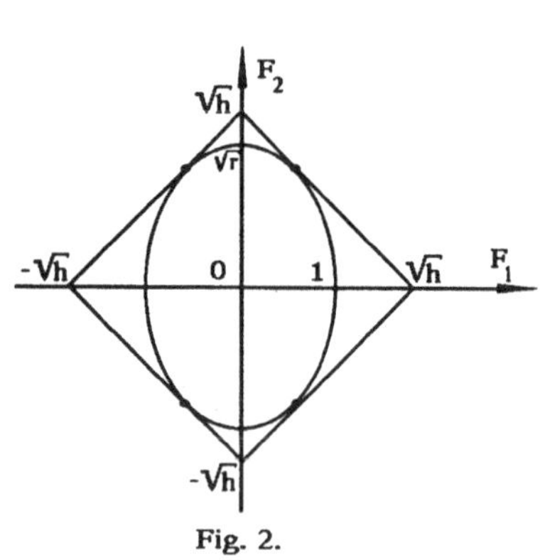

Fig. 2.

Thus, these points are the bifurcation points on the boundary of continuous spectrum in which the flow of the smooth bore type can bifurcate from a piecewise constant flow. It is interesting to note that the mentioned points separate the ellipse arcs consisting of the bifurcation points for internal solitary waves of elevation or depression [5-6]. For $a \neq 0$ the Froude points (3) are outside the spectrum, in this sense the bore is a supercritical flow.

## 3. THE FUNCTION SPACES AND SOLUTION ESTIMATE.

Let us take that $\rho > 0$ and $\alpha > 0$. For problem (4) the function class $E^{\alpha}_{\rho}(\Pi)$ is considered, that is the Banach space of functions $u(x,\psi)$ which are continuous in $\bar{\Pi}$, analytic in $x$ with each $\psi \in [0,h]$ and have the Fourier transform $\hat{u}(z,\psi)$ that is analytic in $z = \xi + i\eta$ within the strip $|\eta| < \alpha$. The norm of $u$ is defined as follows:

$$\|u\|_{\rho,\alpha} = \sup_{|\eta|<\alpha} \int_{-\infty}^{\infty} e^{\rho|\xi|} \sup_{\psi\in[0,h]} |\hat{u}(\xi+i\eta,\psi)|\,d\xi .$$

The functions of this class have an exponential decay

$$|u(x,\psi)| \leqslant \|u\|_{\rho,\alpha} \exp(-\alpha|x|)$$

and the norm possesses a multipilicative property. The subspaces in $E^{\alpha}_{\rho}(\Pi)$ are

a) a class $\overset{0}{E}{}^{\alpha}_{\rho}(\Pi)$ of the functions vanishing at $\psi = 0$ and $\psi = h$;

b) a class $E^{\alpha}_{\rho}(\mathbb{R})$ of the functions $u(x)$ of real variable $x$ with the analogous norm;

c) a class $E^{\alpha}_{\rho,n}(\Pi)$ of the functions belonging to $E^{\alpha}_{\rho}(\Pi)$ together with their derivatives up to the order $n$,

$$\|u\|_{E^{\alpha}_{\rho,n}(\Pi)} = \sum_{0\leqslant m+k\leqslant n} \|D^m_x D^k_{\psi} u\|_{E^{\alpha}_{\rho}(\Pi)}.$$

The space $\overset{0}{H}{}^{\alpha}_{\rho}(\Pi)$ that consists of the same functions $u \in E^{\alpha}_{\rho}(\Pi)$ which restriction to each strip $\Pi_j$ is in $E^{\alpha}_{\rho,2}(\Pi_j)$ is basic for problem (4). The norm in $H^{\alpha}_{\rho}(\Pi)$ is defined as follows:

$$\|u\|_{H^{\alpha}_{\rho}(\Pi)} = \|u\|_{E^{\alpha}_{\rho}(\Pi)} + \max_{j=1,2} \sum_{1\leqslant m+k\leqslant 2} \|D^m_x D^k_{\psi} u\|_{E^{\alpha}_{\rho}(\Pi_j)}.$$

An example (and prototype) of the function of this class is $u_0(x,\psi) = ch^{-2}(x/2)l(\psi)$ when $\alpha < 1$ and $\rho < \pi$.

**LEMMA.** Let us take that $\alpha < 1$. Then for $\varphi \in E^{\alpha}_{\rho,1}(\mathbb{R})$, $f \in E^{\alpha}_{\rho}(\Pi_j)$ $(j = 1, 2)$ and the Froude numbers (3) with $a \in [a_1, a_2]$ (where $-1<a_1<a_2<r$) there exists the solution of the linear problem (4) that is unique in $H^{\alpha}_{\rho}(\Pi)$. This solution is represented as

$$u = \varepsilon^{-2} L_0\langle\varphi,f\rangle + L_{\varepsilon}\langle\varphi,f\rangle \tag{6}$$

where the linear operator $L_0$ acts by the formula

$$L_0\langle\varphi,f\rangle = \frac{3l}{F_1^2 + rF_2^2}\left(D_x^2 - I\right)^{-1}\left[\varphi + \int_0^h F^2 l\, f\, d\psi\right]$$

and for $L_{\varepsilon}\langle\varphi,f\rangle$ with $0 \leqslant m+n \leqslant 2$ the following estimate

$$\|(\varepsilon D_x)^m D^n_{\psi} L_{\varepsilon}\langle\varphi,f\rangle\|_{\rho,\alpha} \leqslant C\left(\|\varphi\|_{E^{\alpha}_{\rho,1}(\mathbb{R})} + \max_{j=1,2}\|f\|_{E^{\alpha}_{\rho}(\Pi_j)}\right)$$

is valid. The constant $C$ is independent here of $a$.

The statement of Lemma follows from the above-obtained explicit formula for $\hat{u}$. The operator $L_0$ corresponds to the main part of the Laurent expansion of the function $1/\Delta(z)$ in the neighbourhood of its poles $z = \pm i\varepsilon$ that are the nearest to the real axis; $L_\varepsilon$ represents a part of solution bounded uniformly in $\varepsilon$. The original idea of this decomposition in the supercritical case may be found in [7].

## 4. NONLINEAR OPERATOR EQUATION.

Returning back to the nonlinear expressions for $f$ and $\varphi$ in (6) gives an equation in the space $H^\alpha_\rho(\Pi)$ with the operator which fixed point is the solution to be found. However, that operator is not contractive, therefore we consider this equation in the equivalent form

$$Au = \Phi(u,a), \tag{7}$$

where

$$Au(x,\psi)= \left(D_x^2 - I\right)u(x,\psi) + \frac{3}{2}\, ch^{-2}(x/2)\; l(\psi)\; u(x,1)$$

is the operator linearized with respect to the approximate solution (with an accuracy up to the $O(|a|)$ order terms). The nonlinear operator $\Phi(\cdot,a)\colon B_\delta \to \overset{0}{E}{}^\alpha_\rho(\Pi)$ satisfies (consequently to Lemma and structure of nonlinearities $f$ and $\varphi$) the Lipschitz condition with the constant proportional to $|a|$ in the closed ball

$$B_\delta = \left\{ u \in H^\alpha_\rho(\Pi) \;\middle|\; \|u\|_{H^\alpha_\rho(\Pi)} \leqslant \delta \right\}$$

with $\alpha < 1$, $\rho < \pi$ and some fixed $\delta > 0$.

The operator $A\colon H^\alpha_\rho(\Pi) \to \overset{0}{E}{}^\alpha_\rho(\Pi)$ with the same $\rho$ and $\alpha$ is the Fredholm operator: it has a one-dimensional kernel formed by the function $u_0(x,\psi)$ (see an example in section 3), and the necessary and sufficient condition for solvability of the nonhomogeneous equation $Au = f$ is the orthogonality of the functions $f(x,1)$ and $u_0(x,1)$ in the scalar product of $L_2(\mathbb{R})$. The fact that the operator $A$ is the Fredholm operator allows the use of the Lyapunov-Schmidt construction, according to which equation (7) is equivalent to the system for a pair $(u,c)$,, $c \in \mathbb{R}$:

$$\begin{aligned} u &= cu_0 + \tilde{A}^{-1}P\,\Phi(u,a), \\ \left[I - P\right]&\Phi(u,a) = 0. \end{aligned} \tag{8}$$

Here $\tilde{A}$ is the restriction of the operator $A$, that acts from a direct complement in $H^{\alpha}_{\rho}(\Pi)$ to the kernel of $\overset{0}{A}$ into the image of $A$, and $P$ is the projection from $E^{\alpha}_{\rho}(\Pi)$ into the same image.

For small $|a|$ in the first equation of (8) there is the contractive operator, therefore this equation has a two-parametric set of solutions $u = u(a,c)$ in $H^{\alpha}_{\rho}(\Pi)$. The second scalar relation still remains, that is the bifurcation equation. In this case the fact that the bifurcation equation turns out to be satisfied identically in $a$ and $c$ is the non-stock effect. To prove this fact, use is made of an analogy of integral (2) that is present in nonlinear eqs.(4), where the additional terms in the right-hand sides are introduced according to projection (8). The parameter $c$ is of a group - theoretical nature, i.e. system (8) describes a set of solutions of the boundary value problem (1) with the fixed main term as $a \to 0$; these solutions are obtained one from the other with the translaion in $x$ by an order of magnitude $O(|a|)$. The areas equality condition fixes unique value of $c$, and as a result, we have

**THEOREM.** For the Froude numbers (3) with small $|a|$ there exists the solution of the problem (1), that describes the bore propagation in a two-layer fluid with the interface

$$y = 1 + \frac{1}{2}a\left[1 + th\,\frac{\varepsilon x}{2}\right] + $$

$$+ a^2\,\frac{1 - r^4}{8r(1 + r^3)}\,\frac{ln(e^{\varepsilon x} + 1) - 1}{ch^2(\varepsilon x/2)} + a^3 v(\varepsilon x,a),$$

where the function $v \in E^{\alpha}_{\rho,2}(\mathbb{R})$ is bounded as $a \to 0$.

It is clearly seen from the specified solution asymptotics presented here that at least with $r \neq 1$ the wave profile is not symmetric respecting its central point as it could be expected if we proceed from the form of the main term.

## REFERENCES.

[1] Kakutani T., Yamasaki N. Solitary waves on the two-layer fluid. J.Phys.Soc.Japan, 45 (1978), pp. 674-679.

[2] Ovsiannikov L.V. et al. Nonlinear problems of surface and internal wave theory. Novosibirsk, Nauka (1985) (in Russian).

[3] Amick C.J., Turner R.E.L. Small internal waves in two-fluid systems. Arch. Rat.Mech.Anal., 1989, v.108, N 2, pp. 111-139.

[4] Benjamin T.B. A unified theory of conjugate flows. Philos.Trans.Roy.Soc. London A, v.269 (1971), pp. 587-643.

[5] Khabakhpasheva T.I. Solitary waves in a two-layer fluid. Dynamics of continuum, N 69, Novosibirsk (1985), pp.96-122 (in Russian).

[6] Amick C.J., Turner R.E.L. A global theory of solitary waves in two fluid systems. Trans. AMS, N 298 (1986), pp. 431-481.

[7] Nalimov V.I. Supercritical flows from under a cover. Journ.Appl.Mech.Techn.Phys, N 2 (1989), pp.77-80 (in Russian).

International Series of Numerical Mathematics, Vol. 106, © 1992 Birkhäuser Verlag Basel 

# NUMERICAL CALCULATION OF MOVABLE FREE AND CONTACT BOUNDARIES IN PROBLEMS OF DYNAMICS DEFORMATION OF VISCOELASTIC BODIES

*L.A. Merzhievsky, A.D. Resnyansky*
*Lavrentyev Institute of Hydrodynamics*
*Novosibirsk 630090, RUSSIA*

Description of evolution of contact and free boundaries is one of problems in numerical simulation of nonstationary high-rate processes. The complex Eulerian-Lagrangian calculation techniques are the most promising for such a description since the purely Eulerian methods require a large number of calculation points to obtain a required approximation, at purely Lagrangian approach the calculation errors increase quickly for large grid deformations. Definition of boundaries is a part of the general problem of solution construction of the initial boundary value problem in continuum mechanics and becomes of current concern for modelling pulse processes in metals. This is due to that in a given class of processes and media inside the continuity regions new free surfaces, such as cracks and fragment boundaries, can appear.

Let us consider the statement of the problem of loading and fracture of metal specimens in a viscoelastic Maxwell-like model [1]. This model comprises the evolution equation of a metric tensor of elastic deformations $\|g_{ij}\|$:

$$\frac{dG}{dt} + G\circ W + W^{*}\circ G = -\Phi, \quad \frac{d}{dt} = \frac{\partial}{\partial t} + \vec{u}\cdot\nabla,$$

and corollaries of the conservation laws of pulse and energy:

$$\frac{d\vec{u}}{dt} - \frac{1}{\rho}\, div\ \sigma = 0,$$

$$\frac{de}{dt} - \frac{1}{\rho}\sigma : W = 0.$$

Here $\Phi$ denotes the prescribed nondifferential right-hand parts which are the nonlinear solution functions simulating relaxation of shear stresses; $\sigma = \|\sigma_{ij}\|$ is the stress tensor; $W = \|\frac{\partial u_i}{\partial x_j}\|$ is the strain rate tensor; $\rho$ is the denisty (function $G$); $e$ is the internal energy related to $G$ and entropy $S$ by the equation of state. The system is closed by the Murnaghan formulae, that follow from the thermodynamic identity, i.e.

$$\sigma_{ij} = -2\rho g_{ik}\frac{\partial e}{\partial g_{kj}}, \quad T = e_s.$$

To simulate fracture, use is made of the time criterion $J > J_0$, where $J$ is found from the equation

$$\frac{dJ}{dt} = (\sigma_e - \sigma_0)^n\, \vartheta(\sigma_e - \sigma_0),$$

Here $\vartheta$ is the Heaviside function; $\sigma_0$, $J_0$, $n$ are the material constants; $\sigma_e$ is the "effective" stress (a certain prescribed function of the stresses state).

The system of equation written was used for a two-dimensional nonstationary case. To solve it, we applied the Godunov method, calculations were made in movable grids, with distinguishing the contact and free boundaries [2]. Thus, a complex Eulerian-Lagrangian approach is used, i.e. mass, pulse and energy can be transferred via the internal boundaries of cells, and the external boundary of the calculation region must be that of a volume of the same particles, namely, it should be Lagrangian.

In this paper, to simulate the fracture, use is made of a "quasibrittle" approach, i.e. it is assumed that a crack always forms at the moment of fracture. The crack is simulated by introducing two free surfaces on the boundaries of two adjacent cells, where the fracture criterion is satisfied. The cracks can propagate with a given velocity and coalesce, thereby producing the fragments.

In this case, a concept of the calculation region structure is of specified sense. The calculation region is a set of all subregions. In this (probably, multiconnected) region, the corresponding initial boundary value problem is solved. In its turn, under the calculation subregion the curvilinear rectange being a region part and the image from a parametric square is meant. Finally, depending on the fracture character, a subregion being the square on a parametric plane contains a substructure that forms in the following way. In accordance with one of two coordinate directions $\xi$ or $\eta$ of the parametric plane (for example, $\eta$) the square is divided into layers consisting of rectangles which do not have the free surfaces (cracks) along the direction $\xi$. Each of these rectangles is divided in its turn into "elementary" rectangles without the "cracks" along $\eta$. Thus, each "elementary" rectangle and the corresponding image in a physical plane $(x,y)$ don't have the "cracks".

In calculations, a real structure of the calculation region can be rather complicated. As an example, Fig. 1 gives results of solution of the plane shock problem in which a steel cylinder 2.3mm in dia. strikes an aluminum target 0.5mm in thickness at 5km/s rate. In this case the structure of the calculation region includes the following elements, i.e. on the whole the calculation region comprises two subregions (striker and target), the first one being structurized for this moment by a set of cracks and contains 21 "elementary" subregions (4 layers of cells, the first and second layers comprise 6 subregions, the third one - 5 subregions and the fourth layer contains 4 subregions). In further calculations, this structure changes in time due to formation of new cracks and possible "healing" old ones.

Thus, a defined structure of the calculation region is required in the first place to build the Eulerian difference grid for each calculation step provided that position of the free and contact boundaries has already been calculated. The problem of grid construction is considered in a great deal of papers, for example, on the use of conformal mappings, etc.

The following necessary and important stage is construction of boundaries of the movable grid which satisfy a condition that the boundary (free or contact) consists on the whole of the same particles. It should be noted that the nodes themselves, which form the boundary, mustn't fulfil this condition, i.e. the nodes may move relatively the medium along the boundary. The boundaries, including those ones that form spallation structure, can be represented as the sides of curvilinear rectanges which compose the "elementary" subregions. Let us assume that velocity of angular points of these "elementary" rectangles is determined in any way, then the object of consideration is an unclosed curve with a given velocity field on this contour and given motion of the contour ends. Mark the nodes $\{\vec{x}_i\}\Big|_{i=1}^{n+1}$ forming this contour. Let us cause them to satisfy the following condition (the arrangement law):

$$s_i/s_{i-1} = t_i, \quad s_i = |\vec{x}_{i-1} - \vec{x}_i|, \quad i = 2,\dots,n+1, \tag{1}$$

the set $\{t_i\}$ has been prescribed. At parametric representation of the contour $\vec{x}(t,l)$ $(0 < l < 1)$ the problem can formally be stated as follows: in accordance with the given velocity field $\vec{u}(t,l)$ and initial position of the contour $\vec{x}(0,l)$ it is necessary to determine $\vec{x}(t,l)$, where the parameter $l$ is chosen to fulfil the arrangement law:

$$\left| \frac{\partial \vec{x}}{\partial l}(t,l) \right| = L(t), \tag{2}$$

where $L(t)$ is the contour length. Let us introduce the "liquid" (Lagrangian) curve $\vec{x}^*(t,l)$:

$$\vec{x}^*(t,l) = \vec{x}(0,l) + \int_0^t \vec{u}(\tau,l)\, d\tau.$$

Due to the gradient of the initial velocity to the tangent line, the "liquid" contour $\vec{x}^*(t,l)$ can't satisfy condition (1). Let us determine the new parameter $r$ which introduction on the "liquid" contour $\vec{x}^*(t,l)$ will allow us to obtain the new contour $\vec{x}(t,l)$ satsifying (2). Incompatibility of the previous parameter $l$ with (1) for $\vec{x}^*(t,l)$ is characterized by the function

$$a_t(l) = \frac{1}{L(t)} \int_0^l \left| \frac{\partial \vec{x}^*}{\partial \xi}(t,\xi) \right| d\xi.$$

By introducing the new parameter $r = a_t(l)$ define that

$$\vec{x}(t,r) = \vec{x}^*(t,a_t^{-1}(r)). \tag{3}$$

Reparametrization with the help of (3) means in essense the rearrangement of the nodes on the "liquid" contour in accordance with condition (2) for $\vec{x}(t,l)$. The velocity of nodes can be found from formula (3):

$$\frac{\partial \vec{x}}{\partial t}(t,l) = \frac{\partial \vec{x}^*}{\partial t}(t,l) + \frac{\partial \vec{x}^*}{\partial l}(t,l)\, \frac{\partial a_t^{-1}}{\partial t}(r).$$

At $t = 0$ we have

$$\frac{\partial a_t^{-1}}{\partial t}(r)\Big|_{t=0} = -\frac{\partial a_t^{-1}}{\partial t}(l)\Big|_{t=0} = \frac{L'(0)}{L(0)}l - \frac{1}{L(0)}\int_0^l \vec{\tau}(\xi)\,\frac{\partial \vec{u}}{\partial \xi}(0,\xi)\,d\xi,$$

where $\vec{\tau}(l) = \frac{\partial \vec{x}}{\partial \xi}(0,l)/L(0)$ is the unit vector that is tangential to the contour at the initial time moment, hence

$$\frac{\partial \vec{x}}{\partial t}(0,l) = \vec{u}(0,l) + \vec{\tau}(l)\left[L'(0)l - \int_0^l \vec{\tau}(\xi)\,\frac{\partial \vec{u}}{\partial \xi}(0,\xi)\,d\xi\right]. \qquad (4)$$

Formula (4) can be interpreted as follows: the velocity of nodes of the "liquid" contour satisfying (2) is obtained as that of particles, which is corrected by an additional velocity, that is tangential to the contour. This tangential component consists of two parts: the first part characterizes the change of the internodal distance due to the contour length change and the second one defines the contribution of the particle rearrangement along the contour owing to flow inhomogeneity (presence of velocity gradient) in the vicinity of the boundary.

To calculate the motion of boundaries of moving grids, use is usually made of the following simple technique comprising two stages: (1) calculation of nodal coordinates of the "liquid" contour and (2) final rearrangement of nodes along the contour enveloped on these nodes (this stage brings the most essential error). The discrete form of formula (4) gives an algorithm which can be more convenient than similar simple techniques for the following reason. The accuracy of the algorithm obtained is higher in cases when the contour curvature and discretization step are significant (the accuracy increases with an appropriate tangential vector $\vec{\tau}$ since the stage of arrangement at the point of considerable contour curvature can bring an additional error into the normal component of the velocity vector). However, the presence of the tangential vector in (4) restricts application of the algorithm obtained to the cases when $\vec{\tau}$ can be determined quite correctly with the least arbitrariness, in this case the sufficient smoothness of the contour is also desirable.

In practical calculations for the cases of significant contour curvature, the mentioned error in the normal boundary velocity can result in a considerable deviation of the contour being calculated from the "liquid" one. The larger the gradient of the tangential velocity component, the more rapid is the error increase. Specially for such cases the following method of boundary calculation is proposed.

A node with number $i$ on the "liquid" contour, that satisfies the given law of arrangement, gets the increment $\vec{h}_i$ at two discrete subsequent time moments. It is suggested to move the $i$-th node in the direction $\vec{h}_i$ with velocity equal to velocity projection onto this direction. Let the contour position at the first moment be defined by the set $\{\vec{x}_i\}\Big|_{i=1}^{n+1}$, and at subsequent moment the unknown coordinates satisfying (1) from the set $\{\vec{y}_i\}\Big|_{i=1}^{n+1}$. Thus, it is required to build such $\{\vec{h}_i\}\Big|_{i=1}^{n+1}$ that the first and last nodes are moved in accordance with the given displacement $\vec{w}_i = \vec{u}_i \Delta t$ $(i = 1, n+1;$ $\Delta t$ is the time step), i.e. for the first and last nodes of the contour we have $\vec{h}_i = \vec{w}_i$, for the other the value of $\vec{h}_i$ is found from the condition $\vec{w}_i \cdot \vec{h}_i = \vec{h}_i \cdot \vec{h}_i$, $i = 2,\dots,n$. We can prove the following assumption.

**ASSUMPTION.** Let us take such field of displacements that $\vec{w}_i \neq 0$, $i = 2,\dots,n$ and $|\vec{w}_i \cdot \vec{n}_i|/|\vec{w}_i| \geqslant a_0 > 0$, where $\vec{n}$ is the arbitrary unit normal that lies in a less of two angles, which are formed by the normals to edges adjacent to the $i$-th node. Construction of the unknown set $\{\vec{h}_i\}\Big|_{i=2}^{n}$ is guaranteed if condition

$$\frac{|\vec{w}_i - \vec{w}_{i-1}|}{|\vec{w}_1|} \leqslant A a_0^k, \quad i = 2,\dots,n.$$

is fulfilled for the field of displacements.

The error value can be characterized by the distance between $\vec{y}_i$ and $\vec{x}_i + \vec{w}_i$, which equals $\rho_i = |\vec{h}_i - \vec{w}_i|$. For $\rho_i$ the estimate

$$\rho_i \leqslant \frac{B|\vec{w}_i - \vec{w}_{i-1}|}{a_0^k}.$$

is valid. Here $A$, $B$, $k$ are the constants dependent on geometrical parameters of the grid and set $\{t_i\}$. From the last estimate it follows that for a smooth velocity field

$$\rho_i \leqslant \frac{B}{a_0^k} \left|\frac{\partial}{\partial s}\vec{w}(s_\vartheta)\right| \Delta t \; |\vec{x}_i - \vec{x}_{i-1}|,$$

where the parameter $s_\vartheta$ marks the point on the edge between the $(i-1)$-th and $i$-th nodes. Thus, for the smooth velocity field the technique proposed doesn't deteriorate the approximation order of the initial difference scheme.

Now let us consider the problem of a slightly conical plate (the angle between the generating line and radial unit vector is $7^\circ$) under pulse loading which gradually propagates over the region from the cone axis to the periphery.

The process of plate material flow is such in this problem that the gradient of the velocity component, which is tangential to the external cone surface, is high near this surface and in the vicinity of axis and small near the internal free cone surface. Use of the simplest algorithm of arrangement to calculate the motion of boundaries results in a rapid increase in error in the vicinity of the loading surface and in catastrophic effect (cell overlapping) illustrated in Fig. 2. With the help of the latter boundary calculation method this problem can be solved correctly. (Fig. 3).

Efficiency of the techniques constructed can be illustrated by numerical solution of the problem of "inverse" cumulation [3]. As the initial configuration, a copper conical shell 2 mm in thickness with a cone angle of $120^\circ$ was taken, to which a conical cylindrical explosive charge was joined. The explosive action was simulated within the framework of stationary detonation. Fig.4 shows the stage of shell loading and formation of a jet and a slug, Fig.5 presents the finite configuration of a freely flying shell with a crack structure in it.

The study performed demonstrates also the fact that the problem of boundary calculation requires special thoroughness in the case when a number of cells is limited by computer capability.

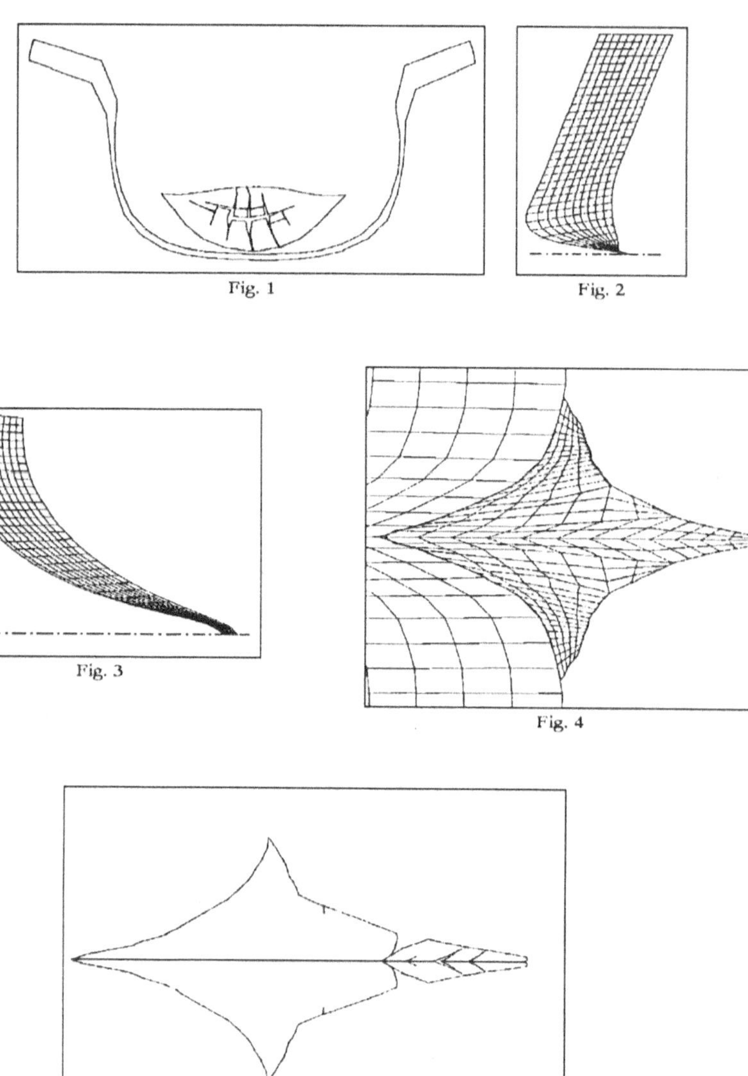

Fig. 1

Fig. 2

Fig. 3

Fig. 4

Fig. 5

## REFERENCES.

[1] Godunov S.K., Elements of continuum mechanics. Moscow, Nauka, 1978, 304 p.

[2] Godunov S.K., Zabrodin A.V., Ivanov M.Ya., Kraiko A.N., Prokopov G.P., Numerical solution of multi-dimensional gas dynamics problem. Moscow, Nauka, 1976, 400 p.

[3] Merzhievsky L.A., Resnyansky A.D., Strength effects in inverse cumulation. DAN SSSR, 1986, vol.290, N 6, pp. 1310-1314.

International Series of Numerical Mathematics, Vol. 106, © 1992 Birkhäuser Verlag Basel 

# ON THE CANONICAL VARIABLES FOR TWO-DIMENSIONAL VORTEX HYDRODYNAMICS OF INCOMPRESSIBLE FLUID

*O.I. Mokhov*
*VNIIFTRI*
*Mendeleevo, 141570 Moscow Region, RUSSIA*

The vorticity equation of two-dimensional hydrodynamics of incompressible fluid is represented in canonical Hamiltonian form by reversible hodograph-type point transformation of physical variables. Analogous transformations determine canonical variables for Rossby waves. A general method of group analysis of Hamiltonian structures of nonlinear equations of mathematical physics is presented and used for those purposes.

Key words: vortex flow, Hamiltonian structure.

Consider a planar vortex flow of incompressible fluid:

$$\frac{\partial u}{\partial t} = -u\frac{\partial u}{\partial x} - v\frac{\partial u}{\partial y} - \frac{\partial P}{\partial x},$$

$$\frac{\partial v}{\partial t} = -u\frac{\partial v}{\partial x} - v\frac{\partial v}{\partial y} - \frac{\partial P}{\partial y},$$

$$\frac{\partial u}{\partial x} + \frac{\partial v}{\partial y} = 0,$$

where the physical variables are: $P$ is the pressure; $u$, $v$ are the components of the fluid velocity. Here the mass density $\rho = 1$.

Accordingly, the vortex $\Omega = v_x - u_y$ satisfies the vorticity equations

$$\frac{\partial \Omega}{\partial t} = \frac{\partial \psi}{\partial x}\frac{\partial \Omega}{\partial y} - \frac{\partial \psi}{\partial y}\frac{\partial \Omega}{\partial x}, \tag{1}$$

where $\psi$ is the stream function, $u = \psi_y$, $v = -\psi_x$, $\Omega = -\Delta\psi$.

It is well-known (see, for example, refs.[1,2]) that a Hamiltonian structure for (1) is given by the noncanonical Poisson bracket

$$\left\{\Omega(x,y),\Omega(x',y')\right\} = \Omega_y\delta_x(x-x',y-y') - \Omega_x\delta_y(x-x',y-y') \qquad (2)$$

The vorticity equation (1) can be represented in a Hamiltonian form as

$$\frac{\partial\Omega}{\partial t} = \left\{\Omega,H\right\} \qquad (3)$$

with the Poisson bracket (2), $H$ is the energy Hamiltonian:

$$H = \frac{1}{2}\int(u^2 + v^2)dx\,dy = -\frac{1}{2}\int \psi\,\Delta\psi\,dx\,dy =$$

$$= -\frac{1}{4\pi}\iint \ln|r-r'|\,\Omega(r)\,P(r')\,d^2r\,d^2r', \qquad (4)$$

$$r = (x,y).$$

In the author's works [3,4] the vorticity equation (1) was reduced to the canonical Hamiltonian form

$$v_t = \frac{d}{dz}\frac{\delta H}{\delta v}, \qquad (5)$$

where $H$ is the usual functional of energy (4).

The corresponding transformation has the form

$$x = v,\ y = z,\ \Omega = w. \qquad (6)$$

This is the classical reversible point transformation of hodograph type with one field variable. One may regard $v(z,w)$ as a new scalar field defined by the relation $x = v(y,\Omega(x,y))$.

Thus, the vorticity equation (1) reduces to the Hamiltonian form (5) with respect to the canonical Poisson bracket of Gardner, Zakharov and Fadeev

$$v_t = \left\{v,H\right\},\quad \left\{v(z,w),v(z',w')\right\} = \delta_z(z-z',w-w').$$

Now, for equation (5) the conjugate canonical variables can

be introduced in just the same way as the well-known Gardner conjugate canonical variables for the Korteweg-de Vries equation [5]:

$$v(z,w) = \sum_{k_1,k_2=-\infty}^{+\infty} r_{k_1k_2} exp\left\{i(k_1z + k_2w)\right\},$$

$$q_k = \frac{r_k}{k_1}, \; p_k = r_{-k}, \; k = (k_1,k_2), \quad k_1 > 0.$$

Now equation (5) is reduced to the canonical system of Hamiltonian:

$$(q_k)_t = \frac{i}{4\pi^2}\frac{\partial H}{\partial p_k}, \quad (p_k)_t = \frac{i}{4\pi^2}\frac{\partial H}{\partial q_k},$$

where $H = H(q_k,p_k)$ is the usual energy Hamiltonian.

Moreover, in the author's works [3,4] it was shown that the generalized vorticity equation describing plasma drift waves and geostrophic fluid flow (see, e.g., ref. [1] and references therein)

$$(c - \Delta)\varphi_t = \varphi_x\left[(c - \Delta)\varphi - \lambda\right]_y - \varphi_y\left[(c - \Delta)\varphi - \lambda\right]_x \tag{7}$$

always can be expressed in the canonical Hamiltonian form. Here, the function $\varphi$ represents the electrostatic potential for drift waves and the stream function for fluid flow, $\lambda$ is a given function of $(x,y)$ related to the local ion cyclotron frequency or Coriolis frequency, and $c = const$. For Rossby waves ($\lambda = \beta y$, $\beta \neq 0$ is the $\beta$-plane approximation in geophysics) the canonical variables were found by Zakharov and Piterbarg [6] (see also refs. [3,4]). We propose an effective general method which is based on group analysis of local Poisson brackets, and was developed in the author's works [7,8].

Consider the most general contact transformation of the variables $(x^i,u,u_i)$, $1 \leqslant i \leqslant n$,

$$x^i = \varphi^i(y^j,v,v_j), \quad 1 \leqslant i, \quad j \leqslant n,$$

$$u = \psi(y^j,v,v_j),$$

$$u_i = \chi_i(y^j,v,v_j).$$

The relations (8) define a contact transformation of the variables $(x^i, u, u_j)$, $1 \leqslant i \leqslant n$ if for every $i$

$$D_i\psi = \chi_k D_i \varphi^k,$$

where $D_i$ is the total derivative in $y^i$, i.e.

$$D_i\varphi = \frac{\partial\varphi}{\partial y^i} + \frac{\partial\varphi}{\partial v}\frac{\partial v}{\partial y^i} + \frac{\partial\varphi}{\partial u_j}\frac{\partial^2 v}{\partial y^i \partial y^j}.$$

Let $F$ be an arbitrary functional of the form

$$F = \int f(x, u, u_k) d^n x, \quad x \in R^n,$$

$$k = (k_1, \ldots, k_n), \quad u_k = \frac{\partial^{|k|} u}{\partial x^k}.$$

Then $F(y, v, v_j) = \int fJ \, d^n x$, where $J = \det(D_j \varphi^i)$ is the Jacobian of the contact transformation.

We have

$$\frac{\delta F}{\delta v} = \rho J \frac{\delta F}{\delta u},$$

where

$$\rho = \frac{\partial\psi}{\partial v} - \chi_k \frac{\partial\varphi^k}{\partial v}.$$

Now consider an arbitrary many-dimensional scalar local Poisson bracket for functionals $H_1$, $H_2$:

$$\left\{H_1, H_2\right\} = \int \frac{\delta H_1}{\delta u} L \frac{\delta H_2}{\delta u} d^n x,$$

where $L$ is the Hamiltonian differential operator, i.e. for arbitrary functionals $H_1$, $H_2$, $H_3$ we have

1) $\left\{H_1, H_2\right\} = -\left\{H_2, H_1\right\}$ (skew-symmetry),

2) $\left\{\left\{H_1, H_2\right\}, H_3\right\} + \left\{\left\{H_2, H_3\right\}, H_1\right\} + \left\{\left\{H_3, H_1\right\}, H_2\right\} = 0$ (the Jacobi identity).

In particular, for the hydrodynamic Poisson bracket (2)

$$L = \Omega_y D_x - \Omega_x D_y .$$

For an arbitrary Hamiltonian differential operator $L$ and an arbitrary contact transformation the differential operator

$$M = \frac{1}{\rho} L \circ \frac{1}{\rho J}$$

is Hamiltonian. Here the operator $L$ must be expressed in the variables $(y,v(y))$.

Thus, the contact transformation defines the automorphism of the set of Hamiltonian differential operators of the fixed order:

$$L \to M = \frac{1}{\rho} L \circ \frac{1}{\rho J}. \tag{8}$$

Here we shall describe the class of two-dimensional Hamiltonian operators generated from the Hamiltonian operator of Gardner, Zakharov and Fadeev $D_{x^1}$ by the contact automorphisms (8). Really, it appears that every two-dimensional scalar differential Hamiltonian operator of the first order belongs to this class.

Consider the most general contact transformation for $x \in R^2$:

$$x^i = \varphi^i(y^1,y^2,v,v_1,v_2),$$

$$u = \psi(y^1,y^2,v,v_1,v_2),$$

$$u_i = \chi_i(y^1,y^2,v,v_1,v_2),$$

where

$$\psi_{y^i} = \chi_k \varphi^k_{y^i}, \quad \varphi_{y^i} = D_{y^i}(\varphi), \quad J = \varphi^1_{y^1}\varphi^2_{y^2} - \varphi^2_{y^1}\varphi^1_{y^2},$$

$$\rho = \frac{\partial\psi}{\partial v} - \chi_k \frac{\partial\varphi^k}{\partial v}, \quad D_{x^1} = \frac{1}{J}\left[\varphi^2_{y^i} D_{y^1} - \varphi^2_{y^1} D_{y^2}\right].$$

The general form of two-dimensional Hamiltonian operator reducible by contact transformation to canonical form $D_{x^1}$ is

$$L = \frac{1}{\rho J}\left[\varphi^2_{y^i} D_{y^1} - \varphi^2_{y^1} D_{y^2}\right] \circ \frac{1}{\rho J} .$$

Consider the equation for Rossby waves in atmosphere of rotating planet:

$$\omega_t + \beta\psi_x = \psi_x\omega_y - \psi_y\omega_x, \tag{9}$$

where $\omega = (c-\Delta)\psi$. Hamiltonian structure of (9) can be described by the linear Poisson bracket (2) with 2-cocycle $-\beta\delta_x(x-x',y-y')$. The corresponding Hamiltonian operator is

$$L = \omega_y D_x - \omega_y D_y - \beta D_x.$$

In [6] Zakharov and Piterbarg guessed the transformation reducing this Hamiltonian operator to the canonical form $D_{y^1}$ if $\beta \neq 0$:

$$\omega(x,y) = v(x,-y/\beta + \omega(x,y)/\beta^2).$$

This transformation in standard form is

$$x = z, \; y = -\beta w + v(z,w)/\beta, \quad \omega = v.$$

It belongs to the group of special contact (and even point) transformations introduced and studied in the author's works [7,8]. We can introduce a parameter in transformation (6) in order that to build the canonical variables for the equation for Rossby waves (9) under any $\beta$:

$$x = v, \; y = z, \; \omega = \beta z + w,$$

$$\omega_x = 1/v_w, \; \omega_y = \beta - v_z/v_w,$$

$$\rho = -1/v_w, \; J = -v_w.$$

One may also regard that $v(z,w)$ is a new scalar field defined by relation

$$x = v(y,\omega(x,y) - \beta y).$$

## REFERENCES.

[1] Weinstein A. Phys. Fluids 26 (1983), p.388.

[2] Zakharov V.E., Kuznetsov E.A. Hamiltonian formalizm for systems of hydrodynamic type. Soviet Scientific Rev., Sect. C, vol. 4 (1984), preprint IAE SO AN SSSR, N 186 (1982).

[3] Mokhov O.I. Teor. Mat. Fiz. 78 (1989), p.136.

[4] Mokhov O.I. Phys. Letters A 139 (1989), p.363.

[5] Gardner C.S. J. Math. Phys. 12 (1971), p.1548.

[6] Zakharov V.E., Piterbarg L.I. Dokl. Akad. Nauk SSSR, 295 (1987), p.86.

[7] Mokhov O.I. Usp. Mat. Nauk 40 (1985), p.257.

[8] Mokhov O.I. Funkt. Anal. Appl. 21 (1987), p.53.

# ABOUT THE METHOD WITH REGULARIZATION FOR SOLVING THE CONTACT PROBLEM IN ELASTICITY

*R.V.Namm*
*Institute of Applied Mathematics*
*Khabarovsk 680003, RUSSIA*

In this paper we consider the dependence of the solution on the parameters in iterative prox-regularization method which is used in a contact problem in elasticity.

Key words: orthogonal projection, convex set, iterative, Lipschitz-continuous boundaries, contact set.

I. Let

$$J(u) = a(u,u) - 2L(u)$$

be a quadratic functional in Hilbert space $H$. Here $a$ is a bilinear, symmetric, non-negative, bounded form on $H \times H$, $L$ is a bounded linear functional on $H$.

We denote

$$R = \left\{v: a(v,v) = 0\right\}, \ \tilde{R} = \left\{z \in R: L(z) = 0\right\}.$$

Let $Q$ be an operator of orthogonal projection from space $H$ on $R$, $P = I - Q$ ($I$ be an identity operator).

We consider the problem

$$\begin{cases} J(u) \to min! \\ u \in G, \end{cases}$$

where $G$ is some closed convex set in $H$.

It is assumed that

a) the form $a$ is weakly coercive, i.e.

$$a(v,v) \geqslant \omega \|Pv\|^2 \quad \forall \ v \in H,$$

where $\omega$ is some positive constant independent on $v$;

b) the kernel $R$ of quadratic form $a$ is finite-dimensional;

c) the set $G^* = \left\{u \in G\colon J(u) = \inf_{v \in G} J(v)\right\}$ is not empty.

Let us consider the iterative algorithm with regularization on each step to solve the problem (1).

1. The sequence of positive numbers $(\varepsilon_k)$ with condition

$$\sum_{k=1}^{\infty} \varepsilon_k^{1/2} < \infty$$

and element $z^0 \in H$ is chosen. Let $m = 1$.

2. Beginning with the point $z^{k-1}$, we define $z^k$ from the criterion

$$J(z^k) + \|z^k - z^{k-1}\|^2 \leqslant \min_{u \in G} \left\{J(u) + \|u - z^{k-1}\|^2\right\} + \varepsilon_k .$$

It was shown in [2] that the sequence $\{z^k\}$ converges to some element $z^* \in G^*$ in $H$.

For any $v^* \in G^*$ and any points $v^k$, $k = 0, 1, 2, \dots,$ such as $v^k \in G$, $\lim_{k \to \infty} v^k = v^*$, it is assumed that

d) if for $s \in \tilde{R}$ it holds that $v^* + s \in G$, then there is a constant $\chi > 0$ for which $v^k + \chi s \in G$ for every $k = 1, 2, \dots$ .

We denote the distance from point $x$ to set $X$ as $\rho(x,X)$. Let for $u^* \in G^*$ the equality

$$\|z^0 - u^*\| = \rho(z^0, G^*)$$

takes place.

**THEOREM 1.** $\|z^* - u^*\| \leqslant \sum_{k=1}^{\infty} \varepsilon_k$ (see [3]).

II. In this section the indicated theorem 1 is considered for the contact problem in elasticity [4].

Let $\Omega \subset \Omega' \cup \Omega'' \subset R^2$, where $\Omega'$, $\Omega''$ are domains with Lipschitz-continuous boundaries $\partial\Omega'$, $\partial\Omega''$ and also $\Omega' \cup \Omega'' = \phi$, $\Gamma_c = \partial\Omega' \cap \partial\Omega''$, $\mathrm{mes}\, \Gamma_c > 0$. For

$$u = (u', u'') \in [W_2^1(\Omega')]^2 \times [W_2^1(\Omega'')]^2$$

we define the strain tensor

$$\varepsilon_{ij}^{N} = \frac{1}{2}\left[\frac{\partial u_i^N}{\partial x_j} + \frac{\partial u_j^N}{\partial x_i}\right], \quad i,\ j = 1,\ 2 \quad (N = {}',\ {}'').$$

Under the fixed decompositions

$$\partial\Omega' = \Gamma_u \cup \Gamma_\tau' \cup \Gamma_c, \quad \partial\Omega'' = \Gamma_\tau'' \cup \Gamma_c \quad (mes\ \Gamma_u > 0)$$

and fixed functions

$$C_{ijpe}^{N} \in L_\infty(\Omega^N) \quad (i,\ j,\ p,\ e = 1,\ 2; \quad C_{ijpe}^{N} = C_{jipe}^{N} = C_{peij}^{N}).$$

$$F = (F', F'') \in [L_2(\Omega')]^2 \times [L_2(\Omega'')]^2,$$

$$T = (T', T'') \in [L_2(\Gamma')]^2 \times [L_2(\Gamma'')]^2.$$

it is required to find the minimum of functional

$$J(u) = a(u,u) - 2L(u)$$

on the set

$$G = \left\{u \in H:\ -u_n' + u_n'' \leqslant 0 \quad \text{on} \quad \Gamma_c\right\},$$

where

$$a(u,u) = \int_{\Omega'} C_{ijpe}'\varepsilon_{ij}'(u)\varepsilon_{pe}'(v)\, d\Omega + \int_{\Omega''} C_{ijpe}''\varepsilon_{ij}''(u)\varepsilon_{pe}''(v)\, d\Omega,$$

$$L(u) = \int_{\Omega'} F_i' u'\, d\Omega + \int_{\Omega''} F_i'' u_i''\, d\Omega + \int_{\Gamma_\tau'} T_i' u_i'\, d\Gamma + \int_{\Gamma_\tau''} T_i'' u_i''\, d\Gamma,$$

$$H = \left\{u = (u', u''):\ u \in [W_2^1(\Omega')]^2 \times [W_2^1(\Omega'')]^2,\ u' = 0 \quad \text{on} \quad \Gamma_u\right\}.$$

It is assumed that the contact set $\Gamma_c = \partial\Omega' \cap \partial\Omega''$ is a straight line segment. For $u^*$ we denote

$$E^* = \left\{x \in \Gamma_c:\ (u_n'')^* - (u_n')^* = 0\right\}.$$

The kernel of quadratic form $a(u,u)$ is

$$R = \left\{s = (s', s''):\ s' \equiv 0,\ s_1''(x) = a_1'' - b'' x_2,\ s_2''(x) = a_2'' + b'' x_1\right\}.$$

It was shown in [1] that the problem has a solution if $L(s) \leqslant 0$ for all $s \in G \cap R$ and $L(s) < 0$ for $s \in G \cap R$, $s''_n - s'_n < 0$ on $\Gamma_c$.

Let the contact set $\Gamma_c$ lie on axis $Ox_1$

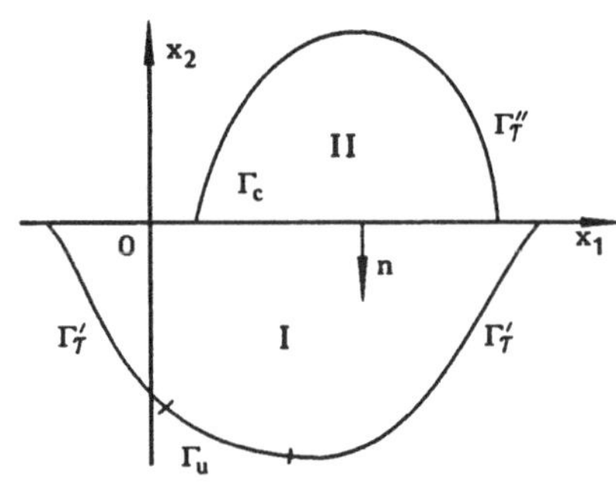


From the solvability condition it follows that for

$$s \in R^* = \left\{P \in G \cap R:\ P''_n = 0 \text{ on } \Gamma_c\right\}$$

the condition $L(s) = 0$ should be realized. The normal $n = (0,-1)$. Therefore the condition $s''_n = 0$ is equivalent to condition $a''_2 + b''x_i$ $\forall\, x_i \in \Gamma_c$, from which $a''_2 \equiv 0$, $b'' \equiv 0$. By the same token, if $s \in R^*$, then $s = (0,0,a''_1,0)$. Since $L(s) = 0$, then

$$\int_{\Omega''} F''_1 a''_1 \, d\Omega + \int_{\Gamma''_\tau} T''_1 a''_1 \, d\Gamma = 0 \quad \text{or}$$

$$\int_{\Omega''} F''_1 \, d\Omega + \int_{\Gamma''_\tau} T''_1 \, d\Gamma = 0.$$

Let us consider $s \in \tilde{R} \backslash R^*$. It can be shown, that

$$L(s) = \int_{\Omega''} F''_1 s''_1 \, d\Omega + \int_{\Gamma''_\tau} T''_1 s''_1 \, d\Gamma = 0.$$

From (2) it follows that

$$a''_2 \left[ \int_{\Omega''} F''_2 \, d\Omega + \int_{\Gamma''_\tau} T''_2 \, d\Gamma \right] -$$

$$- b'' \left[ \int_{\Omega''} (F''_1 x_2 - F''_2 x_1) \, d\Omega + \int_{\Gamma''_\tau} (T''_1 x_2 - T''_2 x_1) \, d\Gamma \right] = 0.$$

From here

$$- \frac{a''_2}{b''} = - \frac{\int\limits_{\Omega''} (F''_1 x_2 - F''_2 x_1)\, d\Omega + \int\limits_{\Omega''} (T''_1 x_2 - T''_2 x_1)\, d\Gamma}{\int\limits_{\Omega''} F''_2 \, d\Omega + \int\limits_{\Gamma''_\tau} T''_2 \, d\Gamma} \equiv const.$$

On the other hand $s''_n = a''_2 + b'' x_1$ on $\Gamma_c$. This means that $s''_n = 0$ for $x^*_1 = -a''_2/b''$. By the same token it is shown that for all $s \in \tilde{R} \backslash R^*$ the corresponding $s''_n$ vanishes at the same point $x^*_1$. From the solvabiliry condition it follows that $x^*_1 \in ri\ \Gamma_c$ (see [1]).

Let us suppose that the set $E^*$ is the union of non-zero measure sets and the scalar function $\varphi = (u''_n)^* - (u'_n)^*$ is continuous on $\Gamma_c$ and differentiable at the point $x^*_1$.

We consider two possible cases:

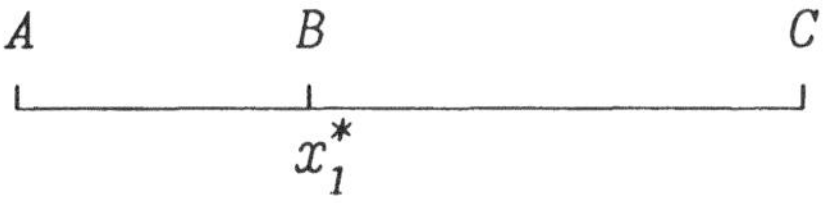


1) $mes\ (E^* \cap [A,B]) > 0$ and $mes\ (E^* \cap [B,C]) > 0$,

2) either $E^* \subseteq [A,B]$ or $E^* \subseteq [B,C]$.

In case 1 it is easily seen that if we have $u^* + s \in G$ for some $s \in \tilde{R}$, then $s \in R^*$. By the same token, the assumption d) is fulfilled.

Let us turn to case 2. We suppose that $x^*_1 \notin E^*$. Then for any $s \in \tilde{R}$ with condition $s''_n < 0$ on $E^*$ there is $a > 0$ for which $\varphi + a s''_n < 0$ on $\Gamma_c$. We take such $s_1 \in R$, that $L(s_1) < 0$ and choose such $\beta > 0$, that $u^* + as - \beta s_1 \in G$. It is easily seen that $J(u^* + as - \beta s_1) < J(u^*)$. This contradicts condition

$$u^* = \arg\min_{u \in G} J(u).$$

Therefore $x^*_1 \in E^*$ and it is an extreme point for $E^*$. Hence, if $u^* + s \in G$ for some $s \in \tilde{R}$, then from the differentiability of $\varphi$ at the point $x^*_1$ it follows that $s \in R^*$.

Thus the assumption d) is fulfilled in both cases.

**Remark.** In case 1 under the condition $\varepsilon_k = q^k$, where $q \in (0,1)$ is some suitably chosen constant, we can prove

**THEOREM 2.** There is a constant $q \in (0,1)$ for which

$\|z^k - z^*\| \leqslant cq_1^k$, $c > 0 - const$ (see [5],[6]).

**REFERENCES.**

[1] Fichera G. Existence theorems in elasticity, Mir, Moscow, (1974);

[2] Kaplan A.A. About the stability of methods to solve the convex programming problems and variational inequalities. Models and methods of optimization, Novosibirsk, Nauka, (1988), pp.132-159;

[3] Namm R.V. About the limited point dependence on the points of generating sequence in iterative prox-regularization method, Preprint of the Institute of Applied Mathematics, Vladivostok, (1990), 15 p.;

[4] Hlavacek I., Haslinger J., Necas I., Lovisek J. Solution of variational inequalities in mechanics, Mir, Moscow, (1986);

[5] Namm R.V. About the convergence rate of algorithm with prox-regularization to solve the variational inequalities with weakly coercive operator, Optimization, Novosibirsk, 46(60), (1988). pp. 49-61;

[6] Kaplan A.A., Namm R.V. About the convergence rate estimate of iterative processes with prox-regularization, Investigation over conditional correctness of mathematical physics problems, Novosibirsk, (1989), pp. 60-77.

# SPACE EVOLUTION OF TORNADO-LIKE VORTEX CORE

*V.V. Nikulin*
*Lavrentyev Institute of Hydrodynamics*
*Novosibirsk 630090, RUSSIA*

In a long-wave approximation, the motion of liquid in a vertical tornado-like vortex core is studied. An extension of the continuous solution to a finite or infinite height is established to be possible. Classification of nonextended solutions is made: the solution doesn't exist either due to vanishing of a vertical velocity component or due to tending of derivatives to infinity (analogously to a "gradient catastrophe" in gas dynamics). For the first case, an analytical example is built, in which the height-finite solution propagates continuously over the whole half-space. A possibility of applying obtained results to calculate the height of a tornado-like vortex, to describe the process of its decay and effects observed in tornados and dust vortices in nature is discussed.

Key words: tornado-like vortex, breakdown.

## 1. STATEMENT OF THE PROBLEM.

A half-space filled in with an inviscid incompressible liquid in the gravity field is examined. The flow is considered to be stationary and rotationally symmetric. A cylindrical system of coordinates $(r,\theta,z)$ is introduced, where $r$ is the radius, $\theta$ the azimuthal angle, $z$ the symmetry axis directed opposite the gravity force. By $r = r_0(z)$, the boundary that separates the vortex core from the outer flow in the region $r > r_0(z)$ is denoted. The liquid density in the outer flow is assumed to be constant. On the core boundary, an abrupt change of the density and tangential velocity component is possible. To change over to non-dimensional values, the scales of the length, velocity and density are introduced. As the length unit, the characteristic scale of changes in the Z-axis is taken; as the velocity unit, the value of rotational velocity component for $r = r_0$, $z = 0$ is adopted; the density of the outer flow is taken as the density unit. In this case, the characteristic pressure and acceleration are equal to unit. The nondimensional value of $r_0$, when $z = 0$, is denoted by $\delta$. Further on, all the values are taken as nondimensional ones (if not specified).

The following notations are introduced: $(u,v,w)$ are the velocity components corresponding to $(r,\theta,z)$, $P$ is the pressure, $\rho$ is the density, $g$ is the gravitational acceleration. The outer flow is assumed to be known and prescribed in the following form that satisfies the equations of motion, i.e.

$$u = w = 0, \; v = \delta/r, \quad P = -\frac{\delta^2}{2r^2} - gz. \tag{1.1}$$

The vortex core flow is studied in a Z-axis long-wave approximation. The coordinates and functions are extended as follows:

$$r^2 \to \delta^2\eta, \; z \to z, \; 2ur \to \delta^2 q, \quad vr \to \delta A, \; w \to w, \; \rho \to \rho, \; P \to P, \; g \to g.$$

The boundary $r_0(z)$ changes over to $\eta_0(z)$. After substitution, the equations of motion and continuity take the form

$$\begin{gathered} q\frac{\partial A}{\partial \eta} + w\frac{\partial F}{\partial z} = 0, \\ \frac{\rho\delta^2}{2}\left[ q\frac{\partial q}{\partial \eta} - \frac{q^2}{2\eta} + w\frac{\partial q}{\partial z} \right] - \frac{\rho A^2}{2} = -2\eta\frac{\partial P}{\partial \eta}, \\ \rho\left[ q\frac{\partial w}{\partial \eta} + w\frac{\partial w}{\partial z} \right] = -\frac{\partial P}{\partial z} - \rho g, \\ \frac{\partial q}{\partial \eta} + \frac{\partial w}{\partial z} = 0, \\ q\frac{\partial P}{\partial \eta} + w\frac{\partial P}{\partial z} = 0. \end{gathered} \tag{1.2}$$

For the symmetry axis and the core boundary, the following boundary conditions

$$q = A = 0, \quad \eta = 0, \tag{1.3}$$

$$P = -\frac{1}{2\eta_0} - gz, \quad \eta = \eta_0, \tag{1.4}$$

$$q = w\left(\frac{\partial \eta_0}{\partial z}\right), \quad \eta = \eta_0. \tag{1.5}$$

are laid out. Condition (1.4) results from (1.1) and a requirement of pressure continuity on the core boundary, (1.5) is the kinematic condition.

It is assumed that $\delta \ll 1$. The summands in (1.2), that are proportional to $\delta^2$, are omitted. The system obtained is transformed analogously [1,2]. New independent variables $z'$, $\nu$; $\nu \in [0,1]$ are introduced according to relations $z = z'$, $\eta = R(z',\nu)$, where $R$ satisfies the equation

$$w\frac{\partial R}{\partial z'} = q \tag{1.6}$$

and boundary conditions

$$R(z',0) = 0, \quad R(z',1) = \eta_0. \tag{1.7}$$

The initial condition $R(0,\nu)$ is the arbitrary single-valued continuous function. In such definition of $R$, the boundary conditions (1.3) (for $q$) and (1.5) are fulfilled automatically. The unknown boundary $\eta_0(z)$ changes to the known one $\nu = 1$. After elimination of $P$ and $q$, system (1.2) takes the following form in the variables $z'$, $\nu$ (further on the prime of $z'$ will be dropped):

$$\rho w\frac{\partial w}{\partial z} = -\frac{1-\rho_1 A_1^2}{2R_1^2}\frac{dR_1}{dz} + \frac{\partial}{\partial z}\int_{\nu}^{1}\frac{1}{2R}\frac{d}{d\nu}\left(\rho A^2\right)d\nu + (1-\rho)g,$$
$$\frac{\partial}{\partial z}\left[w\frac{\partial R}{\partial \nu}\right] = 0, \quad A = A(\nu), \quad \rho = \rho(\nu). \tag{1.8}$$

Here $R_1$, $A_1$, $\rho_1$ are the values of $R$, $A$, $\rho$ for $\nu = 1$ (on the core boundary). System (1.8) is solved with the intial data when $z = 0$. It is taken that $w = w_0(\nu)$, $R = \nu$ for $z = 0$.

**THEOREM.** Let us assume that $A = 0$, $\rho = \rho_1 = const$, $\infty > w_0(v) \geqslant \gamma > 0$, $\gamma = const$,

$$\lambda = \frac{1}{2\rho_1} \int_0^1 \frac{dv}{w_0^2} .$$

Two cases for $\rho_1 < 1$ and $\rho_1 > 1$ are considered.

1. Let us take $\rho_1 < 1$. Then, if $\lambda < 1$, the solution exists with all the values of $z > 0$, $v \in [0,1]$; in this case $r \to 0$, $w \to \infty$, being monotone for $z \to \infty$. If $\lambda > 1$, the solution exists only for $z \leqslant l$,

$$l = \frac{1}{2(1 - \rho_1)g} \left[ \frac{1}{\rho_1} - \gamma^2 - \frac{1}{\rho_1 \int_0^1 w_0 (w_0^2 - \gamma^2)^{-1/2} \, dv} \right], \qquad (1.9)$$

in this case $w = 0$, $\partial R/\partial v = \infty$ when $z = l$ for $v$ prescribed by the equation

$$w_0(v) = \gamma.$$

2. Consider now the case when $\rho_1 > 1$. It is supposed additionally that $(w_0 - \gamma)^{-3/2}$ isn't integrated on [01]. Then for any $\lambda > 0$ the solution exists only for the finite values of $z$, $z \leqslant h$. With $z \to h$, we have $\partial w/\partial z \to \infty$, $\partial r/\partial z \to -\infty$, if $\lambda > 1$ and $\partial w/\partial z \to -\infty$, $\partial R/\partial z \to \infty$, if $\lambda < 1$. When $\lambda > 1$, $w$ increases monotonically and $R$ decreases with $\lambda$, and vice versa when $\lambda < 1$. The value of $h$ is found from the solution of a certain integral equation.

**Proof.** It is assumed that $A = 0$, $\rho = \rho_1$. Equations (1.8) are integrated from $0$ to $z$. The notation $\varphi = w^2 - w_0^2$ is introduced.

We then derive the system

$$\varphi - \frac{1}{\rho_1 R_1} = 2gz(1 - \rho_1)\rho_1^{-1} - \rho_1^{-1}, \quad R = \int_0^v \frac{w_0 \, dv}{(w_0^2 + \varphi)^{1/2}}. \qquad (1.10)$$

From the first equation it follows that $\varphi = \varphi(z)$, its left-hand side is denoted by $f(\varphi)$. Then, taking into account $\varphi(0) = 0$, we obtain $f'(0) = 1 - \lambda$,

$$f''(\varphi) = \frac{3}{4\rho_1 R_1^2}\int_0^1 \frac{w_0\, dv}{(w_0^2 + \varphi)^{5/2}} - \frac{1}{2\rho_1 R_1^3}\left[\int_0^1 \frac{w_0\, dv}{(w_0^2 + \varphi)^{3/2}}\right]^2.$$

The estimation of the second summand by the Bunyakovsky inequality easily shows that $f''(\varphi) > 0$. Hence, $f'(\varphi)$ is the monotonically increasing function. Since

$$\frac{df}{dz} = \frac{df}{d\varphi}\frac{d\varphi}{dz} = 2g(1 - \rho_1)/\rho_1, \tag{1.11}$$

the sign of $d\varphi/dz$ is defined by that of the right-hand part and, by virtue of monotony of $f'(\varphi)$, by the sign of $f'(0)$.

Let us take $\rho_1 < 1$, $\lambda < 1$. Then $\varphi'(z) > 0$. If $\rho_1 < 1$, $\lambda > 1$ then $\varphi'(z) < 0$. From here and (1.10), with the allowance for that expression rooted becomes negative when $\varphi < -\gamma^2$, the statements of theorem follow for the case of $\rho_1 < 1$.

Let us assume that $\rho_1 > 1$. From theorem conditions and expression for $f'(\varphi)$, it is easily established that $f'(\varphi) \to -\infty$ when $\varphi \to -\gamma^2$ and $f'(\varphi) \to 1$ when $\varphi \to \infty$. Thus, $f'(\varphi)$ has the only zero within an interval $(-\gamma^2, \infty)$ and the function $f(\varphi)$ has the only minimum. Since with $\rho_1 > 1$ $df/dz < 0$, the solution exists only for those values of $z$, for which the function $f$ doesn't attain its absolute minimum. From (1.11) it follows that at this point $d\varphi/dz$ tends to $\infty$ or $-\infty$ depending on the sign $f'(0)$. Theorem has been proved.

The results are generalized for the case when directions $g$ and $w$ coincide. Formulation of the theorem isn't changed, with the exception of that the cases of $\rho_1 < 1$ and $\rho_1 > 1$ change places.

## 2. MODEL OF VORTEX BREAKDOWN.

In this section, an analytic example is built for continuous extension of the solution over the whole upper half-space, for the flow which parameters satisfy conditions of the theorem and inequalities $\rho_1 < 1$, $\lambda > 1$. In addition, the value of $w_0(v)$ is supposed to have the only minimum that is equal to $\gamma$ and attained at the point $\gamma = 0$, and the function $(w_0 - \gamma)^{-1/2}$ is assumed to be integrated in the zero point. In this case, according to the theorem, the solution exists only for $z \leqslant l$ (see (1.9)), at the point $z = l$, $v = 0$ the vertical velocity $w$ being equal to zero and $\partial R/\partial v = \infty$. Hence, to extend the solution over the level $z = l$ the line $R = 0$ can be divided into two lines at the point $z = l$, $v = 0$ of an axis plane. Thus, the flow structure in the axis plane is supposed to have the form shown in Figure.

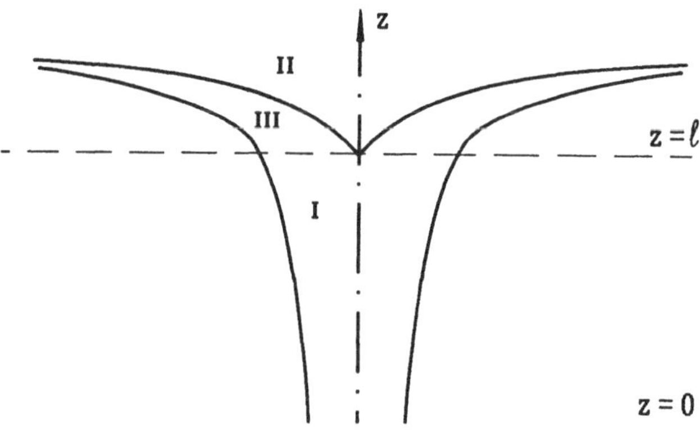


At the interface of regions I and III, the values of pressure, velocity, $R$ and $\rho$ are considered to be continuous. At the interafces of regions II and III, III and outer flow, the pressure is continuous and the kinematic condition is satisfied.

The region II is assumed to be filled in with a quiescent fluid of the constant density $\rho_2$. In III, the flow is described by the same equations as in region I, with $A = 0$, $\rho = \rho_1$. By virtue of equations of hydrostatics in region II and boundary conditions, the restrictions are imposed on the value of $\rho_2$: $\rho_2 = \rho_1$ and the boundaries of region III are determined:

$$R_1 = \left[ \frac{1}{R_{10}} - 2(1 - \rho_1)g(z - l) \right]^{-1},$$

$$R_0 = R_1 - R_{10}.$$

where $R_0$, $R_1$ are the internal and external boundaries of region

III; $R_{10}$ is the value of $R_1$ for $z = l$. The total height of fluid elevation with taking into account the height of region III ($\Delta l$) is restricted in this case and equal to

$$H = l + \Delta l = \frac{\rho_1}{2(1 - \rho_1)g} \left( \frac{1}{\rho_1} - \gamma^2 \right).$$

Since $\lambda > 1$ in this case, it is easily shown that $\gamma^2 < (2\rho_1)^{-1}$. From here the restrictions on $H$ follow, i.e.

$$\left[4(1 - \rho_1)g\right]^{-1} < H < \left[2(1 - \rho_1)g\right]^{-1}.$$

These inequalities allow the value of $H$ to be estimated by an order of magnitude, without knowing $w_0$. For dimensional values, such an estimate has the form

$$H \approx \frac{\rho_0 v_0^2}{2g(\rho_0 - \rho_1)}, \tag{2.1}$$

where $v_0$ is the rotational velocity component on the core boundary for $z = 0$, $\rho_0$ is the density of the outer fluid flow.

## 3. DISCUSSION OF RESULTS, COMPARISON WITH OBSERVATIONS.

In the case of $\rho_1 > 1$, $\lambda > 1$ the vertical and rotational components of velocity increase with $z$. This is, probably, responsible for the effect of fluid suction by tornado [3] and is one of mechanisms of intensification of the tornado core rotation.

The flow structure shown in Figure is of interest in the case of $\rho_1 < 1$, $\lambda > 1$. From the experiments on vortices obtained by heating the underlying surface [4,5] and observations of dust "devils" [6,7], it is known that the vortex core visualized usually by small particles abruptly becomes inviscible at a certain height and disappears. In this case the flow structure changes [4,5]. An abrupt disappearance of the visible core can be explained by that the fluid thickness decreases with spreading over region III, therefore fluid mixes quickly with atmosphere and becomes inviscible. Such a picture of the flow can mainly be a model of the process of vortex breakdown.

Using (2.1), we can obtain the quantitative estimates and

compare them with observations. For calculations, it is convenient to express the density difference by the temperature difference. When $\rho_0 - \rho_1 \ll \rho_0$, we may write that

$$\frac{\rho_0 - \rho_1}{\rho_0} \approx \frac{T_1 - T_0}{T_0},$$

where $T_1$ is the average core temperature, $T_0$ is the atmospheric temperature. Then

$$H \approx \frac{T_0 v_0^2}{2g(T_1 - T_0)}. \tag{2.2}$$

It is assumed that $T_0 = 300^0 K$, $T_1 - T_0 = 20^0 K$, $v_0 = 100\ cm/s$. It is obtained that $H = 75\ cm$. According to [4,5], $H = 45$ and $60\ cm$.

For the dust "devils", it is taken that $T_0 = 300^0 K$, $T_1 - T_0 = 2^0 K$, $v_0 = 10\ m/s$ [8]. We obtain that $H = 750\ m$. On the photo presented [6], the vortex height is approximately $600\ m$.

Thus, in the theoretical model proposed we were a success to study the flow in a core of a tornado-like vortex with taking into account inconstancy of the vertical velocity in a horizontal core section. Owing to this, a possibility of extension of continuous solution to finite or infinite height is established. The nonextended solutions are classified: the solution stops to exist due to either vertical velocity vanishing to zero or tendency of derivatives to infinity. In the first case, an analogy with the boundary layer flow can be shown. Position of the point of stoppage in the boundary layer is due to the beginning of its separation, in the vortex core - due to the beginning of its breakdown. In the second case, we can show an analogy with the gas dynamics flows, i.e. the solution destruction is due, from the mathematical viewpoint, to the same reasons (tendency of derivatives to infinity, i.e. "the gradient catastrophe"). The space evolution of the vortex core is described, and the picture of the flow in the vicinity of its decay is given. A quantitative estimate of the tornado-like vortex height has been obtained. The obtained estimate of an order of magnitude is in agreement with the results of laboratory measurements and observations of dust "devils" in nature.

## REFERENCES.

[1] Teshukov V.M. On hyperbilicity of long wave equations. Dokl.AN SSSR, N 3 (1985), pp. 469-473.

[2] Nikulin V.V. Analogue of equations of vortex shallow water for hollow and tornado-like voretices. Height of stationary tornado-like vortex. PMTF, N 2 (1992).

[3] Nalivkin D.V. Hurricanes, tornados, storms. Leningrad, Nauka, 1969.

[4] Fitzjarrald D.E. A laboratory simulation of convective vortices. J. Atmos. Sci., v.30, N 7 (1973), pp. 894-902.

[5] Mullen J.B., Maxworthy T. A laboratory model of dust devil vortices. Dyn. Atmos. and Oceans. N 1 (1977) 181-214.

[6] Ives R.L. Bahavior of dust devil. Bull. Amer. Meteorol. Soc., v.28, N 4 (1947), pp. 168-174.

[7] Williams N.R. Development of dust wirls. Bull Amer. Meteorol. Soc., v.29, N 3 (1948), pp. 106-117.

[8] Sinclair P.C. The lower structure of dust devils. J. Atmos. Sci., v.30, N 8 (1973), pp. 1599-1619.

International Series of Numerical Mathematics, Vol. 106, © 1992 Birkhäuser Verlag Basel 

# OPTIMAL SHAPE DESIGN FOR PARABOLIC SYSTEM AND TWO-PHASE STEFAN PROBLEM

*S.P.Ohezin*
*Urals State University*
*Department of Mathematical Physics*
*Ekaterinburg 620083, RUSSIA*

In this paper we consider the application of the domain control theory in parabolic systems for the approximation of two-phase one-dimensional Stefan problem [1]. In this approach the role of the control function is carried out by the free-boundary (or interface). We use the shape penalty method [2,3] and investigate the corresponding approximation family of control bilinear systems and prove the theorem about uniform approximation. It is shown that classical solution of Stefan problem is a limit of the approximation solutions.

Key words: optimal control, Stefan problem.

## 1. INTRODUCTION.

We shall consider the classical mathematical model of one-dimensional two-phase Stefan problem [1]:

$$\dot{y}_1(t,x) = a_1^2 y_1''(t,x), \quad 0 < t < T, \quad 0 < x < u(t), \tag{1.1}$$

$$\dot{y}_2(t,x) = a_2^2 y_2''(t,x), \quad 0 < t < T, \quad u(t) < x < X_1, \tag{1.2}$$

$$y_1(0,x) = \varphi_1(x), \; x \in [0,u_0]; \; y_2(0,x) = \varphi_2(x), \quad x \in [u_0,X_1] \tag{1.3}$$

$$y_1(t,0) = 0, \quad y_2(t,X_1) = 0, \tag{1.4}$$

$$y_1(t,u(t)) = y_2(t,u(t)) = 0, \quad 0 \leqslant t \leqslant T, \tag{1.5}$$

$$\dot{u}(t) = æ_2 y_2'(t,u(t)) - æ_1 y_1'(t,u(t)), \quad 0 < t \leqslant T, \tag{1.6}$$

$$u(0) = u_0. \tag{1.7}$$

Here $y_1(t,x)$ $(y_2(t,x))$ is the temperature of "liquid" ("solid") phase at the point $x$ at the instant of time $t$. The function $\varphi_1(x)$ $(\varphi_2(x))$ describes the initial distribution of temperature in "liquid" ("solid") phase. We shall assume that the interface is given by a curve $x = u(t)$ and the temperature $y_i(t,u(t)) = 0$, $i = 1, 2$. The condition (1.6) is a mathematical model of heat balance at the interface. In (1.1), (1.2), (1.6) the point means the differential operator by time $t$, the stroke - by $x$.

We shall investigate the new mathematical model for two-phase Stefan problem. The basic idea is next. One can obtain a priori estimates by the functions $y_i'(t,u(t))$. These estimates allow us to displace the condition (1.6) to the equation

$$\dot{u}(t) = v(t). \tag{1.8}$$

We shall consider $v(\cdot)$ as a control function. Condition (1.6) defines a quality functional given by

$$J(v(\cdot)) = \int_0^T \left[æ_2 y_2'(t,u(t)) - æ_1 y_1'(t,u(t)) - v(t)\right]^2 dt. \tag{1.9}$$

Thus the problem (1.1)-(1.7) is reduced to minimization problem for special functional (1.9). Equations (1.1), (1.2) are approximated by means of the parabolic equations with penalty terms which are defined in a priori known large cylindrical domain [2-4].

## 2. MATHEMATICAL FORMULATION.

Let us assume that functions $\varphi_i(\cdot)$, $i = 1, 2$ may be continued by zero to any interval $[0,X]$ in class $C^{2,\alpha}[0,X]$, $X \geqslant X_1$, $0 < \alpha \leqslant 1$. A priori estimates by the functions $y_i'(t.u(t))$ are known [1], i.e. there exist the real numbers $V_1$ and $V_2$ such that $\dot{u}(t) \in [V_1,V_2]$ for almost every $t \in [0,T]$. We shall also assume that there exists a real number $X \geqslant X_1$ such that for all $v(\cdot) \in V[0,T] = \left\{v(\cdot) \;\middle|\; v(t) \in [V_1,V_2], \text{ for a.e. } t \in [0,T]\right\}$ we have the inclusion $u(t) \in (0,X)$. Here $u(t)$ is the solution of equation (1.8).

Let us consider the approximation system

$$\dot{y}_{1\varepsilon}(t,x) - a_1^2 y_{1\varepsilon}''(t,x) + \varepsilon^{-1} U_{1\varepsilon}(t,x;u(\cdot)) y_{1\varepsilon}(t,x) = 0,$$

$$0 < t \leqslant T, \quad 0 < x < X, \tag{2.1}$$

$$y_{1\varepsilon}(0,x) = \Phi_1(x) = \begin{cases} \varphi_1(x), & x \in [0,u_0] \\ 0, & x \in (u_0,X], \end{cases} \tag{2.2}$$

$$y_{1\varepsilon}(t,x) = 0, \quad (t,x) \in \Sigma = [0,T]\times\partial(0,X), \tag{2.3}$$

$$\dot{y}_{2\varepsilon}(t,x) - a_2^2 y''_{2\varepsilon}(t,x) + \varepsilon^{-1}U_{2\varepsilon}(t,x;u(\cdot))y_{2\varepsilon}(t,x) = 0,$$

$$0 < t \leqslant T, \quad 0 < x < X, \tag{2.4}$$

$$y_{2\varepsilon}(0,x) = \Phi_2(x) = \begin{cases} 0, & x \in [0,u_0), \\ \varphi_2(x), & x \in [u_0,X], \end{cases} \tag{2.5}$$

$$y_{2\varepsilon}(t,x) = 0, \quad (t,x) \in \Sigma, \tag{2.6}$$

$$\dot{u}(t) = v(t), \quad u(0) = u_0. \tag{2.7}$$

Here $\varepsilon > 0$, $v(\cdot)$ is a control function.

$$U_{1\varepsilon}(t,x;u(\cdot)) = \begin{cases} 0, & 0 < x \leqslant u(t) \\ W_{1\varepsilon}(x - u(t) - \varepsilon), & u(t) < x \leqslant u(t) + \varepsilon, \\ 1, & u(t) + \varepsilon < x < X, \end{cases} \tag{2.8}$$

$$W_{1\varepsilon}(x) = \int_{-\varepsilon}^{x} \omega_\varepsilon(|\xi|)\, d\xi, \tag{2.9}$$

$$\omega_\varepsilon(|\xi|) = \begin{cases} 0, & |\xi| \geqslant \varepsilon, \\ C\varepsilon^{-1}exp\left[-\varepsilon^2(\varepsilon^2 - |\xi|^2)^{-1}\right], & |\xi| < \varepsilon. \end{cases} \tag{2.10}$$

The constant $C$ is such that

$$\int_{-\varepsilon}^{x} \omega_\varepsilon(|\xi|)\, d\xi = 1, \tag{2.11}$$

$$U_{2\varepsilon}(t,x;u(\cdot)) = \begin{cases} 1, & 0 < x \leqslant u(t) - \varepsilon, \\ W_{2\varepsilon}(x - u(t) + \varepsilon), & u(t) - \varepsilon < x \leqslant u(t), \\ 0, & u(t) < x < X, \end{cases} \quad (2.12)$$

$$W_{2\varepsilon}(x) = W_{1\varepsilon}(-x). \quad (2.13)$$

Thus we have the optimization problem for the system of parabolic equations which are defined in the cylindrical domain $Q = (0,T)\times(0,X)$ with the following quality functional

$$J(\varepsilon,v(\cdot)) = \int_0^T \left[æ_2 y'_{2\varepsilon}(t,u(t)) - æ_1 y'_{1\varepsilon}(t,u(t)) - v(t)\right]^2 dt. \quad (2.14)$$

It is well known that there exist unique classical solutions of problems (2.1)-(2.6) for all admissible control $v(\cdot) \in V[0,T]$ [1,2].

## 3. THE MAIN RESULTS.

The following statements are true.

**THEOREM 3.1.** The functional $J(\varepsilon,v(\cdot))$ is lower semicontinuous in the weak topology in $L_2(0,T)$.

**THEOREM 3.2.** There exists $v_0^\varepsilon(\cdot) \in V[0,T]$ such that

$$J(\varepsilon,v_0^\varepsilon(\cdot)) = \inf \left\{ J(\varepsilon,v(\cdot)) \,\middle|\, v(\cdot) \in V[0,T] \right\}.$$

We shall use the following designations:
$y_i(t,x;u(\cdot))$ is a classical solution of (1.1)-(1.5) for a given function $u(\cdot)$; $y_{i\varepsilon}(t,x;u(\cdot))$ is a classical solution of (2.1)-(2.6) for a given function $u(\cdot)$, $i = 1, 2$.

With these two notations, we are ready to state our main result in

**THEOREM 3.3.** For any sequence

$$u_\varepsilon(\cdot) \to u_*(\cdot) \quad \text{in} \quad C[0,T], \quad \varepsilon \to 0,$$

we have

$$y'_{i\varepsilon}(t,u_\varepsilon(t);u_\varepsilon(\cdot)) \to y'_i(t,u_*(t);u_*(\cdot)) \quad \text{in} \quad L_2(0,T).$$

**REMARK 3.1.** The proof of Theorem 3.3 is based on the probability representations for solutions of equations (2.1)-(2.6):

$$y_{i\varepsilon}(T - t,x;u(\cdot)) = \tag{3.1}$$

$$= E\left\{\Phi_i(\xi^i_{t,x}(T))\, exp\left[-\varepsilon^{-1}\int_t^T U_{i\varepsilon}(T-s,\xi^i_{t,x}(s);U(\cdot))\, ds\right] \chi\left\{\tau^i_{t,x} \geqslant T\right\}\right\}.$$

Here $E$ denotes the mathematical expectation; $\xi^i_{t,x}(s)$ is the random process given by

$$\xi^i_{t,x}(s) = x + \sqrt{2}\, a_i \int_t^s d\eta, \tag{3.2}$$

in which $\eta$ denotes the Wiener normal process; $\tau^i_{t,x}$ is the random value of the first instant of the leaving time from the domain $Q = (0,T)\times(0,X)$, $\chi_A$ is the characteristic function of the set $A$.

Using representation (3.1), we prove the convergence

$$y_{i\varepsilon}(t,x;u_\varepsilon(\cdot)) \rightarrow y_i(t,x;u_*(\cdot))$$

in Sobolev space $H^{0,2}(Q_i(u_*))$. Then we obtain the result by means of the trace theorem. Here

$$Q_1(u_*) = \left\{ (t,x) \;\middle|\; 0 < t < T,\;\; 0 < x < u(t) \right\},$$

$$Q_2(u_*) = \left\{ (t,x) \;\middle|\; 0 < t < T,\;\; u(t) < x < X \right\}.$$

**REMARK 3.2.** For every $\varepsilon > 0$ there exists the solution of control problem (2.1)-(2.14) and it may be found by means of the special approaches from the control theory. Let us denote the corresponding solution $u_\varepsilon(\cdot)$, $y_{1\varepsilon}(\cdot,\cdot)$, $y_{2\varepsilon}(\cdot,\cdot)$.

The following result may be obtained.

**THEOREM 3.4.** Let $u^0(\cdot)$, $y^0_1(\cdot,\cdot)$, $y^0_2(\cdot,\cdot)$ be a classical solution of two-phase Stefan problem. Then the solution $u_\varepsilon(\cdot)$, $y_{1\varepsilon}(\cdot,\cdot)$, $y_{2\varepsilon}(\cdot,\cdot)$ satisfies

$$u_\varepsilon(\cdot) \rightarrow u^0(\cdot) \quad \text{in} \quad C[0,T],$$

$$y_{i\varepsilon}(\cdot,\cdot) \rightarrow y^0_i(\cdot,\cdot) \quad \text{in} \quad L_2(Q_i(u^0)).$$

## REFERENCES

[1] Meirmanov A.M. The Stefan problem, Novosibirsk: Nauka, 1986 (in Russian).

[2] Lions J.L. Quelques methodes de resolution des problemes aux limites non lineares, Dunod, Paris, 1969.

[3] Ohezin S.P. On the approximation in domain control problem for parabolic system, Prikl. Ma. Mekh., v. 54, N 3, 1990, pp.361-365 (in Russian).

[4] Ohezin S.P. On the mathematical model for Stefan problem. Dif. Uravneniia, v.27, N 6, 1991, pp. 1042-1048 (in Russian).

International Series of Numerical Mathematics, Vol. 106, © 1992 Birkhäuser Verlag Basel


# INCOMPRESSIBLE FLUID FLOWS WITH FREE BOUNDARY AND THE METHODS FOR THEIR RESEARCH

*A.G. Petrov*
*Lomonosov Moscow State University*
*Moscow 119899, RUSSIA*

## 1. VARIATIONAL METHOD.

Thomson W. and Tait P. were the first who applied the Hamilton principle to the fluid [1,2]

$$\int_{t_1}^{t_2} \delta L\, dt = \int_{t_1}^{t_2} dt \int_{\partial\Omega} p\delta_n \cdot dS, \quad \delta_n(t_1) = \delta_n(t_2) = 0, \tag{1.1}$$

where $\delta_n$ is the virtual displacement of fluid flow boundary $\partial\Omega$, the integral represents the work of the force of fluid pressure $p$ on the displacement $\delta_n$, $L$ is the functional of the boundary and the normal velocity (the Lagrange function).

Let $q_1$, $q_2, \ldots$ be the generalized coordinates defining the position of boundary in space, then the variational equation (1.1) is equivalent to the Lagrange equations

$$-\frac{d}{dt}\frac{\partial L}{\partial \dot{q}_i} + \frac{\partial L}{\partial q_i} = Q_i, \tag{1.2}$$

where $Q$ is the generalized force. The successful selection of the generalized coordinates is the key to research problems in question.

a) Potential flows without singularities. Thomson W., Tait P. and Kirchhoff G. presented some interesting examples of application of the method of generalized coordinates to solid body dynamics in liquid [1]. The Lagrange function is equal to the difference of the fluid kinetic energy and the potential energy. Of considerable interest for application is the work of Zukovskii, who began the first investigations of the problems of motion of a solid body with liquid.

Here are some new examples of application of the variational method of investigation of the flow with a free surface.

**Example 1:** The steady motion of a bubble or drop in perfect fluid and stability of motion are considered in works [3-7]. In particular, the stability of bubble motion corresponding to the exact solution of E.B. McLeod (1955) is proved relative to arbitrary perturbation of the cavity shape [7]. The form of the bubble is specified by the conformal mapping $z(\zeta)$ of the exterior of unit circle $|\zeta| > 1$ on the potential flow domain (the exterior of the cavity): $z(\zeta) = 9\zeta - 6\zeta^{-1} - 1/3\ \zeta^{-3}$.

**Example 2:** Variational Ryabushinskii's Principle for the cavitation flow can be obtained from equation (1.1). The Ryabushinskii's functional of the axisymmetric problem of cavitation flow for the small cavitation number $s \ll 1$ can be represented in the following form

$$U = \int_{-1}^{1} \left[ \sigma z(x) + z'^{2}(x) \ln \frac{z(x)}{1 - x^{2}} + z'(x) \int_{-1}^{1} \frac{z'(x) - z'(\zeta)}{|\zeta - x|} d\zeta \right] dx, \tag{1.3}$$

where $z(x)$ is the function defining the cavity boundary $r(x)$ by the formula $r = 2\sqrt{z(x)}$. The function $z(x)$ is determined from the condition $\delta U = 0$, and the drag force $F$ has the following asymptotic expression $F = 3\pi\rho V_{\infty}^{2} l_{x}^{2} U_{0}$, where $U_0$ is the external value of functional (1.3). From (1.3) we can find asymptotic relations for the drag force $F$, length $l_x$ and width $l_y$ of the cavity [8].

**Example 3:** From the variational equation (1.1) the system of Lagrange differential equations for unsteady waves of finite amplitude in [9] is derived. In particular, the steady waves problem is reduced to the problem of the functional extremum

$L(y_1, y_2, \ldots)$.

$$L = \left[ c^2 + \frac{1}{2}\sum_{k=1}^{\infty} k y_k^2 \right] \sum_{k=1}^{\infty} k y_k^2 - \sum_{k=1}^{\infty} y_k^2 - \sum_{k=1}^{\infty} k y_k^2 \sum_{m=1}^{k-1} y_m y_{k-m}.$$

The extremum conditions $\partial L/\partial y_i = 0 \quad (i = 1, 2, \ldots)$ coincide with the equations obtained by Longuet-Higgins (1976).

b) Nonpotential flows with singularities. Variational equation (1.1) can be applied to special vortex flow of perfect fluid in the presence of singularities inside a flow as well as at infinity. In some cases kinetic energy cannot exist, but Lagrange function can be found.

**Case 1.** The flow has a constant vorticity for the plane problem and for the axisymmetric problem the flow has a vorticity, whose intensity decreases proportionally to the distance from the axis of rotation [2, 10].

**Case 2.** Potential flow with multipole singularities of an arbitrary order [2, 11].

In these cases the Lagrange function is obtained as a functional of the boundary flow and the normal velocity. For example, for the plane problem of the body in the potential flow $V_0(t,x)$ with circulation $\Gamma$ the Lagrange function $L$ is as follows:

$$L = \int_{\Omega} \frac{\rho V_0^2}{2}\, d\tau - \int_{V} p_0 \, d\tau - \frac{\rho\Gamma}{2}\, Im(z_0, \bar{z}_0),$$

where $p_0(t,x)$ is the pressure in the stream $V_0(t,x)$, and $\Omega$, $V$ are the regions occupied by fluid and the body, respectively.

This method enables us to write down the ordinary differential equations which determine the dynamics of vortex distribution in a given stream. This approach was used for constructing a model of the Gulf stream ring [12, 13]. The variational formulation of the problems of hydrodynamics provides a possibility of applying straight method of determination of the free boundary or the boundary separating the vortex and potential flows. The dynamic model of a drop is constructed by this method in [10]. In this work the investigation of the stability problem of steady motion of a drop was carried out by the direct Liapunov method.

Variational principle for cavitation flow past bodies in the presence of hydrodynamic singularities is formulated in [14]. With the use of this principle we find asymptotic relations for drag force, length and width of the cavity for axisymmetric and two-dimensional flows with a source.

2. The second method of integral limiting equations for harmonic functions is applied for numerical solutions of steady and unsteady waves on a surface of gravitational fluid.

A numerical algorithm for the determination of the progressive periodic gravity-capillary wave of finite amplitude is presented in [15]. It is demonstrated that the wave profile $Y(x)$ with period $2\pi$ and fixed amplitude $a = 0.5(Y(\pi) - Y(0))$ corresponds to the extremum of the functional $U(y)$ under the following restrictions:

$$\int_0^{2\pi} y\, dx = 0, \quad y(\pi) - y(0) = 2a.$$

A numerical algorithm allows the functional $U$ to be determined through the coordinates of markers with an order of approximation $N^{-4}$, where $N$ is the number of the markers on the boundary.

The numerical method gives a possibility of condensing markers near the points of maximal curvature, which increases the accuracy of computations. The comparison of calculated results with exact solution (Cokelet E.D., 1977) shows that maximum possible accuracy with the single precision ($10^{-6}$) is reached when $N < 32$.

The method is illustrated by following solutions for unsteady waves: breaking waves; Rayleigh-Taylor instability; fluid fluctuation in vessels of different form; waves on a boundary surface between two liquids [16].

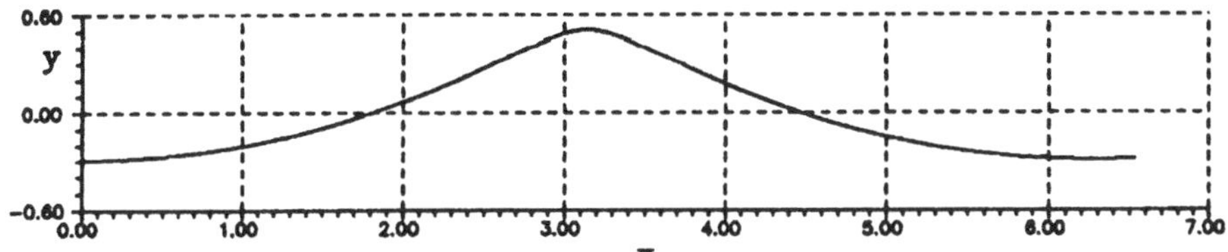


Figure 1. Profile of a steady progressive wave in deep water with amplitude $a = 0.40236$ and wave length $2\pi$.

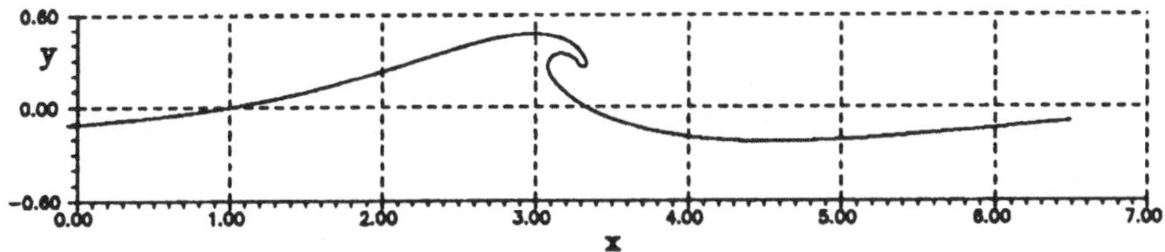


Figure 2. Breaking wave in changing the depth up to $h = 1$.

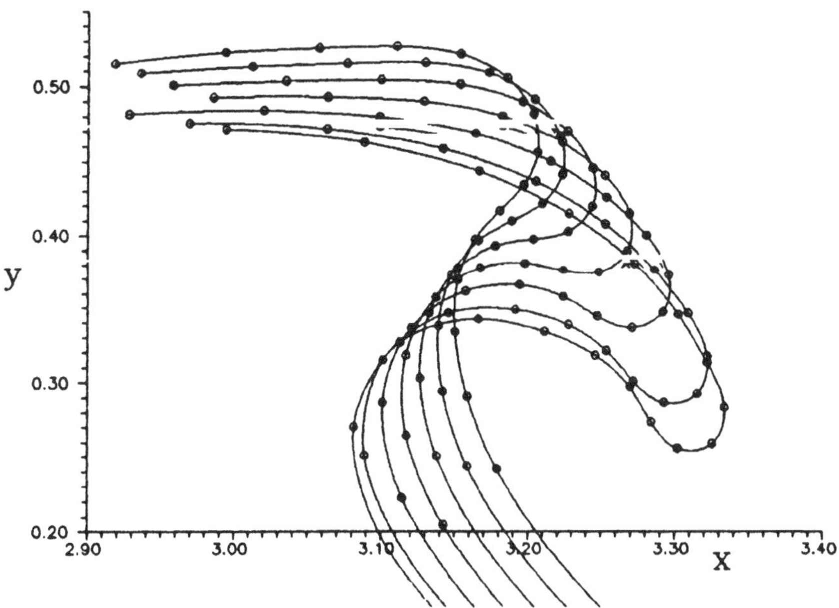


Figure 3. Close-up of the wave-crest at successive times $t = 4.15,\ 4.25,\ 4.45,\ 4.55,\ 4.65,\ 4.75,\ 4.80$.

The numerical results for breaking wave ($N = 48$) are shown in Figs. 1-3. Calculations were carried out by A. Petrov and V. Smolyanin on IBM PC AT-286.

3. A method of duplex estimators of fluid energy dissipation is elaborated to research a viscous fluid flow within a zone on a boundary in which a tangential stress is present. This method is based on the next theorem about an exact upper bound for the dissipation rate of flow with different viscosities [17].

**THEOREM.** In the bounded domain $V$ let $v(Re,x)$ be determined from the solution of the following boundary value problem for the Navier-Stokes equations:

$$(v\nabla)v = -\nabla p + \frac{1}{Re}\nabla^2 v, \quad \nabla v = 0, \quad \vec{x} \in V, \tag{3.1}$$

$$v_i n_i \Big|_{\partial V} = 0, \quad v_i \tau_i \Big|_{\partial V_1} = 0, \quad e_{ij} n_i \Big|_{\partial V_2} = e_i \tau_i + e_n n_i, \tag{3.2}$$

where $\partial V_1$ and $\partial V_2$ are two parts of the domain boundary ($\partial V = \partial V_1 \cup \partial V_2$), $n_i$ and $\tau_i$ are the components of the vectors normal and tangential to the domain boundary $\partial V$, and $e_\tau$ is the function (shear stress) given on the boundary $\partial V$.

The dissipation rate $D(Re)$ is obtained from the solution of the Stokes equations

$$-\nabla p^0 + \frac{1}{Re}\nabla^2 v^0 = 0, \quad \nabla v^0 = 0, \quad \vec{x} \in V \tag{3.3}$$

with the same boundary conditions (2):

$$D(Re) = 2\int_V e_{ij} e_{ij}\, dV \leqslant D(0) = 2\int_V e^0_{ij} e^0_{ij}\, dV,$$
$$e_{ij} = \frac{1}{2}\left(\frac{\partial v_i}{\partial x_j} + \frac{\partial v_j}{\partial x_i}\right), \quad e^0_{ij} = \frac{1}{2}\left(\frac{\partial v^0_i}{\partial x_j} + \frac{\partial v^0_j}{\partial x_i}\right). \tag{3.4}$$

The summation convention is applied.

As $D(Re)$ decreases monotonically, its lower bound will be the limit as $Re \to \infty$. In accordance with the Prandtl-Batchelor theorem, if certain conditions on the boundary streamline are satisfied, the limiting flow as $Re \to \infty$ will have a vortex of constant intensity. The energy dissipation $D(\infty)$ of the flow with the constant intensity vortex is the approximate lower bound of the function $D(Re)$.

The method of duplex estimators is applied for the calculation of a flow inside a falling rain drop at arbitrary Reynolds numbers.

## REFERENCES.

[1] Lamb H. Hydrodynamics, Cambridge (1932).

[2] Petrov A.G. Hamilton's principle and certain problems of dynamics of perfect fluid. PMM, vol.47, N 1(1986), pp.30-36.

[3] Petrov A.G. Curvilinear motion of ellipsoidal bubble. Zh.Prikl.Mekh.Tekh.Fiz., N 3 (1971).

[4] Voinov O.V., Petrov A.G. The Lagrange function of a gas bubble in an inhomogeneous stream. Dokl.Akad.Nauk SSSR, vol.210, N 5 (1973).

[5] Petrov A.G. Stability of the equilibrium shape of a gas bubble in a homogeneous ideal fluid stream. Zh. Prikl. Mekh. Tekh. Fiz., N 5 (1974).

[6] Likhomanov N.I., Petrov A.G. Plane parallel flow around a gas cavity. Izv.Akad.Nauk SSSR, MZhG, N 5 (1975), pp.725-730.

[7] Petrov A.G. Dynamics of a plane cavity in a low viscosity fluid. Izv. Akad. Nauk SSSR, MZhG, N 5 (1973).

[8] Petrov A.G. Drag force acting on an axisymmetric thin body at low cavitation numbers. Dokl.Akad.Nauk SSSR 282 (1985), pp.543-455.

[9] Petrov A.G., Smolyanin V.G. Hamilton's principle and waves on the surface of the liquid. Mekh. Sovremen. Problem, 1987.

[10] Petrov A.G. The Lagrange function for vortex flows and dynamics of deformed drops. PMM, vol.41 (1977), pp.72-86.

[11] Petrov A.G. Reactions for a small body in vortex plane parallel stream. Dokl.Akad.Nauk SSSR, vol. 238 (1978).

[12] Petrov A.G. On the motion of Golfstream rings. Okeanovedenie, vol.20, N 6 (1980).

[13] Perepelkin V.V., Petrov A.G. Dynamics of an elliptic vortex. Izv. Akad. Nauk SSSR, N 4 (1983), pp.539-544.

[14] Petrov A.G., Sotina N.B. Universal, cavity shape-independent relation for cavitation flow with small cavitation numbers. Zh. Prikl. Mekh. Tekh. Fiz., N 5 (1983), pp.758-766.

[15] Petrov A.G., Smolyanin V.G. Calculation of the profile gravity-capillary wave on the surface of the heavy liquid of finite depth. Vestnik MGU, Math., Mech., N 2, 1991.

[16] Petrov A.G., Tshuvashov A.N. On the wave on the boundary separating two flows of heavy perfect liquid. PMM, vol.52, N 4, 1991.

[17] Petrov A.G. Rate of dissipation of the energy of a viscous fluid with a condition for the shear stress on the boundary streamline. Sov.Phys.Dokl. 34(2), 1989, pp.101-103.

[18] Petrov A.G. Circulation inside viscous deformed drops moving in gas with uniform velocity. Zh.Prikl.Mekh.Tekh.Fiz., N 6 (1989).

# ON THE STEFAN PROBLEMS FOR THE SYSTEM OF EQUATIONS ARISING IN THE MODELLING OF LIQUID-PHASE EPITAXY PROCESSES

*A.G.Petrova*
*Altai State University, Barnaul 656099, RUSSIA*

One-dimensional one-phase supercooled Stefan problems for the system of two diffusion equations connected with each other under the free boundary conditions are considered. In the problems of the I-st type called "direct" the values of functions on the free boundary are known while in the problems of the II-nd type called "inverse" these values should be determined with the use of an additional condition. The problems of these two types are considered in the cases of a restricted domain occupied by phase $(0<x<S(t))$ and a semi-infinite one $(S(t)<x<\infty)$. For "direct" problems the possibilities of time-global solvability, finite time disappearance of phase and finite time blow-up are considered; similarity solutions and the asymptotic form of the free boundary $S(t)$ as $t \to \infty$ are investigated. For the "inverse" problems which represent (from mathematical point of view) one phase version of binary alloy crystallization problems similarity solutions and local in time solvability theorems are given.

Key words: free boundary, disappearance of phase, blow-up, similarity solution.

## 1. INTRODUCTION.

The "direct" problem in restricted domain $I$ consists in determination of the functions $S(t)$, $Z_i(x,t)$, $i = 1, 2$ and $Y(t)$ satisfying the following equations and conditions:

$$(Z_i)_t = k_i(Z_i)_{xx} \quad (0 < x < S(t), \quad t > 0), \tag{1.1}$$

$$Z_i = \lambda_i(t) = \mu_i(\theta(t)), \; k_i(Z_i)_x = \dot{S}(t)(\rho_i - Z_i) \quad (x = S(t)), \tag{1.2}$$

$$\rho_1(x) + \rho_2(x) = \rho_1^* Y(x) + \rho_2^*(1-Y(x)), \quad Y = \rho_1/(\rho_1 + \gamma\rho_2),$$

$$0 \leqslant Y \leqslant 1, \tag{1.3}$$

$$Z_i(0,t) = c_i, \quad i = 1,2, \tag{1.4}$$

$$Z_i(x,0) = h_i(x), \ i = 1, 2 \quad (0 < x < 1); \quad S(0) = 1, \tag{1.5}$$

$\lambda_i(t) = \mu_i(\theta(t))$-continuously differentiable and non-increasing for all $t > 0$ functions such that $0 \leqslant \lambda_i(t) \leqslant c_i$ and

$$\lambda_1(0) + \lambda_2(0) < min\{\rho_1^*, \rho_2^*\}, \tag{1.6}$$

$$h_i(x) \geqslant \lambda_i(0) \tag{1.7}$$

are continuous and non-increasing on the interval (0,1) functions,

$$\rho_i^*, \ c_i \ (i = 1,2), \quad \gamma < 1 \tag{1.8}$$

are positive constants.

The model describes the growth of complex binary film from the ternary solution by the method of liquid phase epitaxy ([1]), $Z_i(x,t)$ represents the concentrations of components in liquid phase, $\theta(t)$ is the system temperature, $\rho_i$ are the concentrations in film.

Corresponding problem in semi-infinite domain ($S(t) < x < \infty$), (let us denote it by $I'$) differs from problem $I$ by the conditions (1,4),(1,5) and (1.7) which turns into

$$Z_i(x,t)\Big|_{x\to\infty} = c_i \qquad (i = 1, 2), \tag{1.4$'$}$$

$$Z_i(x,0) = h_i(x), \ i = 1, 2 \qquad (0 < x < \infty); \qquad S(0) = 0, \tag{1.5$'$}$$

$h_i(x) \geqslant \lambda_i(0)$ are continuous and non-decreasing functions (1.7)'

In the "inverse" problems $II$ (for the domain $0 < x < S(t)$) and $II'$ (for $S(t) < x < \infty$) the values $Z_i(S(t),t)$, $i = 1, 2$ are unknown, only the dependences between them are given,( for simplicity let it be linear), but then the distribution of $Y(x)$ is known, therefore, $\rho_1(x)$ and $\rho_2(x)$ are known too. These problems are interpreted as the problems of determination of temperature $\theta(t)$ when the composition of film $Y(x)$ is given. So, the problems $II$ and $II'$ differ from $I$ and $I'$ by the free boundary conditions which are

$$Z_2 = \alpha Z_1 + \delta,\ k_i(Z_i)_x = \dot{S}(\rho_i - Z_i),\ i = 1,\ 2, \tag{1.9}$$

where $\rho_i(x) > 0,\ i = 1,\ 2$ are known functions, $\alpha,\ \delta$ are positive constants.

Note, that if we interpret $Z_1(x,t)$ as the temperature of binary alloy and $Z_2(x,t)$ as the admixture concentration, then (1.9) is the "liquidus" equation and Stefan conditions in the case when the diffusion coefficient in the solid phase is equal to zero.

## 2. "DIRECT" PROBLEM IN RESTRICTED DOMAIN.

Note that in the case when $\lambda_i(t) \equiv c_i,\ i = 1,\ 2,$ the problem (1.1)-(1.8) has a solution only for $h_i(x) \equiv c_i$ and the solution is $S(t) = 1,\ Z_i(x,t) = c_i,\ i = 1,\ 2,\ Y(x)$ is an arbitrary function. Therefore we will consider that $\lambda_i(t) \not\equiv c$ for $i = 1$ or $i = 2$.

**THEOREM 1.** Under the condition

$$c_1 + c_2 < min\ \{\rho_1^*, \rho_2^*\} \tag{2.1}$$

the problem (1.1)-(1.8) has a classical solution on the time interval $(0,T)$ where $T$ is the time of the end of process (i.e. $S(T) = 0$).

The proof of theorem 1 is based on two lemmata.

**LEMMA 1.** Under the corresponding conditions on smoothness of initial functions, the condition $h_1'(0)\cdot h_2'(0) \neq 0$ and also the agreement conditions the problem (1.1)-(1.8) has a solution in Holder classes of functions within a sufficiently small time interval and S(t) is non-decreasing function.

The proof of this lemma is based on applying Schauder fixed-point theorem to the operator

$$\Phi(s) = \int_0^t \frac{k_2 Z_{2x}(S(\tau),\tau)}{\rho_2(S(\tau)) - \lambda_2(\tau)}\, d\tau \qquad (\text{if } h_2'(0) \neq 0).$$

Here $\rho_2(x)$ is a solution of the system which consists of equations (1.3) and equation $\rho_2 - \lambda_2 = L(t)(\rho_1 - \lambda_1)$, where $L(t) = k_1 Z_{1x}(S(t),t)/k_2 Z_{2x}(S(t),t)$ and functions $Z_i$ are the solutions of the initial-boundary problems

$$Z_{it} = k_i Z_{ixx}, \ i = 1, 2 \quad (0 < x < S(t)),$$

$$Z_i\Big|_{x=S(t)} = \lambda_i(t), \quad Z_i\Big|_{x=0} = c_i, \quad Z_i\Big|_{t=0} = h_i(x), \ i = 1, 2.$$

Monotony of $x = S(t)$ is set with the use of maximum principle and the condition (1.6).

**LEMMA** 2. Let

$$h_i(x) \leqslant H_i(1 - x) + \lambda_i(0)x, \ i = 1, 2,$$

if $0 \leqslant x \leqslant 1$; $S_T = \inf\limits_{(0,T)} S(t) > 0$

and there exists the constant $d \in (0, S_T)$ such that

$$c_i > H_i d + \lambda_i(0)(1 - d).$$

Then $|\dot{S}(t)|$ is bounded on $(0,T)$.

For the proof we use the method outlined in [2] and based on the comparison of $Z_i(x,t)$ with

$$W_i(x,t) = \frac{c_i - \lambda_i(t)}{1 - exp(-a_i d)} \Big[1 - exp(a_i(x - S(t))\Big] + \lambda_i(t).$$

in the domain

$$\Omega_\varepsilon = \Big\{(x,t): \ S(t) - d < x < S(t), \ 0 < t < T - \varepsilon\Big\};$$

$a_i$, $i = 1, 2$ are the special chosen constants. In this case unboundedness of $|\dot{S}(t)|$ would contradict the condition (2.1).

For the proof of theorem 1 we should note that for the problem I in principle there exist the next three possibilities:

(A) A classical solution exists for all $t > 0$.

(B) There exists $T_B$ such that $\lim\limits_{t\to T_B-} S(t) = 0$.

(C) There exists $T_C$ such that $\lim\limits_{t\to T_C-} S(t) = 0$ and $\inf\limits_{(0,T_C)} \dot{S}(t) = -\infty$.

Multiplying the equation (1.1) by $x$, then integrating it with respect to $x$ on (0,S(t)) and with respect to $t$ on $(0,t)$ we obtain

$$\int_0^{S(t)} xZ_i(x,t)dx - \int_0^1 xh_i(x)dx = \int_0^t S\dot{S}\rho_i(S(t))dt + k_i\int_0^t (c_i - \lambda_i)dt. \qquad (2.2)$$

The last integral is unbounded as $t \to \infty$ while the rest are bounded, so the case (A) is impossible. From Lemma 2 the case (C) is impossible too.

**THEOREM 2.** If

$$\int_0^1 x(h_1(x) + h_2(x))dx > \max\{\rho_1^*, \rho_2^*\}/2, \qquad (2.3)$$

then for the problem I the case (C) takes place.

**Proof.** Sum up the equations (2.2) and taking into account (2.3) we will have

$$k_1\int_0^t (c_1 - \lambda_1)dt + k_2\int_0^t (c_2 - \lambda_2)dt = - \int_0^1 x(h_1(x) + h_2(x))dx -$$

$$- \int_0^t S\dot{S}(\rho_1 + \rho_2)dt + \int_0^{S(t)} x(Z_1 + Z_2)dx <$$

$$< - \max\{\rho_1^*, \rho_2^*\}/2 - \max\{\rho_1^*, \rho_2^*\} \frac{S^2(t) - 1}{2} + \int_0^{S(t)} x(Z_1 + Z_2)dx.$$

As far as the case (A) is impossible and in the case (B) $S(T_B)=0$, then the last unequality with $t = T_B$ contradicts the condition (1.6).

## SIMILARITY SOLUTIONS OF PROBLEM $I$.

If $\lambda_i(t) = \lambda_i^0 = const$ and

$$c_1 + c_2 < \rho_1^* c_1 /(c_1 + \gamma c_2) + \rho_2^* \gamma c_2/(c_1 + \gamma c_2), \tag{2.4}$$

then with the appropriate initial functions the problem $I$ has an exact solution

$$S(t) = \sqrt{1 - t/T} = \beta \sqrt{T - t},$$

$$Z_i(x,t) = A_i \int_{x/\sqrt{T-t}}^{\beta} \exp(x_i/4k_i)dx + \lambda_i^0, \quad Y(x) = y_0$$

when $0 < t < T$. The constants $\beta$, $A_i$ and $y_0$ are determined from the system of equations

$$\left(\rho_i - \lambda_i^0\right) \frac{\beta}{2k_i} \exp(-\beta^2/4k_i) \int_0^{\beta} \exp(x^2/4k_i)dx = c_i - \lambda_i^0, \quad i = 1, 2$$

$$\rho_1 + \rho_2 = \rho_1^* y_0 + \rho_2^*(1 - y_0), \quad y_0 = \rho_1/(\rho_1 + \gamma\rho_2), \tag{2.5}$$

$$A_i = \left(\rho_i - \lambda_i^0\right) \frac{\beta}{2k_i} \exp(-\beta^2/4k_i).$$

The existence of similarity solution is verified immediately.

To make sure of the solvability of the system (2.5), one must express $\rho_i$ from the first two equations and substitute them into equation connecting $\rho_1$ and $\rho_2$. The equation has the solution $\beta > 0$ under the condition (2.4). Note that this exact solution illustrates finite time $(T = \beta^{-2})$ disappearance of the phase.

## 3. "DIRECT" PROBLEMS IN A SEMI-INFINITE DOMAIN.

**THEOREM 4.** Under the condition (2.1) the problem $I'$ has a solution for $t \in (0,T)$ for every $T > 0$.

Local in time solvability of the problem $I'$ and monotony of $S(t)$ are proved similarly to lemma 1. Also similarly to the case of restricted domain the boundedness of $|\dot{S}(t)|$ for every time interval $(0,T)$ is proved. The condition (2.1) is essential for that.

**THEOREM 5.** If $c_i - h_i(x) \in L_1(0,\infty)$ and the condition

$$c_1 + c_2 > \max\{\rho_1^*, \rho_2^*\} \tag{3.1}$$

is held, then the classical solution of the problem $I'$ can exist only finite time.

For the proof let us integrate the equation

$$(c_i - Z_i)_t = k_i(c_i - Z_i)_{xx}$$

with respect to $x$ from $x = S(t)$ till $x = \infty$ and with respect to $t$ from $0$ till $t$ taking into account the conditions (1.2),(1,4)',(1.5)'. Summing up the obtained equations for $i = 1, 2$ we get

$$\int_{S(t)}^{\infty}(c_1 + c_2 - Z_1 - Z_2)dx - \int_0^{\infty}(c_1 + c_2 - h_1 - h_2)dx =$$

$$= \int_0^t \dot{S}(\rho_1 + \rho_2 - c_1 - c_2)dt$$

The first integral in the left side is nonnegative, the second is finite, hence the condition (3.1) involves the boundedness of $S(t)$ on the hole interval of the existence of solution. If we suppose that the solution exists for all $t > 0$ and carry out Laplace transformation, we'll obtain

$$\int_0^\infty (\dot{S}\rho_i + k_i p\lambda_i)exp(-k_i p^2 t - S(t)p)dt = \int_0^\infty h_i(x)exp(-px)dx,$$

considering $p \to 0$ and taking into account that $\dot{S}(t) \to 0$ as $t \to \infty$ we obtain the contradiction with the conditions on the data.

SIMILARITY SOLUTIONS OF THE PROBLEM $I'$.

If $\lambda_i(t) = \lambda_i^0 = const$ and the condition (2.4) is held, then for $h_i(x) \equiv c_i$ the problem $I'$ has self-similar solution

$$S(t) = \beta\sqrt{t}, \quad Z_i(x,t) = A_i\int_{x/\sqrt{T}}^\infty exp(-x^2/4k_i)dx + c_i,$$

$$i = 1,\ 2,\ Y(x) = y_0,$$

where the constants $\beta$, $A_i$, $i = 1, 2$ and $y_0$ can be determined from the system of equations

$$A_i\int_\beta^\infty exp(-x^2/4k_i)dx + c_i = \lambda_i^0, \quad i = 1,\ 2,$$

$$-A_i k_i exp(-\beta^2/4k_i) = (\rho_i - \lambda_i^0)/2, \quad i = 1,\ 2, \tag{3.3}$$

$$\rho_1 + \rho_2 = \rho_1^* y_0 + \rho_2^*(1 - y_0), \quad y_0 = \rho_1/(\rho_1 + \gamma\rho_2).$$

This statement could be verified similarly to the case of the problem $I$.

If instead of (2.4) the condition

$$c_1 + c_2 = \rho_1^* c_1/(c_1 + \gamma c_2) + \rho_2^* \gamma c_2/(c_1 + \gamma c_2) \qquad (3.4)$$

takes place, then with the appropriate initial functions the problem $I'$ has the solution of the "travelling wave" type with the arbitrary velocity $V$:

$$S(t) = Vt, \quad Z_i(x,t) = (\lambda_i^0 - c_i) exp\left[-(x - Vt)V/k_i\right] + c_i,$$

$$Y(x) = c_1/(c_1 + \gamma c_2).$$

**THEOREM 6.** Let in the problem $I'$ $\lambda_i(t) \to \lambda_i^0$ as $t \to \infty$, where $0 < \lambda_i^0 < c_i$. If the condition (2.4) is held, then

$$S(t) \sim \beta\sqrt{t} \quad \text{as } t \to \infty,$$

where $\beta$ is a constant determined from the system (3.3).

For the proof, following the method from [3] we apply Laplace transformation and tend p to zero.The statement is verified with the use of solution obtained earlier.

## 4."INVERSE" PROBLEMS.

Local in time solvability of the "inverse" problem in semi-infinite domain under the agreement conditions and the compatibility condition $\rho_2(x) \neq \alpha\rho_1(x) + \delta$ was proved in [4]. The existence of the self-similar solution in the case when $\rho_1 = const$, $\delta = 0$ and the condition "$\alpha$ lies between $c_2/c_1$ and $(c_2 - \rho_2)k_1/(c_1 - \rho_1)k_2$" is held was proved in the same paper.

Local in time solvability in the restricted domain $0 < x < S(t)$ can be obtained similarly to the case of semi-infinite domain.The investigation of global in time solvability of the "inverse" problems comes across the same difficulties as in the binary alloy solidification problems.

For the "inverse" problems in the restricted domain there exists the explicit solution which illustrates the finite time disappearance of the phase, too. Indeed, let $\rho_i \equiv const$ and the condition (4.1) be held. Then the direct check-up shows that the solution (with the corresponding initial functions) has a form

$$S(t) = \sqrt{1 - t/T} = \beta \sqrt{T - t}\ ,$$

$$Z_i(x,t) = A_i \int_{x/\sqrt{T-t}}^{B} exp(x^2/4k_i)dx + \lambda_i^0, \qquad i = 1,\ 2,$$

$$\lambda_2 = \alpha\lambda_1 .$$

The constants $A_1$, $A_2$, $\beta$, $\lambda_1$ can be determined from the system

$$k_i A_i\ exp(\beta^2/4k_i) = (\rho_i - \lambda_i)\beta/2 \quad (i = 1,\ 2), \quad \lambda_2 = \alpha\lambda_1 ,$$

$$A_i \int_0^{\beta} exp(x^2/4k_i)dx + \lambda_i = c_i \quad (i = 1,\ 2).$$

Solvability of this system under the condition (4.1) can be proved similarly to [4].

## REFERENCES.

[1] L.G.Badratinova, V.V.Kuznetsov, A.G.Petrova, V.V.Pukhnachov Direct and inverse problems of liquid-phase epitaxy. Proceedings of the Fifth Int. Conf. of the Num. Anal.of Semiconductor Devices and Integrated Circuits. Dublin, Ireland, 1987, pp.136-141.

[2] A.Fasano, M.Primicerio. New results on some classical free-boundary equations, Quart. Apply. Math.,38 (1989), pp.439-466.

[3] J.N.Dewynne, S.D.Howison, J.R.Ockendon, Weiging Xie. Asymptotic behavior of solutions to the Stefan problem with a kinetic condition at the free boundary. J.Austr. Math. Soc. Ser.B 31 (1989), pp.81-96.

[4] Zaltsman B.B., Petrova A.G. Control problem of $A_x B_{1-x} C$ material composition obtained by the method of liquid-phase epitaxy in the weightlessness. Kosmicheskaia nauka i tehnika, Kiev, Naukova dumka 1989, N 4, pp.57-60.

# STEFAN PROBLEM WITH SURFACE TENSION AS A LIMIT OF THE PHASE FIELD MODEL

*P.I.Plotnikov, V.N.Starovoitov*
*Lavrentyev Institute of Hydrodynamics*
*Novosibirsk 630090, RUSSIA*

The quasistationary phase field model equations depending on the small parameter are considered. It is proved that these equations possess a family of the solutions, which converge to the solutions of the Stefan problem with surface tension as the small parameter tends to zero.

Key words: phase field model, Stefan problem with surface tension.

1. In this paper a quasistationary boundary value problem for phase field model equations is considered. It is supposed that continuum medium occupies the bounded domain $\Omega \subset \mathbb{R}^3$. The problem is for a temperature $\theta(\vec{x},t)$ and an order parameter $\varphi(\vec{x},t)$ which satisfy the following equations:

$$\begin{aligned} &\frac{\partial}{\partial t}(\theta + \varphi) = \Delta\theta, && (\Omega\times[0,T]) \\ &-\varepsilon^2\Delta\varphi + W'(\varphi) = \varepsilon\theta && \\ &\theta = 0, \quad \nabla\varphi\cdot\vec{n} = 0 && (\partial\Omega\times[0,T]) \\ &\theta(\vec{x},0) = \theta_0(\vec{x}),\ \varphi(\vec{x},0) = \varphi_0(\vec{x}) && (\Omega). \end{aligned} \tag{1}$$

Here $W(\varphi) = c(\varphi^2 - 1)^2$ is the double-well potential, $\vec{n}$ is the outward normal vector to $\partial\Omega$, $\varepsilon$ is a small parameter. We neglect relaxation effects. Differential equations (1) can not be solved for the time derivatives of unknown functions. Therefore we need compatibility conditions. We shall assume that

$$\begin{aligned} &-\varepsilon^2\Delta\varphi_0 + W'(\varphi_0) = \varepsilon\theta_0 && (\Omega), \\ &\theta_0 = \nabla\varphi_0\cdot\vec{n} = 0 && (\partial\Omega). \end{aligned} \tag{2}$$

Our aim is to prove that problem (1) has the solution which converges as $\varepsilon \to 0$ to the solution of the Stefan problem with surface tension. In this problem $\Omega$ is divided into time varying parts $\Omega^{\pm}(t)$. The interface $\Gamma(t)$ evolves through the motion of its points with the normal velocity $v = [\nabla\theta\cdot\vec{n}]$. The problem involves solving the equation

$$\theta_t = \Delta\theta \quad (\operatorname{int} \Omega^{\pm}(t)) \tag{3}$$

subject to the boundary condition at the interface

$$\theta = k\mu \quad (\Gamma(t)), \tag{4}$$

where $k$ is a mean curvature of $\Gamma$.

Phase field model was suggested by Caginalp [1]. Existence of the generalized solution of Stefan problem with surface tension was proved in the recent paper of Luckhaus [2]. We refer also to [3,4,5].

2. In this section we establish existence of the solution of problem (1).

*Definition.* We say the pair of functions

$$\theta \in L_\infty(0,T;\, L_2(\Omega)) \cap L_2(0,T;\, \overset{0}{H}{}^1(\Omega)),$$

$$\varphi \in L_\infty(0,T;\, H^2(\Omega))$$

is the weak solution of problem (1) if

$$\int_0^T\int_\Omega \left\{(\theta + \varphi)\eta_t - \nabla\theta\cdot\nabla\eta\right\} d\vec{x}\, dt = \int_\Omega (\theta_0 + \varphi_0)\, \eta_0\, d\vec{x}, \tag{5}$$

$$\int_\Omega (\varepsilon^2\, \nabla\varphi\cdot\nabla\zeta + W'(\varphi)\zeta - \varepsilon\theta\zeta)\, d\vec{x} = 0, \tag{6}$$

whenever $\eta, \zeta \in C^\infty(\Omega\times[0,T])$ and $\eta$ vanishes at $\partial\Omega$ and for $t = T$.

We denote by $\langle\cdot,\cdot\rangle$ inner product in $L_2(\Omega)$ and use the notation

$$E_\varepsilon(\varphi) = \int_\Omega \left\{ \frac{\varepsilon}{2}\, |\nabla\varphi|^2 + \frac{1}{\varepsilon}\, W(\varphi) \right\} d\vec{x}.$$

**THEOREM 1.** Let functions $\theta_0 \in L_2(\Omega)$, $\varphi_0 \in H^2(\Omega)$ satisfy the compatibility condition (2). Then problem (1) has at least one solution such that

$$(i)\quad \|\theta\|_{L_\infty(0,T;L_2((\Omega))} + \|\nabla\theta\|_{L_2(\Omega\times[0,T])} \leqslant c,$$

$$\operatorname*{esssup}_{t\in[0,T]} E_\varepsilon(\varphi(t)) \leqslant c,$$

where $c$ does not depend on $\varepsilon$;

(ii) for almost all $t$

$$E_\varepsilon(\varphi(t)) - \langle\varphi(t),\theta(t)\rangle \leqslant E_\varepsilon(v) - \langle v,\theta(t)\rangle + \frac{1}{2}\|\varphi(t) - v\|^2_{L_2(\Omega)}$$

whenever $v \in H^1(\Omega)$.

(iii) $\varphi(t) \to \varphi_0$, $\theta(t) \to \theta_0$ in $L_2(\Omega)$ as $t \to 0$.

**Proof.** We use the semidiscretization method. In order to construct the approximate solution of problem (1), we consider the following scheme:

Find functions $\theta_n$, $\varphi_n$ such that

$$\theta_n(\vec{x},t) = v_k(\vec{x}),\quad \varphi_n(\vec{x},t) = u_k(\vec{x}),\quad (k-1)\tau \leqslant t \leqslant k\tau,$$

$$\Phi_k(u_k) = \inf_{H^1(\Omega)} \Phi_k(u), \tag{7}$$

$$v_k = \frac{1}{\tau} AF_k(u_k).$$

Here the functionals $\Phi_k$ and the functions $F_k$ are given by the formula

$$\Phi_k(u) \equiv E_\varepsilon(u) + 2\tau^{-1}\langle AF_k(u),F_k(u)\rangle,$$

$$F_k(u) \equiv u_{k-1} + \theta_{k-1} - u,\quad \tau = T/n.$$

A continuous positive operator $A\colon H^{s-1} \to H^s(\Omega) \cap \overset{0}{H}{}^1(\Omega)$ is defined by the equalities

$$(-\Delta + \tau^{-1})(Au) = u \quad (\Omega), \quad Au = 0 \quad (\partial\Omega).$$

The functionals $\Phi_k \colon H^1(\Omega) \to \mathbb{R}$ are weak lower semicontinuous and $\Phi_k(u) \to \infty$ as $\|u\|_{H^1(\Omega)} \to \infty$. It follows that $\Phi_k$ has a minimal point $u_k$. Therefore for each $n \geqslant 1$ the problem (1) has at least one approximate solution which satisfies the following inequalities:

$$\|\theta_n\|_{L_\infty(0,T;L_2((\Omega))} + \|\nabla\theta_n\|_{L_2(\Omega\times[0,T])} \leqslant c,$$

$$\operatorname*{esssup}_{t\in[0,T]} E_\varepsilon(\varphi_n(t)) \leqslant c, \tag{8}$$

$$\|\varphi_n\|_{L_\infty(0,T;H^2(\Omega))} \leqslant c(\varepsilon),$$

$$\|(\theta_n + \varphi_n)(t) - (\theta_n + \varphi_n)(t + \Delta t)\|_{L_\infty(0,T;H_{-1}(\Omega))} \leqslant c\,|\Delta t|^{1/2}.$$

This can be verified by a straightforward computation based on (7) and a priori estimates for the solutions of elliptic partial differential equations.

Now, our goal is to prove that approximate solutions converge to a generalized solution of problem (1). It follows from (8) that the sequence of approximate solutions $\varphi_n$, $\theta_n$ contains a subsequence $\theta_k$, $\varphi_k$ such that

$$D_x^\beta\,\theta_k \to D^\beta\theta, \quad D_x^\alpha\,\varphi_k \to D_x^\alpha\varphi \quad \text{weak in} \quad L_2(\Omega\times[0,T]),$$

$$U_k = \varphi_k + \theta_k \longrightarrow U \quad \text{strong in } L_2(\Omega\times[0,T]),$$

$$|\beta| \leqslant 1, \quad |\alpha| \leqslant 2.$$

Here $\theta$, $\varphi$, $U$ are some functions belonging to the Banach spaces $L_\infty(0,T;\ L_2(\Omega)) \cap L_2(0,T;\ \overset{0}{H}{}^1(\Omega))$, $L_\infty(0,T;\ H^2(\Omega))$, $L_2(0,T;\ H^1(\Omega))$ respectively.

In order to pass to the limit in equation (6) we need to prove that the functions $\varphi_n$ converge to $\varphi$ in $L_p(\Omega\times[0,T])$. This fact is based on the following compactness lemma.

Let $X \subset Y$ be Banach spaces with compact embedding $X \longrightarrow Y$

and $K_t$, $t \in [0,T]$ be a family of relatively compact subsets of $Y$. It is supposed that the following condition is fulfilled:

*(k)* If $f$, $g$ are the limit points of $K_t$ and $f - g \in X$, then $f = g$.

We consider a sequence of functions

$\varphi_n: [0,T] \to Y$, $\theta_n: [0,T] \to X$ such that

(a) $\|\varphi_n\|_{L_\infty(0,T;Y)} \leqslant c$, $\|\theta_n\|_{L_p(0,T;X)} \leqslant c$;

(b) $\varphi_n(t) \in K_t$ for almost all $t \in [0,t]$;

(c) functions $U_n = \varphi_n + \theta_n$ converge as $n \to \infty$ in $L_p(0,T;Y)$, $p > 1$, to some function $U$.

**LEMMA** 1. Under the above assumptions there exist $\varphi \in L_\infty(0,T;Y)$, $\theta \in L_p(0,T;X)$ such that $\varphi_n \to \varphi$, $\theta_n \to \theta$ in $L_p(0,T;Y)$.

In order to prove strong convergence of the approximate solutions of problem (1) in $L_2(0,T; L_p(\Omega))$ we choose $X = \overset{0}{H}{}^1(\Omega)$, $Y = L_p(\Omega)$, $K_t = \{\varphi_n(t),\ n \geqslant 1\}$

Let $f_j$, $j = 1, 2$, be limit points of $K_t$. It follows from (7) that functions $f_j$ satisfy the following equation:

$$-\varepsilon^2 \Delta f_j + W'(f_j) + \varepsilon f_j = \varepsilon U$$

and $f = f_1 - f_2$ is a solution of the second order linear homogeneous elliptic equation

$$-\varepsilon^2 \Delta f + af = 0, \quad a \in C^\alpha(\Omega).$$

If $f \in \overset{0}{H}{}^1(\Omega)$, then $f$ and $\nabla f \cdot \vec{n}$ vanish at the boundary of the region $\Omega$. Since $f = 0$ is a unique solution of the Cauchy problem for elliptic equation then $K_t$ satisfies the condition *(k)*. Therefore, approximate solutions satisfy all assumptions of Lemma 1 and converge as $n \to \infty$ to some weak solution of problem (1). The variational principle *(ii)*, the relation *(iii)* follow from (7) and Lemma 1.

## 3. LIMITING PASSAGE TO PROBLEM (3)-(4).

Denote by $BVC(\Omega)$ the set of functions from $BV(\Omega)$ which take only two values: ± 1. Let $\chi \in BVC(\Omega)$, then $\nabla\chi$ is a measure and $|\nabla\chi|$ is its total variation. Since the measure $\nabla\chi$ is absolutely continuous with respect to $|\nabla\chi|$, then there exists $|\nabla\chi|$-measurable function $\vec{v}$ such that

$$\int_A \nabla\chi \; (d\vec{x}) = \int_A \vec{v} \; d|\nabla\chi|$$

for every Borel set $A \subset \Omega$. This follows from the theorem of Radon-Nykodim. It is not difficult to show that $|\vec{v}| = 1$. We call the function $\vec{v}$ the normal vector to the surface $S = \partial E \cap \Omega$, where $E = \left\{ \vec{x} \in \Omega \;\middle|\; \chi(\vec{x}) = 1 \right\}$.

*Definition.* The pair of functions

$$\theta \in L_2(0,T; \overset{0}{H}{}^1(\Omega)) \cap L_\infty(0,T; L_2(\Omega)),$$

$$\chi \in L_\infty(0,T; BVC(\Omega))$$

is a generalized solution of problem (3)-(4) if

$$\int_0^\Omega \int_\Omega \left[(\theta + \chi)\eta_t - \nabla\theta\cdot\nabla\eta\right] d\vec{x}\, dt = \int_\Omega (\theta_0 + \varphi_0)\, \eta_0 \, d\vec{x}, \tag{9}$$

$$\frac{1}{\mu}\int_\Omega \chi \; div\; (\theta\vec{\phi})\; d\vec{x} = \int_\Omega \left[\sum_{i,j=1}^{3} \frac{\partial\phi_i}{\partial x_j} v_i v_j - div\; \vec{\phi}\right] d|\nabla\chi|(\vec{x}) \tag{10}$$

for almost all $t \in [0,T]$. Here $\eta$, $\vec{\phi}$ are any functions from $C^\infty(\Omega\times[0,T])$ such that $\eta$ vanishes at $\partial\Omega$ and for $t = T$.

Equality (10) is a generalized formulation of the condition (4).

**THEOREM 2.** Let $\theta_0 \in L_2(\Omega)$, $\chi_0 \in BVC(\Omega)$. Then there exists the generalized solution of problem (3)-(4) satisfying the following variational principle:

$$\mu\int_\Omega d|\nabla\chi|(\vec{x}) - \langle\theta,\chi\rangle \leqslant \mu\int_\Omega d|\nabla\chi|(\vec{x}) - \langle\theta,v\rangle + \frac{1}{2}\|\chi - v\|^2_{L_2(\Omega)} \quad (11)$$

for every function $v \in BVC(\Omega)$ and almost all $t \in [0,T]$.

**Proof.** Denote by $\theta_\varepsilon$, $\varphi_\varepsilon$ the generalized solution of problem (1). Further we shall construct the solution of (3)-(4) as a limit of $\theta_\varepsilon$, $\varphi_\varepsilon$.

The following estimate is a consequence of *(i)*:

$$\operatorname{esssup}_t \left[ E_\varepsilon(\varphi_\varepsilon) + \int_\Omega \theta_\varepsilon^2 \, d\vec{x} \right] + \int_0^T\int_\Omega (\nabla\theta_\varepsilon)^2 \, d\vec{x}\, dt \leqslant \quad (12)$$

$$\leqslant E_\varepsilon(\varphi_{0,\varepsilon}) + \int_\Omega \theta_{0,\varepsilon}^2 \, d\vec{x}.$$

Assume that $\theta_{0,\varepsilon} = \theta_0$, $\varphi_0$ tends to $\chi_0$ in $L_2(\Omega)$ and $E_\varepsilon(\varphi_{0,\varepsilon}) \leqslant C$, where constant $C$ does not depend on $\varepsilon$.

Note that the last inequality implies that $\chi_0 \in BVC(\Omega)$. Really, define ([6])

$$\zeta(u) = \int_0^u W^{1/2}(\xi)\, d\xi \sqrt{2},$$

$$\mu = \int_{-1}^{-1} W^{1/2}(\xi)\, d\xi \sqrt{2}.$$

Then

$$\mu\int_\Omega d|\nabla\chi| = \int_\Omega d|\nabla\zeta(\chi_0)| \leqslant \varliminf_{\varepsilon\to 0} \int_\Omega d|\nabla\zeta(\varphi_{0,\varepsilon})| \leqslant$$

$$\leqslant \varliminf_{\varepsilon\to 0} \int_\Omega |W^{1/2}(\varphi_{0,\varepsilon})\cdot\nabla\varphi_{0,\varepsilon}| \, d\vec{x} \leqslant E_\varepsilon(\varphi_{0,\varepsilon}).$$

Using the analogous reasoning, we can obtain from (12) the estimates which do not depend on $\varepsilon$:

$$\|\varphi_\varepsilon\|_{L_\infty(0,T;BV(\Omega))} \leqslant C,$$

$$\|\theta_\varepsilon\|_{L_\infty(0,T;L_2(\Omega))} + \|\nabla\theta_\varepsilon\|_{L_2(\Omega\times[0,T])} \leqslant C.$$

Now it is easy to show that sequences $\varphi_\varepsilon$, $\theta_\varepsilon$ are compact in $L_2(\Omega\times[0,T])$. Let in Lemma 1

$$X = \overset{0}{H}{}^1(\Omega),\ Y = L_2(\Omega),\ K_t = \left\{ \varphi_\varepsilon(t),\ \varepsilon > 0 \right\}.$$

If $f$, $g$ are limiting points of $K_t$ in $L_2(\Omega)$, then $f, g \in BVC(\Omega)$. Hence, if $f - g \in \overset{0}{H}{}^1(\Omega)$, then $f = g$. It remains to apply Lemma 1.

The limiting passage as $\varepsilon \to 0$ in equation (6) can be carried out in the same manner as in [6].

**REMARK.** With the help of the variational principle (11) and results of U.Massari [7] we can obtain that for almost all $t \in [0,T]$ the interface is a hypersurface of class $C^{1+1/8}$ if $\Omega \subset \mathbb{R}^3$ and $C^{1+\gamma}$, $\gamma < 1/4$, if $\Omega \subset \mathbb{R}^2$.

**REFERENCES.**

[1] Caginalp G. An analysis of a phase field model of a free boundary. Arch. Rat. Mech. Anal. 92 (1986), pp. 205-245.

[2] Luckhaus S. Solutions for the two-phase Stefan problem with the Gibbs-Thomson Law for the melting temperature. European Journal of Appl. Math., 1990, v.1, No.2, pp.101-111.

[3] Gurtin M. On the two-phase Stefan problem with interfacial energy and entropy. Arch. Rat. Math. Anal. 96 (1986), 199-241.

[4] Caginalp G. Stefan and Hele-Shaw type models as asymptotic limits of the phase field equations. Phys. Review A39 (1989), pp. 5887-5896.

[5] Elliott C.M. The Cahn-Hilliard model for the kinetics of phase separation. Math. Models for Phase Change Problems (ed. J.F. Rodrigues), International Series of Numerical Math., v. 86 (1989).

[6] Luckhaus S., Modica L. The Gibbs-Thomson relation within the Gradient Theory of Phase Transitions. Arch. Rat. Mech. Anal. (1988), pp. 71-83.

[7] Massari U. Frontiere orientate di curvatura media assegnata... Rend. Sem. Mat. Univ. Padova, v. 53, 1975, pp.37-52.

International Series of Numerical Mathematics, Vol. 106, © 1992 Birkhäuser Verlag Basel


# THE MODELIZATION OF TRANSFORMATION PHASE VIA THE RESOLUTION OF AN INCLUSION PROBLEM WITH MOVING BOUNDARY

*H.Sabar, M.Buisson, M.Berveiller*
*Laboratoire de Physique et Mecanique des Materiaux*
*Universite de Metz, Ile du Saulcy, 57045 Metz Cedex, FRANCE*

*U.R.A. - C.N.R.S. n$^{o}$1215*

The field-equations in statics of polycrystalline materials constitued, for example, of martensitic and austenitic grains are described by use of Green's functions taking into account inelastic and elastic heterogeneities. Now, question arises when looking at the evolution of inelasticity (phase transformation, ...) induced by external loading. The main feature is that we observe experimentally movements of interfaces with velocities which are different than the particles, a natural concept is bringing out: the usual static field-equations are derived in time with respect to the proper velocities of the interfaces instead of the particles velocities. This technique gives us explicitly a simultaneous evolution of inelasticity with associated moving-boundaries.

## 1. INTRODUCTION.

Experimental studies related to phase-transformation (austenite - martensite ...) or some others metallurgical phenomena (recrystallization...) have improved significantly during the last years; for example, in case of granular (or multiphased) polycrystals, much attention is focused on a modification of the granular structure and a simultaneous change of the orientations of grains associated with their morphological change. In order to go with these metal physics, we propose a micromechanical formulation taking simultaneously account of the growth and interactions of phases in a heterogeneous material when these phenomena are closely related with *moving boundaries* between the constituents. Here the method and certain suggested applications will be described in the scope of the linear elastoquasistatics associated with a time-varying distribution of some inelastic strain (e.g. plastic-, thermal-, or stress free phase transformation -strain); details are given in [1],[2],[3].

This new resolution method is called *Inclusion-problem with moving boundary* in the sense that it can be considered as a natural *extension* of the well-known *Eshelby-Kroner static inclusion-problem*

## 2. STATICS-FUNDAMENTAL EQUATIONS AND INTEGRAL FORMULATION.

### 2-1. Statement of the problem.

Let us consider an unbounded inhomogeneous anisotropic elastic medium with the elastic moduli tensor $C(x)$ and inelastic strain field $\varepsilon^p(x)$ at point $x$. A fixed rectangular Cartesian coordinate system with coordinate axes $x_i$ $(i = 1, 2, 3)$ is used, by convention only repeated Latin - indices are summed.

The governing equations are given by the Hooke's law where the components $\sigma_{ij}(x)$ of the stress tensor are related to the elastic part $\varepsilon^e(x)$ of the total local linearized strain-tensor $\varepsilon(x)$:

$$\sigma_{ij}(x) = C_{ijkl}(x) \left[\varepsilon_{kl}(x) - \varepsilon^p_{kl}(x)\right] \tag{1}$$

with

$$\varepsilon^e(x) = \varepsilon(x) - \varepsilon^p(x) \tag{2}$$

and the quasi-static equilibrium equations in the absence of body forces

$$\sigma_{ij,j}(x) = 0, \tag{3}$$

where $,j$ denotes partial differentiation with respect to the $x_j$ coordinate.

Assuming uniform given informations at infinity: $\sigma(\infty) = \sigma^0$, $C(\infty) = C^0$, $\varepsilon^p(\infty) = \varepsilon^{p_0}$, the first problem is to find the resulting total displacement-gradients $u_{i,j}(x)$ of the material particles throughout the body by requiring that the total strain must be compatible

$$\varepsilon_{ij}(x) = \frac{1}{2} \left[u_{i,j}(x) + u_{j,i}(x)\right]. \tag{4}$$

## 2-2. Integral formulation-Application of Green's tensor.

The above partial differential equations taken in the sense of generalized functions are transformed into integral equations by introducing the Green's functions $G_{ij}(x)$ for the infinitely extended homogeneous medium with elasticity. These functions satisfy [4],[5];

$$C^{0}_{ijkl}G_{km,lj}(x - x') + \delta_{im}\delta(x - x') = 0,$$
$$G_{ij}(x) = 0 \quad \text{at infinity,} \tag{5}$$

where $\delta(x - x')$ is the three-dimensional distribution of Dirac and $\delta_{im}$ is the Kronecker's symbol. Denoting by $(\ldots)'$ the following perturbations:

$$C(x) = C^{0} + C'(x), \tag{6}$$

$$u_{i,j}(x) = u^{0}_{i,j} + u'_{i,j}(x), \tag{7}$$

where $u^{0}_{i,j}$ is uniform displacement gradient deduced from the stress and strain-boundary information at infinity (with usual symmetries of $C_{ijkl}$):

$$\sigma^{0}_{ij} = C^{0}_{ijkl}\left(u^{0}_{l,k} - \varepsilon^{p_0}_{kl}\right) \tag{8}$$

and with notation of the pseudo-stresses:

$$\sigma^{p}_{ij}(x) = C_{ijkl}(x)\varepsilon^{p}_{kl}(x),$$
$$\sigma^{p_0}_{ij} = C^{0}_{ijkl}\varepsilon^{p_0}_{kl}, \tag{9}$$
$$\sigma^{p'}_{ij} = \sigma^{p}_{ij}(x) - \sigma^{p_0}_{ij},$$

we get partial differential equations analogous to the Navier equations of elasticity

$$C^{0}_{ijkl}u'_{l,kj} + \left[C'_{ijkl}\varepsilon_{kl} - \sigma^{p'}_{ij}\right]_{,j} = 0, \tag{10}$$

from which we obtain the displacement-gradient in the actual material and by the integral equations:

$$u_{ij}(x)=u^{0}_{i,j} + \iiint_{-\infty}^{+\infty} G_{ki,lj}(x-x')\left[C'_{klmn}(x')u_{n,m}(x')-\sigma^{p'}_{kl}(x')\right]dx'. \tag{11}$$

Such calculations were used, for example, in self-consistent methods [6]) or in other cases related to heterogeneous materials ([7],[8]). For information, let us note that interactions between all the constituents may be investigated if we proceed to an adequate discretization of the infinite volume [9],[10].

However, in order to avoid complicated calculations and promote an original method of investigation, we first reduce the topology of this problem to that one of an ellipsoidal inclusion (uniform anisotropic elasticity $C^{I}$ and inelastic strain $\varepsilon^{pI}$, total strain $\varepsilon^{I}$, volume $V$) embedded in a homogeneous infinite matrix (uniform anisotropic elasticity $C^{0}$, uniform inelastic strain $\varepsilon^{p_0}$).

This topology corresponds to the Eshelby-Kroner inclusion's problem. By this topology, it appears, that the following terms reduce to zero outside the inclusion

$$\begin{aligned} &C'(x) = 0 \quad \text{if} \quad x \text{ located in the Matrix,} \\ &\sigma^{p'}(x) = 0 \quad \text{if} \quad x \text{ located in the Matrix,} \end{aligned} \tag{12}$$

and the preceding integral equation (11) becomes at each time $t$:

$$u_{i,j}(x)=u^{0}_{i,j} + \iiint_{V(t)} G_{ki,lj}(x-x')dx'\left[C'_{klmn}\varepsilon^{I}_{mn} - \sigma^{p'}_{kl}\right]. \tag{13}$$

In the following, we observe the evolution of the displacement-gradients which are going simultaneously with the evolution of properties $\varepsilon^{p_0}(t)$ and $\varepsilon^{pI}(t)$ and morphological change $V(t)$.

## 3. INCLUSION PROBLEM WITH MOVING BOUNDARY.

### 3-1. Formulation.

The question arises when looking at the evolution of the displacement-gradients inside and outside the inclusion. According to experimental observations (phase transformation, recrystallisation...) the objective of the method is to include

a) the evolution of the inelasticity via time dependent functions

$$\varepsilon^{p0}(t) \text{ and } \varepsilon^{pI}(t),$$

b) the change of the inclusion's shape by considering in (13) a time varying integration domain $V(t)$.

Nevertheless, for simplicity we neglet here time-variations of the elastic-properties $C^0(t)$ and $C^I(t)$ which may be induced by some transformation-phase or other evolution phenomena.

It is crucial to point out that the volume $V(t)$ is in fact the geometrical zone filled by the eigen-strain $\varepsilon^{pI}$: the Eshel-by-Kroner inclusion is considered as a geometric volume and the movement of his boundary involves a velocity which in most cases is different from that of the particles.

It leads us naturally to the mathematical technique corresponding to time derivation with respect to the variety defined by the geometric volumes of the inclusions instead of the particle-velocity field. Such a technique gives the following differentiations ([11]):

$$\frac{\delta}{\delta t}\left(\ldots\right) = \frac{\partial}{\partial t}\left(\ldots\right) + W_p(x,t)\left(\ldots\right)_{,p},$$

$$\frac{\delta}{\delta t}\iiint_{V(t)} F(x,t)\,dx = \iiint_{V(t)} \frac{\partial F}{\partial t}(x,t)\,dx + \iint_{\partial V(t)} F(y,t)\,\vec{W}(y,t)\cdot\vec{n}(y,t)\,dy,$$

or, using the divergence theorem

$$\frac{\delta}{\delta t}\iiint_{V(t)} F(x,t)\,dx =$$

$$= \iiint_{V(t)} \left[\frac{\partial F}{\partial t}(x,t) + F_{,p}(x,t)W_p(x,t) + F(x,t)W_{p,p}(x,t)\right] dx,$$

where respectively $V(t)$ denotes the volume of the inclusion, $\partial V$

its boundary with $n$ as unit external normal vector, $y$ some arbitrary point of the boundary endowed with a geometric velocity $W(y,t)$, $F$ some arbitrary tensorial or scalar valued function, $\delta/\delta t$ time-derivation with respect to the velocity field, $W(x,t)$ and $\partial/\partial t$ partial derivation with respect to time.

According to (14) we obtain from (13):

$$\frac{\delta}{\delta t} u_{i,j}(x,t) - \frac{\delta u^0_{i,j}(t)}{\delta t} = \tag{17}$$

$$= \iiint\limits_{V(t)} G_{ki,lj}(x-x')dx' \left[ (C^I - C^0)_{klmn} \frac{\partial \varepsilon^I_{mn}}{\partial t} - C^I_{klmn} \frac{\partial \varepsilon^I_{mn}}{\partial t} + C^0_{klmn} \frac{\partial \varepsilon^{p_0}_{mn}}{\partial t} \right] +$$

$$+ \frac{\delta}{\delta t} \left\{ \iiint\limits_{V(t)} G_{ki,lj}(x-x')dx' \right\} \left[ (C^I - C^0)_{klmn} \varepsilon^I_{mn} - C^I_{klmn} \varepsilon^{pI}_{mn} + C^0_{klmn} \varepsilon^{p_0}_{mn} \right]$$

and in agreement with (15-16):

$$\frac{\delta}{\delta t} \iiint\limits_{V(t)} G_{ki,lj}(x-x')dx' = \tag{18}$$

$$= \sum_{\alpha=1}^{3} \iiint\limits_{V(t)} \left[ G_{ki,lj\alpha}(x-x') W_{\alpha'}(x',t) + G_{ki,lj}(x-x') W_{\alpha',\alpha'}(x',t) \right] dx'.$$

Consequently the only integral terms to be calculated are successively

$$\iiint\limits_{V(t)} G_{ki,lj}(x-x')dx', \quad \sum_{\alpha=1}^{3} \iiint\limits_{V(t)} G_{ki,lj\alpha}(x-x') W_{\alpha'}(x',t)dx'$$

and $$\sum_{\alpha=1}^{3} \iiint\limits_{V(t)} \left[ G_{ki,lj}(x-x') W_{\alpha,\alpha}(x',t)dx' \right.$$

The first integral has been already appeared in classical

problems involving Eshelby-Kroner inclusion, the last two terms are rather new and must be calculated in order to get clearly from (17) both the variations of $V(t)$ and the eigen-displacement of its boundary associated with the evolution of inelastic strains and external loading.

Globally, we have in (17) a localization formula related to rates of displacement-gradient taking explicitly and simultaneously into account external agency, plastic-strain rates, inhomogeneous elasticity, and morphological change. It will be an easy task to obtain the rates of strain and rotation (respectively $\delta\varepsilon_{ij}/\delta t$, $\delta w_{ij}/\delta t = 1/2\ \delta(u_{i,j} - u_{j,i})/\delta t$ by symmetrization and antisymmetrization with respect to $i$ and $j$.

## 4. SOME APPLICATIONS.

In order to obtain applications we investigate the case of an ellipsoidal embedded in a matrix and transforming into another non necessarily homothetic ellipsoid.

### 4-1. Ellipsoidal growing.

The ellipsoidal growing is represented by three functions $a_\alpha(t)$ $(\alpha = 1, 2, 3)$ corresponding to the semi-axis of the ellipsoid. The coordinates $x_i$ $(i = 1, 2, 3)$ are attributed to the principal orthogonal axis; for sake of clarity, we do not consider any geometric rotation of the ellipsoid.

In such a case the velocity vector $W(x,t)$ is given by the coordinate $W_\alpha(x,t)$ $(\alpha = 1, 2, 3)$ without summation on Greek letter $\alpha$:

$$W_\alpha(x,t) = \frac{\dot{a}_\alpha(t)}{a_\alpha(t)}\, x_\alpha, \tag{20}$$

where $\dot{a}_\alpha = \partial a_\alpha/\partial t$ and for the divergence

$$W_{i,i}(x,t) = \sum_{i=1}^{3} \frac{\dot{a}_\alpha}{a_\alpha} \tag{21}$$

so that (18) becomes:

$$\frac{\delta}{\delta t} T_{ijkl}(x,t) = \sum_{\alpha=1}^{3} \frac{\dot{a}_\alpha}{a_\alpha} \left\{ T_{ijkl}(x,t) - T^{\alpha}_{ijkl}(x,t) \right\} \tag{22}$$

with

$$T_{ijkl}(x,t) = - \iiint_{V(t)} G_{ki,lj}(x-x')dx' \tag{23}$$

and

$$T^{\alpha}_{ijkl}(x,t) = \iiint_{V(t)} G_{ki,lj\alpha'}(x-x')x'_\alpha dx' . \tag{24}$$

Intergals are calculated from Fourier's transform. Numerical results are obtained [2] in case of anisotropic elasticity (Copper and Zinc). For information we give analytical results in case of isotropic elasticity with the uniform and time independent Lame's coefficients $\lambda^0$ and $\mu^0$,

$$C^0_{ijkl} - \lambda^0 \delta_{ij}\delta_{kl} + \mu^0(\delta_{ik}\delta_{jl} + \delta_{il}\delta_{jk}) \tag{25}$$

when the boundary-movement corresponds to the flattening of a sphere $(a_1(0) = a_2(0) = a_3(0))$ into an ellipsoid $(a_1(0)+\delta a_1, a_2(0)+\delta a_2, a_3(0)+\delta a_3)$. For located inside the ellipsoid $T$ and $T^{\alpha}$ are uniform [2] and independant of the initial radius:

$$T_{ijkl} = \left\{(9-10\nu^0)\delta_{ik}\delta_{jl} - (\delta_{ij}\delta_{kl} + \delta_{jk}\delta_{il})\right\} / \left[30\mu^0(1 - \nu^0)\right], \tag{26}$$

$$T^{\alpha}_{ijkl} = \left\{(13-14\nu^0)(\delta_{ik}\delta_{jl}+2\delta_{ik}\delta_{j\alpha}\delta_{l\alpha})-(\delta_{ij}\delta_{kl}+\delta_{kj}\delta_{il})- \right. \tag{27}$$

$$\left. -2(\delta_{jk}\delta_{i\alpha}\delta_{l\alpha}+\delta_{ij}\delta_{k\alpha}\delta_{l\alpha}+\delta_{il}\delta_{k\alpha}\delta_{j\alpha}+\delta_{kl}\delta_{i\alpha}\delta_{j\alpha}+\delta_{jl}\delta_{i\alpha}\delta_{k\alpha})\right\} / \left[70\mu^0(1-\nu^0)\right]$$

(without summation on $\alpha$)

where $\nu^0 = \lambda^0/(2\mu^0 + 2\lambda^0)$.

Let us precise that a great variety of ellipsoidal kinematics may be investigated. Such results may be considered as tabulated data in this sense: we observe that (17) becomes for $x$ inside the ellipsoid

$$\frac{\delta}{\delta t}u_{i,j}(t) = \frac{\delta}{\delta t}u^{0}_{i,j}(t) + \tag{28}$$

$$+ T_{ijkl}(t)\left[C^{I}_{klmn}\frac{\partial\varepsilon^{pI}_{mn}}{\partial t} - C^{0}_{klmn}\frac{\partial\varepsilon^{p_0}_{mn}}{\partial t} - (C^{I}_{klmn} - C^{0}_{klmn})\frac{\partial\varepsilon^{I}_{mn}}{\partial t}\right] +$$

$$+ \frac{\delta}{\delta t}T_{ijkl}(t)\left[C^{I}_{klmn}\varepsilon^{pI}_{mn} - C^{0}_{klmn}\varepsilon^{p_0}_{mn} - (C^{I}_{klmn} - C^{0}_{klmn})\varepsilon^{I}_{mn}\right],$$

it appears consequently in (28) that the coefficient of $\dot{a}_\alpha/a_\alpha$ (movements of the boundary) $\partial\varepsilon^{pI}/\partial t$ and $\partial\varepsilon^{p_0}/\partial t$ (evolution of inelasticity) are given in the corresponding $T$ and $T^{\alpha}$ of the morphological change have been preestablished via formulas (22,23,24).

Nevertheless complete simulation of the displacement-gradients involves the knowledge of both the evolution law of in-elastic strains (e.g. flow rule for plastic-strain, heat conduction for thermal-strain, yield transformation for phase-change...) and the intrinsic behaviour of the boundary (for example, some relation between velocity $W$ and the generalized applied force on it, see Eshelby [12] and Hill[13]).

The knowledge of these evolution relations depends on the physics of the investigated problem and this question is outside the aim of this work, nevertheles we suggested in the following abstact a simple study related to reorientation of martensitic variants (detail in [13]) where both the concept of moving boundary and his associated intrinsic behaviour are involved.

### 4-2. Reorientation of martensitic variants.

We propose a formulation of constitutive equations dealing with phase-transformation where interface-movements are observed between martensitic variants. Analytically, we describe simply the topology and kinematics of a representative volume constituted of several compatible variants whose interfaces have a velocity differing from those of the particles. The driving force deduced from the energy-momentum concept is compared with the critical threshold from which we define a reorientation criterion. We obtain the macroscopic flow rule by using this criterion as a plastic potential.

### 4-3. Effect of grain size associated with dislocations.

In order to take into account the fine structure of dislocations at grain-boundaries, we use a very simple theory of dislocations piles up [14]. The stress field for a plastic inclusion with mobile interface is calculated in order to bring out the effects of grain size associated to the fine structure of interfacial dislocations.

## 5.CONCLUSION.

This method may be applied to a wide range of unnecessary homothetic ellipsoidal growing and generalized, as indicated, to arbitrary morphological changes. The calculation technique does not present supplementary difficulty compared with the static inclusion problems previously solved by Eshelby and Kroner. Moving boundaries and evolution of inelasticity are explicitly and simultaneously associated in the description of instantaneous local fields; in this way the method appears as an effective mean to progress in physics and micromechanics of moving boundary problems.

## REFERENCES.

[1] Sabar H., Berveiller M., Buisson M. "Probleme d'inclusion a frontiere mobile", Comptes Rendus de l'Academie des Sciences, Serie II, (1990) t.310, pp. 477-452, Paris.

[2] Sabar H., Buisson M., Berveiller M. The Inhomogeneous and Plastic Inclusion Problem with Moving boundariy, in press in Inter.J. of Plasticity.

[3] Buisson M., Patoor E., Berveiller M. Comportement global associe aux mouvements d'interafces entre variantes de martensites, submit. in C.R.Acad.Sci.II.

[4] Lifshits I.M., Rosentsveig L.N. Construction of the Green's tensor for the basic equation of the theory of elasticity for the case of an infinite elastic-anisotropic medium, (1947)

ZhETF, v. 17, N 9.

[5] Indenbom V.L., Orlov S.S. Construciton of the Green's functions in terms of Green's function of lower dimension. (1968) J. Appl. Math. and Mech. v. 32, N 3, pp.414-420.

[6] Berveiller M, Zaoui A. Journal de Mechanique (1980).

[7] Mura T. Micromechanics of defects in Solids, Martinus Nijhoff Publishers, Dordreht. (1987)

[8] Levin B.M. Contraintes thermoelasticues dans les milieux composites. Prikl. Mat. Meh., (1982) v. 46, N 3, pp.502-506.

[9] Berveiller M., Fassi-Fehri O, Hihi A. Multiply Site Self Consistent Scheme, Int.J.Engng.Sci. (1897) 25, 681.

[10] Buisson M., Molinari A., Berveiller M. On a constitutive relation for heterogeneous thermoelastic media. Arch.of Mech., v. 42-2 (1990), Warszawa.

[11] Germain P., "Mecanique," Ellipses, Ecole Polytechnique, (1986), Paris, 1.

[12] Eshelby J.D., Inelastic Behaviour od Solids.ed. M.F.Kanninen, W.F.Adler, A.R.Rosenfield, R.I.Joffee, (1970) p. 77, Mac Craq Hill, New York.

[13] Hill R. Energy-Momentum Tensors in Elastostatics: some reflections on the general theory. J.Mech.Phys.Solids, v. 34, N 3, (1986) pp. 305317.

[14] Berveiller M., Sabar H. Cont. Mod. and Discret Systems, Dijon II,ed Maugin, (1989).

International Series of Numerical Mathematics, Vol. 106, © 1992 Birkhäuser Verlag Basel 

# TO THE PROBLEM OF CONSTRUCTING WEAK SOLUTIONS IN DYNAMIC ELASTOPLASTICITY

*V.M.Sadovskii*

*Computing Center of Academy of Sciences*
*Krasnoyarsk, 660036, RUSSIA*

On the basis of formulation of elastic-plastic flow theory with isotropic and kinematic hardening in terms of variational inequality the integral generalization has been obtained which allows one to study the class of weak solutions. In the problem of shift waves propagation the comparison of discontinuous solutions for different hardenings has been carried out.

Key words. Elastoplasticity, flow theory, hardening, variational inequality, integral generalization, discontinuity relationships.

## INTRODUCTION.

The questions of constructing weak solutions for dynamic problems in Prandtl-Reiss flow theory was firstly considered by J.Mandel [1], who made a wrong conclusion on the ambiguity of velocity and stress fronts discontinuity in this theory. The full system of strong discontinuity relationships was obtained [2] by means of the principle of maximal plastic dissipation of energy in a front, for linear isotropic and kinematic hardening model.

It was shown [3] that Prandtl-Reiss system of quasi-linear equations which corresponds to the elastic perfectly-plastic model can't be reduced to divergent form. Thus, this system can't be generalized as a full number of integral conservation laws, and a construction of discontinuous solutions is impossible similar to models of ideal continua [4].

In the given paper the integral formulation which is equivalent to the equations of flow theory with an arbitrary hardening curve has been made and the system of discontinuity relationships has been obtained on the basis of this formulation exclusively.

1. A geometrically linear model of elasto-plastic solid can

be presented as a system of motion equations, Hook law and a principle of maximum speed of plastic energy dissipation:

$$\rho\, v_{i,t} = \sigma_{ij,j}\,, \qquad e^{0}_{ij} = a_{ijkl}\,\sigma_{kl,t}\,,$$

$$(\bar{\sigma}_{ij} - \sigma_{ij})\, e^{p}_{ij} \leqslant 0\,. \tag{1}$$

$$\frac{1}{2}\,(v_{i,j} + v_{j,i}) = e^{0}_{ij} + e^{p}_{ij}\,.$$

Here $\rho$ is density, $v_i$ - vector of velocity in Cartesian coordinate system, $a_{ijkl}$- elastic coefficients tensor, possessing the property of symmetry and positively defined; $e^{0}_{ij}$ , $e^{p}_{ij}$ - are elastic and plastic parts of strain rate tensor. The maximum principle is satisfied for an arbitrary variation of stress tensor constrained by:

$$f(\sigma_{ij} - \tau_{ij}) \leqslant \theta\,, \tag{2}$$

where $f = f(\sigma_{ij})$ is a convex positively defined homogeneous yield function of undeformed material; $\tau_{ij}$ - a symmetrical micro-stress tensor, $\theta$ is variable yield limit.

The inequality (2) describes classical variants of hardening: the case $\tau_{ij} = 0$ corresponds to isotropic, $\theta = 1$ - to kinematic hardening. For elastic perfectly-plastic continua $\tau_{ij} = 0$, $\theta = 1$. In the general case it's necessary to add to (1),(2) the system of evolution for $\tau_{ij}$ and $\theta$:

$$\xi_{ij,t} = e^{p}_{ij}\,, \qquad \theta\,\eta_{,t} = (\sigma_{ij} - \tau_{ij})\, e^{p}_{ij}\,. \tag{3}$$

In these equations the plastic strain rate tensor $\xi_{ij}$ and scalar coefficient $\eta$ are the specified functions of hardening parameters so that

$$\xi_{ij} = \partial\Phi_0/\partial\tau_{ij}, \quad \eta = \partial\Phi_0/\partial\theta,$$

where $\Phi = \Phi\,(\tau_{ij}, \theta)$ is Gibbs plastic potential.

The system of relationships (1),(3) is equivalent to the inequality

$$(v_i^* - v_i)(\rho v_{i,t} - \sigma_{ij,j}) + (\sigma_{ij}^* - \sigma_{ij})(a_{ijkl}\,\sigma_{kl,t} - v_{i,j}) +$$
$$+ (\tau_{ij}^* - \tau_{ij})\,\xi_{ij,t} + (\theta^* - \theta)\,\eta_{,t} \geqslant 0, \tag{4}$$

in which a variation of the velocity vector is arbitrary, and variations of stress tensor and hardening parameters satisfy the constraint (2). Really, when choosing in (4) the acceptable variations by certain means, one could obtain all the relationships (1),(3). For example, when $\sigma_{ij}^* = \bar{\sigma}_{ij}$, $\tau_{ij}^* = \tau_{ij}$, $\theta^* = \theta$, there follows (1). When $\sigma_{ij}^* = \sigma_{ij} + \gamma_{ij}$, $\tau_{ij}^* = \tau_{ij} + \gamma_{ij}$, where $\gamma_{ij}$ is arbit rary symmetrical tensor, the system of evolution for $\tau_{ij}$ can be obtained. Acceptable variations $\sigma_{ij}^* = \tau_{ij}^* = \theta^* = 0$ and $\sigma_{ij}^* = 2\sigma_{ij}$, $\tau_{ij}^* = 2\tau_{ij}$, $\theta^* = 2\theta$ give the rest equation (3). On the other hand, putting in (1)

$$\bar{\sigma}_{ij} = (\sigma_{ij}^* - \tau_{ij}^*)\,\frac{\theta}{\theta^*} + \tau_{ij}\,,$$

one could derive (4) with respect to (3).

The inequality (4), which represents one of the equivalent formulation of the flow theory, is a special case of hyperbolic variational inequality with non-linear operator:

$$(u^* - u)\,(N\langle u\rangle - g) \geqslant 0 \qquad u\,,\,u^* \in K \tag{5}$$

$$N\langle u\rangle = \partial\varphi/\partial t - \sum_{s=1}^{n} \partial\psi_s/\partial x_s\,.$$

Here $u = u(t,x)$ - is the unknown $m$ - dimensional vector-function, $K = K(t,x)$ - a convex set of acceptable variations of solution,

$$\varphi = \partial\Phi/\partial u, \qquad \psi_s = \partial\Psi_s/\partial u\,,$$

where $\Phi = \Phi(t,x,u)$ and $\Psi_s = \Psi_s(t,x,u)$ - are specified scalar potentials; $g = g(t,x,u)$ - is a specified vector function. Notations $\partial/\partial t$ and $\partial/\partial x_s$ denote total derivatives with respect to the corresponding variables. Below $\varphi$ and $\psi_s$ are assumed to be continuously differentiable and $K$, $g$ - continuous functions of their arguments.

In the model under consideration $u$ includes the components of the velocity vector, stress tensor and hardening parameters;

$$\Phi = \Phi_0 + \frac{1}{2}\rho\, v_i v_i + \frac{1}{2} a_{ijkl}\sigma_{ij}\sigma_{kl}, \quad \Psi_s = \sigma_{is}\, v_i,$$

vector $g$ is equal to zero, and the set $K$ is defined by the constraint (2).

In the case the function $\Phi$ is rigorously convex, and $g$ satisfies Lipschitz condition for the variable $u$, for the inequality (5) a'priori estimates are valid, which generalize the solution estimates of quasi-linear hyperbolic systems in characteristic conoids. These estimates prove a uniqueness and continuous dependence on the initial data "in the small in time" of Cauchy problem solution:

$$u\Big|_{t=0} = u_0(x)$$

and mixed problems with dissipative boundary conditions, and, also, the limitedness of the solutions dependence area (a finiteness of the disturbance propagation velocity in corresponding models) [5].

The formulation (5) also allow us to construct the effective numerical methods of the problems solution. For this the approximation of the differential operator $N\langle u\rangle$ and the constraint $u \in K$ is made, after that the problem is induced to the discrete variational inequality, the numerical realization algorithm of which represents a generalization of the "stress correction" procedure by M.L.Wilkins [6].

2. A variational inequality of the general form can be written in a divergent form

$$u^* N\langle u\rangle - (u^* - u)\, g \geqslant \partial(u\,\varphi - \Phi)/\partial t - \sum_{s=1}^{n} \partial(u\,\psi_s - \Psi_s)/\partial x_s + h,$$
$$h(t,x,u) = \Phi_{,t} - \sum_{s=1}^{n} \Psi_{s,s}$$

and, thus, it has an integral generalization, equivalent to it on smooth solutions:

$$\iint_G \{- \varphi\, (\chi\, u^*)_{,t} + \sum_{s=1}^{n} \psi_s\, (\chi\, u^*)_{,s} - (u^* - u)\, \chi\, g\}\, d\,\omega_x dt \geqslant$$
$$\geqslant \iint_G \{- (u\,\varphi - \Phi)\, \chi_{,t} + (u\,\psi_s - \Psi_s)\, \chi_{,s} + h\,\}d\,\omega_x dt\ . \tag{6}$$

Here $\chi \in C^{\infty}(G)$- is an arbitrary finite in $G$ non-negative function, $u^* = u^*(t,x) \in K$ - is a continuously differentiable vector function.

The integral inequality defines the set of weak solutions, which includes the various vector-functions $u \in L_2(G)$, which satisfy (6) with every acceptable $u^*$. This set includes the solutions with a strong discontinuity, which have a discontinuity of the first order on some hyper-surface $S$ and continuously differentiable in the rest part of $G$.

Applying Green formula to the integrals on subdomains of such solution continuity, one can show, with respect to $\chi$ arbitrariness, that at every point of hyper-surface $S$

$$u^*\, R \;\geqslant c\; [u\,\varphi - \Phi] + \sum_{s=1}^{n} [u\,\psi_s - \Psi_s]\ ; \qquad R = c\; [\varphi] + \sum_{s=1}^{n} [\psi_s]\, \nu_s\ ,$$

where $c \geqslant 0$ - is a velocity of discontinuity front motion, which represents a section $S$ by hyper-plane $t = const$, in the direction of its normal $\nu_s$; brackets denote a function shock.

The latter inequality is rewritten in the equivalent form

$$(u^* - u^0)\, R \;\geqslant d \equiv c\; (\varphi^0\; [u] - [\Phi]) + \sum_{s=1}^{n} (\psi_s^0\; [u] - [\Psi_s]), \tag{7}$$

Here $u^0 = (u^+ + u^-)\,/2$ ($u^{\pm}$ are one-sided limits of the solution on $S$), and values $\varphi$, $\psi_s$ are determined by analogy. Putting in (7) $u^* = u^0 \in K$, one could obtain a condition of discontinuity realizability: $d \leqslant 0$, which is for the variational inequality (5) of the same importance, as a condition of entropy non-decrease in the models of ideal media.

By the expansion of shocks $[\Phi]$ and $[\Psi_s]$ into Teilor series, it's easy to show, that $d\,/[u]^2 \to 0$ with $[u] \to 0$, i.e. a value

$d$ has a third order infinitesimal as compared with $[u]$.

Further two types of solutions with a strong discontinuity will be considered: regular solutions, for which at the points of hyper-surface $S$ the left side of inequality (7) is non-negative with every $u^* \in K$, and irregular ones, for which such vector $u' \in K$ exists, that $(u' - u^0)R < 0$ at some point $S$.

The following statement is valid: on the front of the regular solution discontinuity for every $\lambda : |\lambda| \leqslant 1/2$ (in this case $u^\lambda = (\lambda + 1/2)\, u^+ + (1/2 - \lambda)\, u^- \in K$ )

$$(u^* - u^\lambda)\, R \geqslant 0 \; . \tag{8}$$

Really, by the definition of the regular solution the inequality (8) is satisfied with $\lambda = 0$. Assuming in it as $u^*$ firstly vector $u^+$, then $u^-$, and summing inequalities obtained, one can state $[u]\; R = 0$. This results in (8) with every $\lambda$ with respect to $u^\lambda = u^0 + \lambda\, [u]$.

From geometrical point of view the proved statement means, that if some point $u^\lambda$ of the section in $m$ - dimensional space with ends $u^+$, $u^-$ is disposed strictly inside the set $K$, then due to the arbitrariness of variation $u^0 - u^\lambda$, included into (8), the equality $R = 0$ is satisfied. In the case the whole section belongs to the boundary $K$, vector $R$ is directed along its inner normal at any point of the section.

Thus, the regular solution in elasto-plastic flow theory can contain discontinuities of two types: elastic waves, defined by the system $R = 0$, and plastic waves, satisfying gradientality condition with respect to the yield surface. Note, that in the case of elastic perfectly-plastic medium functions $\Phi$ and $\Psi_s$ are quadratic, therefore ($d = 0$) any discontinuous solution is regular. The full system of strong discontinuity relationships for the elastic perfectly-plastic model, which follows from (8), was obtained by the method considered here in [7,8].

3. In a closed form the weak solutions with a strong discontinuity can be constructed in one-dimensional problem on propagation of plane shear waves, caused by application of tangential stress $\sigma_{13} = q$ to the surface of semi-space $x_1 \geqslant 0$. In this

problem the variational inequality (8) and the constraint on variations take the following form ($\mu$ is shear modulus, $\tau_s$ is yield limit):

$$(\sigma_{13}^{*} - \sigma_{13}^{\lambda})\left(\frac{1}{\mu} - \frac{1}{\rho c^2}\right)[\sigma_{13}] + (\tau_{13}^{*} - \tau_{13}^{\lambda})\,[\xi_{13}] + \\ + (\theta^{*} - \theta)\,[\eta] \geqslant 0\ ; \qquad \sigma_{13} - \tau_{13} \leqslant \tau_s\,\theta\ . \tag{9}$$

For the case of the active pure shift the hardening parameters $\tau_{13}$ and $\theta$ are proved to be dependent, i.e. they are connected by the equality $\eta = \tau_s\,\xi_{13}$, the comparison of which with a diagram of material deformation: $\xi_{13} + \sigma_{13}/\mu = F(\sigma_{13})$ allows us to determine a value of plastic potential

$$\Phi_0 = \int \xi_{13}\, d\tau_{13} + \eta\, d\theta = \int_0^{\tau_{13}+\tau_s\theta} \{\, F(\sigma) - \sigma/\mu\ \}\, d\sigma\ .$$

With a constant value of $q > \tau_s$ the regular weak solution of the problem includes two discontinuity lines: the elastic precursor $x_1 = c_s\, t$, propagating with a velocity of transverse waves $c_s = \sqrt{\mu/\rho}$, and plastic wave $x_1 = c_\tau\, t$. In the domain between discontinuities $\sigma_{13} = \tau_s$, behind the front of plastic wave $\sigma_{13} = q$. Its velocity can be found from (9) with account of Lagrangian factors rule

$$c_\tau = \left[\frac{1}{\rho}\,\frac{q - \tau_s}{F(q) - F(\tau_s)}\right]^{1/2} .$$

The condition of the plastic discontinuity realizability leads to the inequality

$$\int_{\tau_s}^{q} \sigma\, d\,F(\sigma) \leqslant \{F\,(q) - F\,(\tau_s)\}\,(q + \tau_s)\,/2\ ,$$

which is automatically satisfied for the down convex diagram of deformation. In the case of up convex diagram a self-similar problem solution is realized, depending on $x_1/t$, with singular discontinuity line: $x_1 = c_s t$.

More often the real diagram has down convex part of material flow and up convex part of active hardening. In this case the realizability condition does not allow a weak solution to include two plastic shock waves (Fig. 1), as far as two coinciding shocks, too. For sufficiently large value of $q$ the line of plastic discontinuity corresponds to tangent point $\tau_c$ on diagram, and $c_\tau$ is calculated from the same formula after substituting $q$ by $\tau_c$. For $q > \tau_c$ the central simple wave joins this line as shown in Fig. 2.

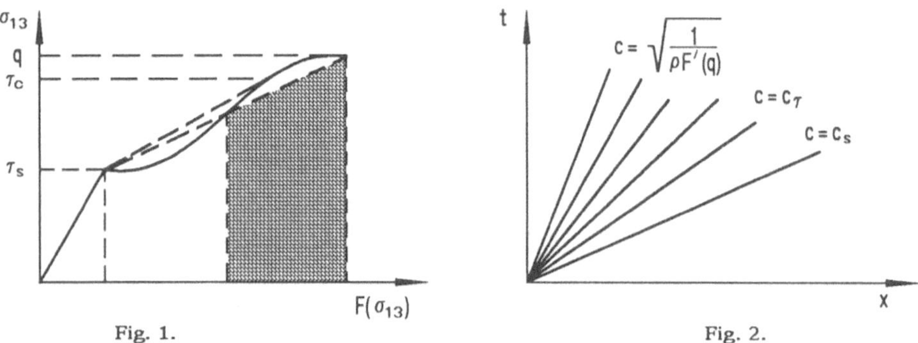


Fig. 1. Fig. 2.

Thus, the solution, describing a propagation of the shear loading wave, is defined only by a concrete form of the function $F$ and doesn't depend on the character of the material hardening.

## REFERENCES

[1] Mandel J. Plactic waves in infinity 3D medium. Mekhanika: Sb. perevodov, 1963, N 5, pp. 119-141.

[2] Bykovtsev G.I., Kretova L.D. On shock wave propagation in elastoplastic media. Prikl. Matemetika i mekhanika, 1972, v. 36, vyp. 2, pp. 106-116.

[3] Kukudzhanov V.N. On investigation of equations in dinamics of elastoplastis bodies under finite deformations. In: Nonlenear waves of deformation. Tallinn, 1978, pp. 102-105.

[4] Sedov L.I. Continuum mechanics. Part I, Moscow, Nauka, 1976, 536 p.

[5] Sadovskii V.M. Variational and quasivariational inequalities

in elastoplastic dynamics problems. Krasnoyarsk, 1990, 44 p., Dep. VINITI 3.4.90, N 1780 - V90.

[6] Wilkins M.L. Calculation of elastoplastic flows. In: Calculation methods in hydrodynamics. Moscow, Mir, 1967, pp. 212-263.

[7] Sadovskii V.M. On dynamic correctness of theory of elastic ideally plastis flow. In: Dinamika Sploshnoi Sredy, Novosibirsk, 1983, vyp. 63, pp. 147-151.

[8] Drugan W.J., Shen Yinong. Restrictions on dynamically propagating surfaces of strong discontinuity in elastic-plastic solids. - J. Mech. and Phys. Solids, 1987. V. 35, N 6. pp.771 - 787.

# THE JUSTIFICATION OF THE CONJUGATE CONDITIONS FOR THE EULER'S AND DARCY'S EQUATIONS

*V.V. Shelukhin*
*Lavrentyev Institute of Hydrodynamics*
*Novosibirsk 630090, RUSSIA*

The surface and subsoil waters joint motion problem is considered. Conjugate conditions for the Euler's and Darcy's equations are imposed. The boundary value problem is reduced to a system of integro-differential equations. A local existence and uniqueness theorem is proved in a scale of Banach spaces of analytic functions.

Key words: Euler's equations, Darcy's equations, conjugate conditions

## 1. INTRODUCTION.

In making the ecological examination of hydrotechnical projects the question arises about poisoning the open reservoir water with an unhealthy subsoil substance. This paper is concerned with a poisoning model in which the mass exchange of subsoil and surface waters is taken into account. We assume that surface flow is vortex-free and is governed by the Euler's equations. The filtration law is taken in the Darcy's form. The conjugate conditions for the two systems are formulated and their mathematical justification is given in two-dimensional problem.

Naturally, the flow in the open reservoir requires a free boundary problem approach. For the nonstationary Euler's equations this problem known as the Cauchy-Poisson one has been treated in [1]. Since the conjugate condition problem for the Euler's and Darcy's equations has never been solved before, we first investigate the flow under a solid roof because it is simpler than the free boundary problem. Here we show that the conjugate condition problem for the flow under the roof can be solved by the method applied to the Cauchy-Poisson problem [1]. So this paper is a step in developing the technique for solving the free boundary problem.

## 2. PROBLEM STATEMENT.

The reservoir water is assumed as occupying the layer

$$Q_1 = \left\{ x,y\colon x \in R,\ H(x) < y < h \right\}.$$

When in this layer, the admixture concentration $c$ varies according to equations

$$c_t + \bar{v}\cdot\nabla c = \nu_1 \Delta c, \tag{$1_c$}$$

$$\rho(\bar{v}_t + \bar{v}\nabla\bar{v}) = -\nabla p + \rho\bar{g},\quad div\ \bar{v} = 0,\ rot\ \bar{v} = 0. \tag{1}$$

Here $\bar{v}$ represents the velocity of flow, $p$ is the pressure, $\rho$ is the density.

The layer

$$Q_2 = \left\{ x,\ y\colon x \in R,\ -h < y < H(x) \right\}$$

is occupied by a porous medium. In this layer the admixture propagation is influenced by the rate of filtration $\bar{u} = m\bar{v}$ as follows:

$$c_t + \bar{v}\cdot\nabla c = \nu_2 \Delta c, \tag{$2_c$}$$

$$\bar{u} = -\frac{k}{m}\left( \nabla p + \rho\bar{g} \right),\quad div\ \bar{v} = 0. \tag{2}$$

The coefficients of porosity $(m)$, permeability $(k)$, viscosity $(\mu)$, diffusion $(\nu_i)$ and the density $\rho$ are considered to be constant.

We impose the conjugate conditions for systems $(1_c)$, (1) and $(2_c)$, (2) on the bottom

$$\Gamma = \left\{ x,\ y\colon x \in R,\ y = H(x) \right\}$$

of the reservoir as follows:

$$[c] = 0,\quad \left(c\bar{v} - \nu_1\nabla c\right)\bar{n}\Big|_{\Gamma_1} = \left(c\bar{u} - m\nu_2\nabla c\right)\bar{n}\Big|_{\Gamma_2}, \tag{$3_c$}$$

$$\bar{v}\cdot\bar{n}\Big|_{\Gamma_1} = \bar{u}\cdot\bar{n}\Big|_{\Gamma_2},\quad [p] = \lambda\bar{v}\cdot\bar{n}\Big|_{\Gamma_1}. \tag{3}$$

Here $\bar{n}$ is the normal vector to $\Gamma$ pointing to $Q_2$, $f\big|_{\Gamma_i}$ stands for the trace of the function $f: Q_i \to R$ on $\Gamma \subset \partial Q_i$,

$$[f] = f\big|_{\Gamma_1} - f\big|_{\Gamma_2}.$$

Formulas (3) imply that the mass flux has no jump across the line $\Gamma$ and the pressure jump is proportional to the flux, the parameter $\lambda$ is to be defined elsewhere.

We treat the lines $y = -h$, $y = h$ as a solid bottom and roof, so we put

$$\frac{\partial c}{\partial n}\bigg|_{y=-h} = \frac{\partial c}{\partial n}\bigg|_{y=h} = 0, \tag{$4_c$}$$

$$\bar{u}\cdot\bar{n}\big|_{y=-h} = \bar{v}\cdot\bar{n}\big|_{y=h} = 0. \tag{4}$$

The initial values of the functions $c$, $\bar{v}$ should be specified as

$$c\big|_{t=0} = c_0(x,y) \left[ c_0\big|_{Q_1} = 0 \right], \; \bar{v}\big|_{t=0} = v_0(x,y), \tag{$5_c$, 5}$$

The model proposed has an advantage. Had the vectors $\bar{u}$, $\bar{v}$ been known, the admixture concentration $c$ might have been found from relations $(1_c)-(5_c)$ which constitute the well-known diffraction problem for $c$ [2]. With this in mind we confine ourselves to problem (1)-(5) only.

Let $\varphi$ be the potential function for velocity, $\bar{v} = \nabla\varphi$, and let us introduce the function $q = p - y\rho g$. Then, due to the Cauchy-Lagrange integral, problem (1)-(5) can be rewritten as follows:

$$\nabla\varphi\big|_{Q_1} = 0, \; \Delta q\big|_{Q_2} = 0, \; \left(\frac{\partial\varphi}{\partial n} + \frac{k}{\mu}\frac{\partial\varphi}{\partial n}\right)\bigg|_{\Gamma} = 0,$$

$$\left\{\rho\left[\varphi_t + \frac{1}{2}|\nabla\varphi|^2 + 2gy\right] + q + \lambda\frac{\partial\varphi}{\partial n}\right\}\bigg|_{\Gamma} = 0, \tag{6}$$

$$\frac{\partial\varphi}{\partial n}\bigg|_{y=h} = \frac{\partial q}{\partial n}\bigg|_{y=-h} = 0,$$

$$\varphi \Big|_{t=0} = \Phi(x,y), \ \Delta\Phi \Big|_{Q_1} = 0.$$

We are concerned with the existence of periodical solutions, so we suggest that $\Phi(x,y)$ is $2\pi$ - periodical in $x$.

## 3. OPERATOR EQUATION.

Let $D_2$ be the trace operator, $D_2(\gamma) = v|_\Gamma$, where $v$ is the $2\pi$ - periodical in $x$ solution of the problem

$$\Delta v \Big|_{Q_2} = 0, \ \frac{\partial v}{\partial n} \Big|_{y=-h} = 0, \ \frac{\partial v}{\partial n} \Big|_\Gamma = \gamma, \ v(0,H(0)) = 0.$$

Here $\bar{n}\big|_\Gamma$ points to $Q_1$. The solvability condition is

$$\int_{\Gamma\cap(r\leqslant x\leqslant r+2\pi)} \gamma \, ds = 0, \quad r \in R. \tag{7}$$

We define the normal derivative operator $N_1$ as follows: $N_1(\sigma) = \frac{\partial w}{\partial n}\Big|_\Gamma$, where

$$\Delta w \Big|_{Q_1} = 0, \ \frac{\partial w}{\partial n} \Big|_{y=h} = 0, \ w \Big|_\Gamma = \sigma.$$

The operators $D_2$, $N_1$ enable us to transfer problem (6) to the line $\Gamma$. The vector function

$$\zeta(x,t) = (\xi,\eta), \ \xi = \varphi(x,H(x),t), \ \eta = \xi_x$$

satisfies the Cauchy problem $\dot{\zeta} = F(\zeta)$, $\zeta(0) = \zeta_0$, i.e.

$$\dot{\xi} = -\frac{\eta^2}{2(1+H_x^2)} - \frac{1}{2}\left[N_1(\xi)\right]^2 - 2gH - \frac{\mu}{k\rho} D_2 N_1(\xi) - \frac{\lambda}{\rho} N_1(\xi) \equiv G,$$

$$\dot{\eta} = -\frac{\partial}{\partial x} G, \tag{8}$$

$$\zeta(0) = (\xi_0,\eta_0) = \left[\Phi(x,H(x)), \frac{d}{dx}\Phi(x,H(x))\right].$$

We remark that the function $N_1(\xi)$ satisfies condition (7), so the operator $D_2 \circ N_1$ is well defined.

To tackle Cauchy problem (8) we introduce the scale $E = \cup E^2_\rho$ $(\rho > 0)$ of the Banach spaces $E^2_\rho = E_\rho \times E_\rho$, where $E_\rho$ is the space of analytic $2\pi$ - periodical functions $\xi(x)$ with the norm

$$\|\xi\|_\rho = \sum_{-\infty}^{\infty} |\xi_n| \, e^{\rho|n|}, \quad \xi = \sum_{-\infty}^{\infty} \xi_n \, e^{inx}.$$

The norm in $E^2_\rho$ is defined as follows: $|\zeta|_\rho = \|\xi\|_\rho + \|\eta\|_\rho$, $\zeta = (\xi,\eta)$.

It is easy to establish the following properties of the norm $\|\xi\|_\rho$:

N1) $\|\xi\|_\sigma \leqslant \|\xi\|_\rho, \quad \sigma \leqslant \rho,$

N2) $|\xi(x)| \leqslant \|\xi\|_\rho,$

N3) $\|\xi\eta\|_\rho \leqslant \|\xi\|_\rho \|\eta\|_\rho,$

N4) $\left\|\dfrac{\partial^k}{\partial x^k} \xi\right\|_\rho = \dfrac{\partial^k}{\partial x^k} \|\xi\|_\rho,$

N5) $\dfrac{\partial}{\partial\rho} \|\xi + \eta\|_\rho \leqslant \dfrac{\partial}{\partial\rho} \|\xi\|_\rho + \dfrac{\partial}{\partial\rho} \|\eta\|_\rho,$

N6) $\dfrac{\partial^k}{\partial\rho^k} \|\xi\|_\rho \leqslant k! \, (\sigma - \rho)^{-k} \, \|\xi\|_\sigma, \quad \sigma > \rho,$

N7) $\dfrac{\partial^2}{\partial\rho^2} \|\xi\|_\rho \geqslant 0.$

Given positive numbers $\rho_0$ and $r$, we introduce the sets

$$O_\rho(r) = \left\{ \zeta\colon \, |\zeta|_\rho \leqslant r \right\}, \quad O(r,\rho_0) = \bigcup_{0<\rho\leqslant\rho_0} O_\rho(r).$$

As proved by Ovsiannikov [1], the solvability of Cauchy problem (8) results from the following properties of the operator $F\colon O(r,\rho_0) \to E$:

$F_1$) $\zeta \in \mathcal{O}_\rho(r) \Rightarrow F(\zeta) \in E^2_\sigma, \quad \sigma < \rho \leqslant \rho_0,$

$F_2$) There is a positive number $L$ such that

$$|F(\zeta_1) - F(\zeta_2)|_\sigma \leqslant L \left[1 + \frac{\partial}{\partial\sigma}\right]\left\{ \left(1 + |\zeta_1|_\sigma + |\zeta_2|_\sigma\right)|\zeta_1 - \zeta_2|_\sigma\right\},$$

$\zeta_1, \zeta_2 \in \mathcal{O}_\rho(r), \quad \sigma < \rho \leqslant \rho_0.$

If the operator F satisfies these conditions and the $E_\rho$ - norm satisfies the condition N1, N5, N7, then the following theorem holds.

**THEOREM 1.** ([1]) Let $\zeta_0 \in \mathcal{O}_{\rho_0}(r)$ and

$$\int_0^{\rho_0} |F(0)|_\sigma \, d\sigma = M < \infty.$$

Then there is a unique solution to problem (8) in the scale $E$, and $\zeta(t) \in E^2_\rho$ when $t \geqslant 0$, $\rho > 0$, $\rho + at < \rho_0$, where

$$a = 4M + (1 + 2r)(1 + 4r \exp \rho_0)L.$$

Moreover, the solution satisfies the estimate

$$|\zeta(t) - \zeta_0|_\rho \leqslant \frac{1}{L}\int_\rho^{\rho+at} e^{\sigma-\rho}|F(\zeta_0)|_\sigma d\sigma.$$

## 4. ESTIMATES OF THE OPERATORS $N_1$ AND $D_2$.

To prove the existence and uniqueness of a solution to problem (8) it suffices to establish the properties $F_1$ and $F_2$ for the operators $N_1$ and $D_2$.

These operators can be defined implicitly as the solutions of linear integral equations. As far as the operator $N_1$ is concerned, the equation can be derived as follows. Given a harmonic function $w$ in $Q_1$, the analytic function $V(z) = w_x - iw_y$ $(z = x + iy)$ has an analytic extension on the strip

$$Q_1^* = \left\{ x \in R,\ H(x) < y < 2h - H(x) \right\}.$$

By the Cauchy theorem

$$\pi i V(z) = v.p.\int_{\partial Q_1^*} \frac{V(\zeta)}{\zeta - z}\, d\zeta, \quad z = x + iH(x). \tag{10}$$

Since

$$(1 + iH_x)V \Big|_\Gamma = w_s(x) + iw_N(x).$$

where $w_s = \frac{d}{dx}w(x,H(x))$, $w_N \sqrt{1 + H_x^2}\, \frac{\partial w}{\partial n}\Big|_\Gamma$ ($\bar{n}$ points to $Q_2$), then it follows from formula (10) that

$$\pi w_N(x) = \int_{-\infty}^{\infty} w_N(\xi)P(x,\xi)\, d\xi + \int_{-\infty}^{\infty} w_s(\xi)Q(x,\xi)\, d\xi,$$

where

$$P = \frac{H'(x)(\xi - x) - (2h - H(\xi) - H(x))}{(\xi - x)^2 + (2h - H(\xi) - H(x))^2} - \frac{H(\xi) - H(x) - H'(x)(\xi - x)}{(\xi - x)^2 + (H(\xi) - H(x)^2},$$

$$Q = \frac{\xi - x + H'(x)(2h - H(\xi) - H(x))}{(\xi - x)^2 + (2h - H(\xi) - H(x))^2} - \frac{\xi - x + H'(x)(H(\xi) - H(x))}{(\xi - x)^2 + (H(\xi) - H(x)^2},$$

An analogous integral equation holds for the operator $D_2$.

**LEMMA.** For any $h > 0$ there is a number $\delta(h) \in (0,1)$ and a function $m(h, \|H\|_\rho, \|H_x\|_\rho)$ such that the following estimates hold

$$\|N_1(\sigma)\|_\rho \leqslant m\|\sigma_x\|_\rho, \quad \|\frac{\partial}{\partial x} N_1(\sigma)\|_\rho \leqslant \frac{\partial}{\partial \rho}\left[m\|\sigma_x\|_\rho\right],$$

$$\|D_2(\gamma)\|_\rho + \|\frac{\partial}{\partial x} D_2(\gamma)\|_\rho \leqslant m\left[\|\gamma\|_\rho + \|\gamma_x\|_\rho\right],$$

when $\|H\|_\rho \leqslant \delta$, $\|H_x\|_\rho \leqslant \delta$.

**Remark.** The function $m$ and the number $\delta$ can be described explicitly.

The lemma is a crusial point. To prove it, we'v used the ideas from [1].

It follows from the lemma that the operators $N_1$ and $D_2$ satisfy the conditions $F_1$, $F_2$.

We formulate the main result.

**THEOREM 2.** Let $\rho_0$, $r$ be given positive numbers. Let the functions $H(x)$ and $\Phi(x,y)$ satisfy the conditions

$$\|H\|_{\rho_0} \leqslant \delta, \quad \|H_x\|_{\rho_0} \leqslant \delta, \quad \left( \Phi(x,H(x)), \frac{d}{dx} \Phi(x,H(x)) \right) \in O_{\rho_0}(r)$$

(here the number $\delta$ is chosen in the lemma). Then problem (8) is uniquely solvable in the scale $E$ and the solution $\zeta(t)$ belongs to the space $E^2_\rho$ if $\rho + ta < \rho_0$ ($a$ is defined in theorem 1, where $M$ is assumed to be equal to $4\rho_0 g\delta$). The estimate of type (9) holds for the solution $\zeta$ .

**REFERENCES.**

[1] Ovsiannikov L.V., Makarenko N.I., Nalimov V.I., et al. Nonlinear problems of surface and internal wave theory. Novosibirsk, Nauka, 1985 (in Russian).

[2] Ladyzhenskaya O.A., Solonnikov V.A., Ural'tseva N.N. Linear and quasilinear equations of parabolic type. Trans.Math. Monographs 23, Amer. Math. Soc., Providence, R.I., 1968.

# ON AN EVOLUTION PROBLEM OF THERMOCAPILLARY CONVECTION

*V.A.Solonnikov*
*LOMI, Fontanka 27*
*St.Petersbourg 191011, RUSSIA*

We consider a free boundary problem of incompressible viscous flow governing the motion of an isolated liquid mass. The liquid is subjected to capillary forces at the boundary, and the coefficient of the surface tension depends on the temperature satisfying the heat equation with convection and dissipation terms. It is shown that if the initial data are close to the rest state, i.e. the velocities and the temperature are small and the domain occupied by the liquid is close to a ball, then the problem possesses a unique classical solution which is determined for all positive values of time.

Key words: free boundary problem, Navier-Stokes equations, thermocapillary convection.

## 1. INTRODUCTION.

We consider the following free boundary problem: find a bounded domain $\Omega_t \subset R^3$, $t > 0$, vector field $\vec{v}(x,t) = (v_1, v_2, v_3)$ and functions $p(x,t)$ and $\theta(x,t)$ satisfying in $\Omega_t$ the following equations, initial and boundary conditions:

$$\vec{v}_t + (\vec{v}\cdot\nabla)\vec{v} - \nabla T = 0, \quad \nabla\cdot\vec{v} = 0,$$

$$\theta_t + (\vec{v}\cdot\nabla)\theta - \nabla\cdot\varkappa\nabla\theta = \lambda|S(\vec{v})|^2 \quad (x \in \Omega_t,\ t > 0),$$

$$v_t \Big|_{t=0} = \vec{v}_0(x), \quad \theta \Big|_{t=0} = \theta_0(x) \quad (x \in \Omega_0), \tag{1}$$

$$\frac{\partial\theta}{\partial n} + \beta\theta = 0,$$

$$T\vec{n} - \sigma(\theta)\, H\vec{n} = \nabla_\tau\sigma(\theta) \quad (x \in \Gamma_t \equiv \partial\vec{\Omega},\ t > 0).$$

Here $S(\vec{v})$ is the strain tensor with the elements

$$S_{ij} = \frac{\partial v_i}{\partial x_j} + \frac{\partial v_j}{\partial x_i},$$

$T(\vec{v},p) = -pI + \nu S(\vec{v})$ is the stress tensor, $\beta = const > 0$, $\lambda$, $æ$ and $\sigma$ are smooth positive functions of the temperature $\theta$, $\nabla = \left(\partial/\partial x_1, \partial/\partial x_2, \partial/\partial x_3\right)$, $\nabla_\tau \sigma = \nabla\sigma - \vec{n}\,\frac{\partial\sigma}{\partial n}$ is the gradient on the surface $\Gamma_t$, $\vec{n}$ is an exterior normal to $\Gamma_t$ and $H(x,t)$ is the twice mean curvature of $\Gamma_t$ which is negative if $\Omega_t$ is convex in the neighborhood of $x$. We suppose that $\Gamma_t$ is given by the equation

$$|x| = R(\omega,t), \quad \omega = \frac{\vec{x}}{|x|}, \tag{2}$$

and $R$ satisfies the kinematic boundary condition

$$RR_t = Rv_r - \vec{v}_\omega \cdot \nabla_\omega R, \tag{3}$$

where $v_r = v_\omega \cdot \frac{\vec{x}}{|x|}$ and $\vec{v}_\omega = \vec{v} - \frac{\vec{x}}{|x|}\, v_r$ are radial and angular component of $\vec{v}$, respectively.

In recent years problems of thermocapillary convection were studied by many authors (see the reference article by V.V.Pukhnachov [5] and the literature cited there.) M.V.Lagunova and the present author [2] proved the solvability of problem (1)-(3) in a finite time interval under the assumption that $\nu=const$, $æ=const$ and that there is no dissipation term $\lambda|S(\vec{v})|^2$ in the equation for $\theta$. ( In general case, the proof of a local existence theorem is essentially the same.) Under these assumptions on the equations (1) the solvability of problem (1) with initial data close to the rest state was established by the present author [10]. In the present communication we extend this result to a general system (1).

By $C^l(\Omega)$ with a positive non-integer $l$ we mean standard Holder spaces of functions (or vector fields) defined in the domain $\Omega \subset R^n$. The norm $C^l(\Omega)$ has a form

$$|u|_\Omega^{(l)} = \sum_{|j|<l} |D^j u|_\Omega + \sum_{|j|=[l]} \langle D^j u\rangle_\Omega^{(l-[l])},$$

where

$$|u|_\Omega = \sup_{x\in\Omega} |u(x)|, \quad D^j u = \frac{\partial^{|j|} u}{\partial x_1^{j_1} \partial x_2^{j_2} \partial x_3^{j_3}}, \quad |j| = j_1 + j_2 + j_3,$$

$$\langle v\rangle_{\Omega}^{(\lambda)} = \sup_{x,y\in\Omega} |x-y|^{-\lambda}|v(x)-v(y)| \quad (0<\lambda<1).$$

We also introduce anisotropic spaces

$$C^{l_1,l_2}(Q_T) = C(0,T;C^{l_1}(\Omega))\cap C(\Omega;C^{l_2}(0,T)), \quad Q_T = \Omega\times(0,T)$$

mainly in the case $l_1 = 2l_2 = l$. The norm in $C^{l,l/2}(Q_T)$ may be defined by the formula

$$|u|_{Q_T}^{(l,l/2)} = \sum_{|a|+2b\leqslant[l]} |D_x^a D_t^b u|_{Q_T} + \sum_{|a|+2b=[l]} \langle D_x^a D_t^b u\rangle_{Q_T}^{(l-[l],\frac{l-[l]}{2})} +$$

$$+ \sum_{|a|+2b=[l]-1} \langle D_x^a D_t^b u\rangle_{t,Q_T}^{(1+l-[l])/2}$$

with

$$\langle v\rangle_{Q_T}^{(\lambda,\lambda/2)} = \langle v\rangle_{x,Q_T}^{(\lambda)} + \langle v\rangle_{t,Q_T}^{(\lambda/2)} \equiv$$

$$= \sup_{t\in(0,T)} \langle v(\cdot,t)\rangle_{\Omega}^{(\lambda)} + \sup_{x\in\Omega} \langle v(x,\cdot)\rangle_{(0,T)}^{(\lambda/2)}.$$

For $0<\alpha,\ \beta<1$ we define the norm in $C^{\alpha,\beta}(Q_T)$ in the following ways:

$$\langle v\rangle_{Q_T}^{(\alpha,\beta)} = |u|_{Q_T} + \langle u\rangle_{x,Q_T}^{(\alpha)} + \langle u\rangle_{t,Q_T}^{(\beta)}.$$

We also introduce Sobolev spaces $W_2^k(\Omega)$ with the norm

$$\|u\|_{W_2^k(\Omega)}^2 = \sum_{|j|\leqslant k} \int_\Omega |D^j u|^2 dx \equiv \sum_{|j|\leqslant k} \|D^j u\|_{L_2(\Omega)}^2.$$

Holder and Sobolev spaces of functions defined on manifolds (for example, on $\Gamma_t$, $S_T = \Gamma_0\times(0,T)$ etc.) are defined in a similar way, by means of local maps and partition of unity.

The main result of the paper is as follows:

**THEOREM 1.** Suppose that the domain is defined by inequality $|x| < R(\omega,0)$, $\omega = \frac{x}{|x|}$ with $R\in C^{3+\alpha}(S_1)$, $\alpha\in(0,1)$,

$$S_1 = \{|\omega| = 1\}.$$

For arbitrary $\vec{v}_0 \in C^{2+\alpha}(\Omega_0)$, $\theta_0 \in C^{2+\alpha}(\Omega_0)$ satisfying natural compatibility conditions $\nabla\cdot\vec{v}_0 = 0$,

$$\nu(\theta_0)\Big[S(\vec{v}_0)\vec{n}_0 - \vec{n}_0(\vec{n}_0\cdot S(\vec{v}_0)\vec{n}_0)\Big] - \nabla_\tau\sigma(\theta_0)\Big|_{\Gamma_0} = 0, \quad \frac{\partial\theta_0}{\partial n} + \beta\theta_0\Big|_{\Gamma_0} = 0,$$

where $\vec{n}_0$ is exterior normal to $\Gamma_0$, and the condition

$$|\vec{v}_0|^{(2+\alpha)}_{\Omega_0} + |\theta_0|^{(2+\alpha)}_{\Omega_0} + |R(\cdot,0) - R_0|^{(3+\alpha)}_{S_1} \leqslant \varepsilon_0 \tag{4}$$

with $R_0 = (3|\Omega_0|/4\pi)^{1/3}$ and with a small $\varepsilon_0 > 0$, there exists a unique solution of (1), (3) with the following properties:

$$R \in C^{3+\alpha,(3+\alpha)/2}(S_1\times(0,\infty)),\ R_t \in C^{2+\alpha,(2+\alpha)/2}(S_1\times(0,\infty)),$$

$$\vec{v} \in C^{2+\alpha,(2+\alpha)/2}(Q^\infty), \quad \nabla p \in C^{\alpha,\alpha/2}(Q^\infty), \quad p \in C^{1+\alpha,(1+\alpha)/2}(\Sigma^\infty),$$

where $Q^\infty = \Big\{x \in Q_t,\ t > 0\Big\}$, $\Sigma^\infty = \Big\{x \in \Gamma_t,\ t > 0\Big\}$. Moreover,

$$|R(\cdot,t) - R_0|^{(3+\alpha)}_{S_1} + |\vec{v}(\cdot,t)|^{(2+\alpha)}_{\Omega_t} + |\theta(\cdot,t)|^{(2+\alpha)}_{Q_t} \leqslant c\varepsilon_0. \tag{5}$$

**REMARK.** Condition $\vec{v} \in C^{2+\alpha,(2+\alpha)/2}(Q^\infty)$ is understood in the following sense: in view of $|R(\cdot,t) - R_0|^{(3+\alpha)}_{S_1} \leqslant c\varepsilon_0$, there exists a smooth invertible mapping of $Q^\infty$ onto $Q_\infty = \Omega_0\times(0,\infty)$ and $\vec{v} \in C^{2+\alpha,(2+\alpha)/2}(Q_\infty)$.

## 2. PROOF OF THEOREM 1.

Theorem 1 is proved in several steps.

**Step 1. Conservation laws.** The solution of problem (1)-(3) is subjected to the following conservation laws:

$$|\Omega_t| = |\Omega_0| = \frac{4}{3}\pi R_0^3 \quad \text{(conservation of mass)},$$

$$\int_{\Omega_t} \vec{v}(x,t)\,dx = \int_{\Omega_0} \vec{v}_0(x)\,dx \quad \text{(conservation of momentum)}$$

$$\int_{\Omega_t} [\vec{v}\times\vec{x}]\, dx = \int_{\Omega_0} [\vec{v}_0\times\vec{x}]\, dx \quad \text{(conservation of angular momentum)}$$

$$\frac{1}{2}\int_{\Omega_t} |\theta(x,t)|^2 dx + \int_0^t\int_{\Omega_\tau} æ|\nabla\theta|^2 dx\, d\tau + \int_0^t\int_{\Gamma_\tau} \beta|\theta|^2 dx\, d\tau =$$

$$= \int_0^t\int_{\Omega_\tau} \lambda|S(\vec{v})|^2\,\theta\, dx\, d\tau + \frac{1}{2}\int_{\Omega_0} |\nabla\theta_0|^2 dx, \tag{6}$$

$$\frac{1}{2}\int_{\Omega_t} |\vec{v}(x,t)|^2 dx + \frac{1}{2}\int_0^t\int_{\Omega_\tau} \nu\, |S(\vec{v})|^2 dx\, d\tau + \sigma(0)\Big(|\Gamma_t| - 4\pi R_0^2\Big) +$$

$$+ \int_0^t I(\tau)\, d\tau = \frac{1}{2}\int_{\Omega_0} |\vec{v}_0(x)|^2 dx + \sigma(0)\Big(|\Gamma_0| - 4\pi R_0^2\Big) \tag{7}$$

(conservation of energy). Here $|\Gamma_t|$ is the surface area of $\Gamma_t$,

$$I(t) = \int_{\Gamma_t} \Big(\sigma(\theta) - \sigma(0)\Big)\frac{1}{\sqrt{g}}\frac{\partial}{\partial t}\sqrt{g}\, dS =$$

$$= \int_{S_1} \Big(\sigma(\theta) - \sigma(0)\Big)\frac{\partial}{\partial t}\left[R(\omega,t)\sqrt{R^2(\omega,t) + |\nabla_\omega R|^2}\right] d\omega$$

and $g$ is the determinant of the metric tensor of $\Gamma$. The relationship $|\Omega_t| = |\Omega_0|$ is a consequence of the incompressibility of the liquid (i.e. of the equation $\nabla\cdot\vec{v} = 0$), and other conservation laws are obtained essentially by the same arguments as in [7], theorem 5.

Introducing, if necessary, a new coordinate system, we may satisfy the equations

$$\int_{\Omega_t} \vec{v}(x,t)^2 dx = 0, \tag{8}$$

$$\int_{\Omega_t} \vec{x}\, dx = 0, \tag{9}$$

$$\int_{\Omega_t} [\vec{v}\times\vec{x}]\, dx = (0,0,\mu).$$

The first two equations mean that the center of mass of the liquid does not move and it coincides with the origins of the coordinate system $\{x\}$. The third equation shows that the total angular momentum is parallel to $x_3$-axis. Under conditions (8), (9) the relationship (7) implies

$$\frac{1}{2}\int_{\Omega_t} |\vec{v}(x,t)|^2 dx + \frac{1}{2}\int_0^t\int_{\Omega_\tau} \nu\, |S(\vec{v})|^2 dx\, d\tau +$$

$$+ C_1\sigma(0) \int_{S_1} \Big(|R - R_0|^2 + |\nabla_\omega R|^2\Big) d\omega + \Big| \int_0^t I(\tau)\, d\tau \Big| \leqslant \tag{10}$$

$$\leqslant \frac{1}{2}\int_{\Omega_0} |\vec{v}_0(x)|^2 dx + C_2\sigma(0)\int_{S_1} \Big(|R(\omega,0) - R_0|^2 + |\nabla_\omega R(\omega,0)|^2\Big) d\omega,$$

provided that

$$|R(\omega,\tau) - R_0| + |\nabla_\omega R(\omega,\tau)| \leqslant \delta R_0 \quad (0 \leqslant \tau \leqslant t) \tag{11}$$

with sufficiently small $\delta > 0$.

**Step 2. Transformation of problem (1)-(3).** As in the case $\sigma = const$, the limiting regime in the problem (1)-(3) (as $t \to \infty$) corresponds to the rotation of the liquid as a rigid body about $x_3$-axis with a total angular momentum $\mu$, i.e.

$$\vec{v}(x,\infty) = \vec{v}_\infty(x) = \alpha\vec{\eta}_3,\ \vec{\eta}_3(x) = (x_2,-x_1,0),$$

$$p_\infty(x) = \alpha^2(x_1^2 + x_2^2)/2 + p_1,\ p_1 = const,\quad \theta_\infty(x) = 0.$$

The domain $\Omega_t$ will tend to an equilibrium figure $\Omega_\infty$ of a rotating fluid whose surface $\Gamma_\infty$ is defined by the equation $|x| = r\left(\frac{x}{|x|}\right)$, where $r(\omega)$ is a rotationally symmetric function on a unit sphere $S_1$ (i.e. independent of the angle $\varphi$, $tg\,\varphi = x_2/x_1$), even with respect to $\omega_3$ and satisfying the equation

$$\sigma(0)H_\infty + \frac{\alpha^2}{2}\left(\omega_1^2 + \omega_2^2\right)r^2 + p_1 = 0, \quad \omega \in S_1, \tag{12}$$

where $H_\infty$ is the twice mean curvature of $\Gamma_\infty$ and

$$\alpha = \mu\left[\int_{\Omega_\infty}(x_1^2 + x_2^2)\,dx\right]^{-1}.$$

As $t \to \infty$, the temperature $\theta$ tends to $0$. It is proved in [9], theorem 5.1, that for small $\mu$ the equation (12) has a unique solution $(r,p_1)$ in the class described above such that $|\Omega_\infty| = 4\pi R_0^3/3$.

Let us introduce new functions $\vec{w} = \vec{v} - \vec{v}_\infty$, $s = p - p_\infty$. They satisfy the following equations and boundary conditions:

$$\vec{w}_t + (\vec{v}\cdot\nabla)\vec{w} - \nabla T(\vec{w},s) = -\alpha(\vec{w}\cdot\nabla)\vec{\eta}_3(x), \; \nabla\cdot\vec{w} = 0 \quad (x \in \Omega_t), \tag{13}$$

$$T(\vec{w},s)\vec{n} - \sigma(\theta)\,H\vec{n} = \nabla_\tau\sigma(\theta) + \left(\alpha^2(x_1^2 + x_2^2)/2 + p_1\right)\vec{n} \quad (x \in \Gamma_t).$$

Since $\vec{\eta}_3\cdot\nabla_\omega r = 0$, the equation (3) may be rewritten in the form

$$R_t = w_r - \vec{w}_\omega\frac{\nabla_\omega R}{R} - \frac{\alpha}{R}\vec{\eta}_3\cdot\nabla_\omega(R - r).$$

Using this formula we can show that

$$|I(t)| \leqslant C_3\|\theta\|_{L_2(\Gamma_t)}\left(\|\vec{w}\|_{W_2^2(\Gamma_t)} + |\mu|\,\|R - r\|_{W_2(S_1)}\right),$$

provided that (11) holds and $|D_\omega^2 R(\omega,t)| \leqslant C_4 R_0$.

**3. Step 3. Local estimates of $(\vec{w},s,\theta)$ (with respect to time).**

Suppose that the solution of the problem (1)-(3) is defined for $t \in (0,t_0 + 1)$, and let us estimate higher order derivatives of $\vec{w}$, $s$, $\theta$ in the interval $(t_0 + 1/2, t_0 + 1)$ in terms of lower order norms of the same functions in a larger interval $(t_0,t_0 + 1)$. To this end, we write (13) and the equation for $\theta$ in (1) in Lagrangian coordinates $(\xi,t)$, $\xi \in \Omega_{t_0}$ taking $t_0$ as the initial moment of time. As shown in [9], § 6, these coordinates are related to Eulerian coordinates by the formula

$$\vec{x} = Z(t)\vec{y}, \quad \vec{y} = \vec{\xi} + \int_0^t Z^{-1}(\tau)\, \vec{w}(\xi,\tau)\, d\tau,$$

where

$$Z(t) = \begin{pmatrix} \cos\alpha(t - t_0) & \sin\alpha(t - t_0) & 0 \\ -\sin\alpha(t - t_0) & \cos\alpha(t - t_0) & 0 \\ 0 & 0 & 1 \end{pmatrix}.$$

We also transform the boundary condition in (13) using the well-known formula

$$H\vec{n} = \Delta(t)\vec{x} = Z(t)\, \Delta(t)\vec{y}$$

($\Delta(t)$ is the Laplace-Beltrami operator on $\Gamma_t$) and separating the normal and tangential parts in this condition. After elementary calculations (see also [7], theorem 6, [9], theorem 4.1 and [10], § 4 we can rewrite the above relationships in the form

$$\frac{\partial\hat{\theta}}{\partial t} - \hat{\nabla}\cdot\varkappa(\hat{\theta})\hat{\nabla}\hat{\theta} = \lambda(\hat{\theta})|S(\hat{\vec{\omega}})|^2,$$

$$\frac{\partial\hat{\vec{w}}}{\partial t} - \hat{\nabla}\nu\, \hat{S}(\hat{\vec{w}})\vec{n} + \hat{\nabla}\hat{s} = -\alpha(\hat{\vec{w}}\cdot\hat{\nabla})\vec{\eta}_3(Zy), \quad \nabla\cdot\hat{\vec{w}} = 0,$$

$$\nu(\theta)\Pi_0\Pi\hat{S}(\hat{\vec{w}})\vec{n} = \sigma'_\theta(\hat{\theta})\Pi_0\Pi\hat{\nabla}\hat{\theta},$$

$$Z(t)\vec{n}_0\cdot\hat{T}(\hat{\vec{w}},\hat{s})\vec{n} - \sigma(\hat{\theta})\vec{n}_0\cdot\Delta(t)\vec{y} = Z(t)\vec{n}_0\cdot\sigma'_\theta(\hat{\theta})\Pi\hat{\nabla}\hat{\theta},$$

where

$\hat{\theta}(\xi,t) = \theta(x(\xi,t),t)$, $\hat{\vec{w}}(\xi,t) = \vec{w}(x(\xi,t),t)$, $\hat{s}(\xi,t) = s(x(\xi,t),t)$,

$$\hat{\nabla} = \left[\frac{\partial\xi}{\partial x}\right]^T\nabla = Z(t)\left[\frac{\partial\xi}{\partial y}\right]^T\nabla, \quad \left[\hat{S}(\vec{w})\right]_{ij} = \sum_{m=1}^{3}\left[A_{mi}\frac{\partial w_j}{\partial\xi_m} + A_{mj}\frac{\partial w_i}{\partial\xi_m}\right],$$

$$A_{mi} = \frac{\partial\xi_m}{\partial x_i}, \quad \hat{T}(\vec{w},s) = sI + \nu\hat{S}(\vec{w}),$$

$\Pi\vec{\omega} = \vec{\omega} - \vec{n}(\vec{\omega}\cdot\vec{n})$ and $\Pi_0\vec{\omega} = \vec{\omega} - \vec{n}_0(\vec{\omega}\cdot\vec{n}_0)$ are projections to tangent planes to $\Gamma_t$ and $\Gamma_{t_0}$, respectively.

Let $\zeta(t)$ be a smooth monotone function vanishing for $t < t_0 + 1/4$ and equal to one for $t > t_0 + 1/2$. Rearranging terms in boundary conditions it is not hard to verify that $\vec{w}' = \hat{\vec{w}}\zeta$, $s' = \hat{s}\zeta$, $\theta' = \hat{\theta}\zeta$ may be considered as solutions of the following initial-boundary value problem in $Q = \Omega_{t_0}\times(t_0,t_0+1)$:

$$\frac{\partial\theta'}{\partial t} - \hat{\nabla}\cdot\varkappa(\hat{\theta})\hat{\nabla}\theta' = \lambda(\hat{\theta})\, S(\hat{\vec{\omega}}) :\hat{S}(\vec{w}') + \hat{\theta}\zeta_t' \equiv f(\xi,t), \tag{14}$$

$$\theta'\Big|_{t=t_0} = 0, \quad \vec{n}\cdot\hat{\nabla}\theta' + \beta\theta'\Big|_{\xi\in\Gamma_{t_0}} = 0,$$

$$\frac{\partial\vec{w}}{\partial t} - \hat{\nabla}\nu(\hat{\theta})\,\hat{S}(\vec{w}) + \hat{\nabla}s' = -\alpha(\vec{\omega}'\cdot\hat{\nabla})\vec{\eta}_3(x(\xi,t)) + \hat{\vec{w}}\zeta_t' \equiv \vec{f}(\xi,t),$$

$$\hat{\nabla}\cdot\vec{w}' = 0, \tag{15}$$

$$\vec{w}'\Big|_{t=t_0} = 0,$$

$$\nu(\hat{\theta})\Pi_0\Pi\hat{S}(\vec{w}')\vec{n}\Big|_{\xi\in\Gamma_{t_0}} = \sigma_\theta'(\hat{\theta})\Pi_0\Pi\hat{\nabla}\theta'\Big|_{\xi\in\Gamma_{t_0}} \equiv \vec{d}(\xi,t),$$

$$\vec{n}_0\cdot\hat{T}(\vec{\omega}',s')\vec{n} - \sigma(0)\vec{n}_0\cdot\Delta(t)\int_{t_0}^{t}\vec{w}'\,d\tau\Big|_{\xi\in\Gamma_{t_0}} = b + \int_{t_0}^{t} B\,d\tau.$$

Here

$$b(\xi,t) = (\vec{n}_0 - Z(t)\vec{n}_0)\cdot\hat{T}(\vec{w}',s')\vec{n} + \sigma'_\theta(\hat{\theta})(Z(t)\vec{n}_0 - \vec{n})\Pi\hat{\nabla}\theta',$$

$$B(\xi,t) = -\sigma(0)\vec{n}_0\cdot\dot{\Delta}(t)\int_{t_0}^{t}\vec{w}'\,d\tau +$$

$$+ \left[\sigma(\vec{\theta}) - \sigma(0)\right]\vec{n}_0\cdot Z^{-1}(t)\Delta(t)\vec{w}' + \sigma(0)\vec{n}_0\cdot\left[Z^{-1}(t) - I\right]\Delta(t)\vec{w}' +$$

$$+ \zeta'_t Z(t)\vec{n}_0\cdot\hat{T}(\vec{w}',s')\vec{n} + \sigma'_\theta(\hat{\theta})\left[\vec{n}_0\cdot\Delta(t)\vec{y}\right]\left[\frac{\partial\theta'}{\partial t} - \hat{\theta}\zeta'_t\right] +$$

$$+ \zeta(t)\sigma(\hat{\theta})\left[\vec{n}_0\cdot\dot{\Delta}(t)\vec{y}\right] + \zeta\frac{\partial}{\partial t}\left[\left[\frac{\alpha^2}{2}\left(x_1^2 + x_2^2\right) + p_1\right]\left(Z\vec{n}_0\cdot\vec{n}\right)\right] -$$

$$- \zeta'_t\left(Z\vec{n}_0-\vec{n}\right)\Pi\hat{\nabla}\sigma(\hat{\theta}), \tag{16}$$

$$\dot{\Delta}(t) = \frac{\partial}{\partial t}\Delta(t).$$

Suppose that $\hat{\theta}$ and $\vec{u}(\xi,t) = \alpha\vec{\eta}_3(x(\xi,t)) + \vec{\hat{w}}(\xi,t) = \vec{v}(x(\xi,t),t)$ satisfy the inequality

$$|\vec{u}|_Q^{(2+\alpha,1+\alpha/2)} + |\hat{Q}|_Q^{(1+\alpha,(1+\alpha)/2)} \leqslant \delta_1 \tag{17}$$

with a small $\delta_1 > 0$. Then we may consider (14) as linear initial-boundary value problems for equations with variable coefficients; in virtue of (17), these coefficients are close to constants and have uniformly bounded Hölder norms. It follows from the theory of parabolic initial-boundary value problems (see [6], and [1], ch.VII) that

$$|\theta'|_Q^{(2+\alpha,1+\alpha/2)} \leqslant c_4|f|_Q^{(\alpha,\alpha/2)} \leqslant$$

$$\leqslant c_5\left[|\hat{\theta}|_Q^{(\alpha,\alpha/2)} + |\vec{w}'|_Q^{(1+\alpha,(1+\alpha)/2)}\right]. \tag{18}$$

The solution of problem (15) is estimated in [3,4] (in particular, the case of variable coefficient of viscosity $\nu$ is considered in [3]. It is shown that

$$|\vec{w}'|_Q^{(2+\alpha,1+\alpha/2)} + |\nabla s'|_Q^{(\alpha,\alpha/2)} + |s'|_S^{(1+\alpha,(1+\alpha)/2)} \leqslant$$
$$\leqslant c_7\Big(|\vec{f}|_Q^{(\alpha,(\varepsilon+\alpha)/2)} + |\vec{d}|_S^{(1+\alpha,(1+\alpha)/2)} +$$
$$+ |b|_S^{(1+\alpha,(1+\alpha)/2)} + |B|_S^{(\alpha,\alpha/2)}\Big) \tag{19}$$

with arbitrary $\varepsilon \in (0,1-\alpha)$; $S = \Gamma_{t_0}\times(t_0,t_0+1)$.

Functions $\vec{f}$, $\vec{d}$, $b$, $B$ contain higher order derivatives of $\vec{w}'$ and $s'$ only with small coefficients (such as $Z(t)-I$, $\vec{n}-\vec{n}_0$, $\sigma(\hat{\theta})-\sigma(0)$ etc.) which have an order of magnitude $\delta_1+|\mu|$. Therefore after elementary but lengthy calculations involving estimates of every term in 9150, 9160 and application of interpolation inequalities of the type

$$|\vec{w}|_Q^{(l,l/2)} \leqslant \varepsilon|\vec{w}|_Q^{(2+\alpha,1+\alpha/2)} + c_8(\varepsilon)\|\vec{w}\|_{L_2(Q)} \quad (0 < l < 2+\alpha)$$

we can deduce from (18), (19) the estimate

$$X'^2 \leqslant c_9\Big(\delta_1 + |\mu| + \varepsilon\Big)^2 X^2 + c_{10}(\varepsilon)\Big(\|\hat{\vec{w}}\|^2_{L_2(Q)} + \|\hat{\theta}\|^2_{L_2(Q)} + \|\hat{s}\|^2_{L_2(S)}\Big), \tag{20}$$

where

$$X = |\hat{\vec{w}}|_Q^{(2+\alpha,1+\alpha/2)} + |\nabla\hat{s}|_Q^{(\alpha,\alpha/2)} + |\hat{s}|_S^{(1+\alpha,(1+\alpha)/2)} + |\hat{\theta}|_Q^{(2+\alpha,1+\alpha/2)},$$

$$X' = |\hat{\vec{w}}|_{Q'}^{(2+\alpha,1+\alpha/2)} + |\nabla\hat{s}|_{Q'}^{(\alpha,\alpha/2)} + |\hat{s}|_{S'}^{(1+\alpha,(1+\alpha)/2)} + |\hat{\theta}|_{Q'}^{(2+\alpha,1+\alpha/2)},$$

$$Q' = \Omega_{t_0}\times(t_0+1/2,\ t_0+1),\quad S' = \Gamma_{t_0}\times(t_0+1/2,\ t_0+1).$$

**Step 4. The completion of the estimate.**

The normal component of the boundary condition in (13) has a form

$$\sigma(\theta)H + \frac{\alpha^2}{2}R^2(\omega_1^2+\omega_2^2) + p_1 = \vec{n}\cdot T(\vec{w}',s)\vec{n}\,\Big|_{|x|=R(\omega,t)} \tag{21}$$

Taking the difference of (21) and (12) multiplied by $R$ and $r$, respectively, we obtain

$$\sigma(\theta)\left[RH[R] - rH[r]\right] + \frac{\alpha^2}{2}\left[R^3 - r^3\right](\omega_1^2 + \omega_2^2) + p_1(R - r) =$$

$$= R\vec{n}\cdot T(\vec{w}', s')\vec{n} + (\sigma(0) - \sigma(\theta))rH[r], \tag{22}$$

where

$$H[R] = \frac{1}{R}\left[\frac{\Delta_{S_1} R}{\sqrt{R^2 + |\nabla_\omega R|^2}} + \frac{\Delta_\omega R}{\sqrt{R^2 + |\nabla_\omega R|^2}}\cdot\nabla_\omega R - \frac{2R}{\sqrt{R^2 + |\nabla_\omega R|^2}}\right]$$

is the twice mean curvature of the surface $\Gamma_t$: $|x| = R(\omega, t)$ expressed in terms of $R$ and its derivatives. Equation (22) can be written as an elliptic equation for $R - r$ on $S_1$ with the principal part $\sigma(\theta)(R^2 + |\nabla_\omega R|^2)^{-1/2}\Delta_{S_1}(R - r)$ ($\Delta_{S_1}$ is the Laplacian on $S_1$). It was proved in [9], [10] on the basis of equations (15), (21), (22) that $R - r$ and $\hat{s}$ satisfy the estimate

$$\|R - r\|^2_{W_2(S_1)} + \|\hat{s}\|^2_{L_2(\Gamma_{t_0})} \leqslant$$

$$\leqslant c_{11}\left[\|\hat{\vec{w}}\|^2_{L_2(\Omega_{t_0})} + \|\hat{\vec{w}}\|^2_{W_2^2(\Omega_{t_0})} + \|\hat{\theta}\|^2_{L_2(\Gamma_{t_0})}\right] \tag{23}$$

(the last norm appears due to the presence of the term $(\sigma(\theta) - \sigma(0))rH[r]$ in (22). We use (23) to evaluate the norm of $\hat{s}$ in (20) and we also apply the interpolation inequality

$$\|\hat{\vec{w}}\|^2_{W_2^{2,1}(Q)} + \|\hat{\theta}\|^2_{L_2(S)} \leqslant \varepsilon X^2 + c_{12}(\varepsilon)\left[\|\hat{\vec{w}}\|^2_{L_2(Q)} + \|\hat{\theta}\|^2_{L_2(Q)}\right].$$

This leads to

$$X'^2 \leqslant c_{13}\left[\delta_1 + |\mu| + \varepsilon\right]^2 X^2 + c_{14}(\varepsilon)\left[\|\hat{\vec{w}}\|^2_{L_2(Q)} + \|\hat{\theta}\|^2_{L_2(Q)}\right].$$

The norm of $\hat{\vec{w}}$ can be estimated by the Korn's inequality in the form presented in [8], namely,

$$\|\vec{w}\|^2_{L_2(\Omega_t)} \leqslant c_{15}\left\{\sum_{i=1}^{3}\left[\int_{\Omega_t}\vec{w}\cdot\vec{\eta}_i dx\right]^2 + \|S(\vec{w})\|^2_{L_2(\Omega_t)}\right\},$$

where

$$\vec{\eta}_1 = (0,x_3,-x_2),\ \vec{\eta}_2 = (-x_3,0,x_1),\ \vec{\eta}_3 = (x_2,-x_1,0).$$

The scalar products $\int_{\Omega_t}\vec{w}\cdot\vec{\eta}_i dx$ were evaluated in [9], § 6, by $c_{16}|\mu|\,\|R-r\|_{L_2(S_1)}$ so, applying (23) again we arrive at

$$X'^2 \leqslant c_{17}\left(\delta_1+|\mu|+\varepsilon\right)^2X^2 + c_{19}(\varepsilon)\int_{t_0}^{t_0+1}\left[\|S(\vec{w})\|^2_{L_2(\Omega_t)} + \|\theta\|^2_{L_2(\Omega_t)}\right]dt. \tag{24}$$

Let us introduce the following notations:

$$t_k = \frac{k}{2},\ Q_k = \Omega_{t_k}\times(t_k,t_k+1),\ Q'_k = \Omega_{t_k}\times(t_k+1/2,t_k+1),$$

$$S_k = \Gamma_{t_k}\times(t_k,t_k+1),\ S'_k = \Gamma_{t_k}\times(t_k+1/2,t_k+1)\quad (k = 0,\ 1,\ \ldots),$$

$$X_k = |\hat{\vec{w}}_k|^{(2+\alpha,1+\alpha/2)}_{Q_k} + |\nabla\hat{s}_k|^{(\alpha,\alpha/2)}_{Q_k} + |\hat{s}_k|^{(1+\alpha,(1+\alpha)/2)}_{S_k} + |\hat{Q}_k|^{(2+\alpha,1+\alpha/2)}_{Q_k},$$

$$X'_k = |\hat{\vec{w}}_k|^{(2+\alpha,1+\alpha/2)}_{Q'_k} + |\nabla\hat{s}_k|^{(\alpha,\alpha/2)}_{Q'_k} + |\hat{s}_k|^{(1+\alpha,(1+\alpha)/2)}_{S'_k} + |\hat{Q}_k|^{(2+\alpha,1+\alpha/2)}_{Q'_k},$$

where $\hat{\vec{w}}_k$, $\hat{s}_k$, $\hat{\theta}$ are $\vec{w}$, $s$, $\theta$ expressed as functions of the Lagrangian coordinates $\xi \in \Omega_{t_k}$. Suppose that the solution of the problem (1)-(3) is defined for $t \leqslant t_N + 1$ and that conditions

(11) and (17) are satisfied for all $t \leqslant t_N + 1$ and $Q_k$, $k \leqslant N$, respectively. Then inequality (24) holds for all $Q_k$ with the same constants $c_{17}$, $\varepsilon$, $c_{19}$. Moreover, since Lagrangian coordinates $\xi^{(k)} \in \Omega_{t_k}$ and $\xi^{(k+1)} \in \Omega_{t_{k+1}}$ are related to each other by an invertible transformation

$$\vec{\xi}^{(k+1)} = \vec{\xi}^{(k)} + \int_{t_k}^{t_{k+1}} \vec{v}(x(\xi^{(k)},t),t)dt,$$

the norms $\left(\sum_1^N X_k^2\right)^{1/2}$ and $\left(\sum_0^N X_k'^2\right)^{1/2}$ are equivalent, and from (24) we easily obtain

$$\left(\sum_0^N X_k'^2\right)^{1/2} \leqslant c_{20} \left(\delta_1 + |\mu| + \varepsilon\right)^2 \left(\sum_0^N X_k'^2 + X_0^2\right) +$$

$$+ c_{19}(\varepsilon) \int_0^{t_N+1} \left(\|S(\vec{w})\|^2_{L_2(\Omega_t)} + \|\theta\|^2_{L_2(\Omega_t)}\right) dt.$$

To estimate the integral term we apply inequalities (6) and (7). This gives

$$\int_0^{t_N+1} \|\theta\|^2_{L_2(\Omega_t)}dt \leqslant c_{12}\left[\|\theta_0\|^2_{L_2(\Omega_0)} + \delta_1 \int_0^{t_N+1} \|S(\vec{w})\|^2_{L_2(\Omega_t)}dt\right],$$

$$\int_0^{t_N+1} \|S(\vec{w})\|^2_{L_2(\Omega_t)}dt \leqslant c_{22}\left[F^2 + \int_0^{t_N+1} |I(t)|\, dt\right] \leqslant$$

$$\leqslant c_{22}F^2 + \varepsilon^2 \int_0^{t_N+1} \left(\|\vec{w}_t\|^2_{L_2(\Omega_t)}dt + \|\vec{w}\|^2_{W_2^2(\Omega_t)}\right) dt +$$

$$+ c_{23}\varepsilon^{-2} \int_0^{t_N+1} \|\theta\|^2_{L_2(\Omega_t)}dt,$$

so finally

$$\sum_{k=0}^{N} X_k'^2 \leqslant \left[c_{24}\left(\delta_1 + |\mu| + \varepsilon\right)^2 + c_{25}(\varepsilon)\delta\right] \sum_{k=0}^{N} X_k'^2 + c_{26}\left(F^2 + X_0^2\right),$$

where

$$F^2 = \|\vec{v}_0\|^2_{L_2(\Omega_0)} + \|\theta_0\|^2_{L_2(\Omega_0)} + \|R(\cdot,0) - R_0\|^2_{W_2^1(S_1)}.$$

Suppose that $\delta$, $\mu$ are so small and choose $\varepsilon$ so small that

$$c_{24}(\delta_1 + |\mu| + \varepsilon)^2 + c_{25}(\varepsilon)\delta_1 \leqslant 1/2.$$

Then we can get a uniform estimate

$$\sum_{k=0}^{N} X_k'^2 \leqslant 2c_{26}\left(F^2 + X_0^2\right). \tag{25}$$

**Step 5. Completion of the proof of theorem 1.**

According to local existence theorem (for the case of constant $\nu$, $\varkappa$ and $\lambda = 0$ it is established in [2]), problem (1)-(3) is solvable in a finite time interval, say, in $(0,1)$ and

$$X_0 = |\hat{\vec{w}}|^{(2+\alpha,1+\alpha/2)}_{Q_0} + |\nabla\hat{S}|^{(\alpha,\alpha/2)}_{Q_0} + |\hat{S}|^{(1+\alpha,(1+\alpha)/2)}_{S_0} +$$

$$+ |\hat{\theta}|^{(2+\alpha,1+\alpha/2)}_{Q_0} + \leqslant c_{27}\left(|\vec{w}(\cdot,0)|^{(2+\alpha)}_{\Omega_0} + |\theta_0|^{(2+\alpha)}_{\Omega_0}\right) \leqslant$$

$$\leqslant c_{28}\left(|\vec{v}_0|^{(2+\alpha)}_{\Omega_0} + |\theta_0|^{(2+\alpha)}_{\Omega_0}\right).$$

Since $\Gamma_t$ is given by $\vec{x} = \vec{\xi} + \int_0^t \hat{\vec{v}}(\xi,\tau)\, d\tau$, $\xi \in \Gamma_0$, it's not hard to verify that conditions (11), (17) hold if $\varepsilon_0$ in (5) is small. Moreover, it follows from a coersive estimate for satisfying an elliptic equation (22) that

$$|R - r|^{(3+\alpha)}_{S_1} \leqslant c_{29}\left(|S|^{(1+\alpha)}_{\Gamma_t} + |\vec{w}|^{(2+\alpha)}_{\Omega_t} + |\theta|^{(1+\alpha)}_{\Omega_t} + \|R-r\|_{L_2(S_1)}\right) \tag{26}$$

so $|R-r|_{S_1}^{(3+\alpha)}$ is also small. Now, if the solution is defined for $t \in (0,t_N+1)$, and the smallness conditions (11), (17) are satisfied, then the estimate (25) gives a bound $c_{30}\varepsilon_0$ for $|\vec{v}(\cdot,t_N+1)|_{\Omega_{t_N+1}}^{(2+\alpha)} + |\theta(\cdot,t_N+1)|_{\Omega_{t_N+1}}^{(2+\alpha)}$, and we can make one step further and extend our solution to the interval $(t_N+1,t_N+3/2)$. Since our estimates are uniform, we can continue this procedure and establish the solvability of problem (1)-(3) for all $t > 0$. The inequality (25) holds for $N = \infty$, which means that $X_k \to 0$ as $k \to \infty$, i.e. $\vec{w}$, $s$, $\theta$ tend to zero. In virtue of (23), (26), $R - r$ also tends to zero. The proof of theorem 1 is completed.

## REFERENCES.


[1] Ladyzhenskaya O.A, Solonnikov V.A., Uraltseva N.N. Linear and quasilinear equations of parabolic type. Moscow, Nauka, (1967);

[2] Lagunova M.V., Solonnikov V.A. Nonstationary problem of thermocapillary convection. Stability and Appl.Anal.Conf.Media, vol.1, (1991), p.47-72.

[3] Mogilevskii I.Sh. Estimates in $C^{21,1}$, for solutions of a boundary value problem for the nonstationary Stokes system with the surface tension in boundary condition. To appear in Proceedings of the conference on Navier-Stokes equations (Oberwolfach, 1991).

[4] Mogilevskii I.Sh., Solonnikov V.A. On the solvability of a free boundary problem for the Navier-Stokes equations in the Hölder space of functions. - Nonlinear Analysis, a tribute in honour of G.Prodi, Pisa, 1991, p.257-271.

[5] Pukhnachov V.V. Thermocapillary convection under low gravity, Fluid dyn. Transactions, vol.14, (1989), p.145-204.

[6] Solonnikov V.A. On boundary value problems for general linear parabolic systems of differential equations, Proceedings of Steklov Math.Inst., vol.83,(1965), p.3-162.

[7] Solonnikov V.A. On a nonstationary motion of a finite liquid

mass bounded by a free surface. Zapiski Nauchn.Semin.LOMI, vol.152 (1986), p. 137-157.

[8] Solonnikov V.A. On the transient motion of an isolated volume of viscous incompressible fluid, Izvestia Acad.Sci.USSR, Ser.Mathem., vol.51 (1987), p.1065-1087.

[9] Solonnikov V.A. On a nonstationary motion of a finite mass of selfgravitating fluid. Algebra and analysis, vol.1 (1989), p.207-249.

[10] Solonnikov V.A. Solvability of evolution problem of thermocapillary convection in an infinite time interval, IMA preprint (1991).

# FRONT TRACKING METHODS FOR ONE-DIMENSIONAL MOVING BOUNDARY PROBLEMS

*Uwe Streit*
*Technical University Chemnitz*
*P.O. Box 964, 9010 Chemnitz, GERMANY*

## 1. INTRODUCTION.

In a series of papers (cf.Section 6) we showed how a combination of the two main attitudes towards MBPs (moving boundary problems), i.e. with or without explicit reference to the moving boundaries, may yield efficient numerical schemes in the case of one space dimension. In fact, classical formulations often suggest a simple front tracking procedure, especially if the equations restricted to the subdomains are linear, while weak formulations, based on nonlinear, possibly degenerate or singular equations in the whole domain (with fixed boundaries), provide a convenient tool for the error analysis. Though there is a large number of numerical methods for MBPs, it should be noted that many of them, especially those of front tracking type, suffer from a lack of rigorous error analysis. There is no place for a review on error estimates here. Instead, we refer to Nochetto & Verdi (1988).

The general pattern of our finite difference based front tracking methods is as follows:choose a fixed space-time-grid such that the discrete interface moves within one time step is *not greater* than the width of the space mesh. At the current time level determine a nearest grid point to the moving interface and solve the (linear) discrete field equations in each subdomain separately. Locate the moving boundary at the next time level from an explicit formula.

To obtain an error estimate for the field variable, the front tracking scheme is shaped into a scheme for a corresponding fixed domain formulation, for which error estimates are easier available. In most cases the moving boundary error can be estimated using the classical formulation instead of introducing level sets for the weak solution.

In Section 2-5 of the present paper we illustrate our idea by a simple model problem.

## 2. MODEL PROBLEM.

The nonlinear boundary value problem

$$\begin{cases} (u + \lambda v)_t - u_{xx} = 1, \quad v \in \theta(u) & (x \in \Omega,\ t > 0), \\ -u_x + u + 1 = 0 & (x = 0,\ t > 0), \\ u_x = 0 & (x = 1,\ t > 0), \\ u = -1,\ v = 0 & (x \in \Omega,\ t = 0). \end{cases} \tag{1}$$

over the fixed domain $\Omega = \{x\colon 0 < x < 1\}$ describes the melting, caused by body heating, of a pure solid substance with latent heat $\lambda > 0$. The dependence of the concentration $v(x,t)$ of the liquid phase on the temperature $u(x,t)$ is given by the Heaviside graph: $\theta(u) = 0 (u < 0)$, $\theta(u) = 1 \quad (u > 0)$, $\theta(0) = [0,1]$, which means phase transition at zero temperature.

As a mathematical model of the spot welding process problem (1) was analysed by Atthey (1974). Primicerio (1983) and Meirmanov (1986) gave a formulation in terms of classical MBPs: for small times problem (1) is linear since $v \equiv 0$. At some time $T_0 > 0$ the temperature at $x = 1$ reaches zero and phase change will start with formation of a mushy region (characterized by $u = 0$) separated from the solid region ($u < 0$) by a moving boundary $x = s(t)$. At time $t = T_0 + \lambda$ a liquid region ($u > 0$) begins to form together with a second moving boundary. At the end the mushy region disappears, only solid and liquid remain.

Our concern here is only that part of the process between $t = T_0$ and $t = T_0 + \lambda$. Shifting the time scale by an amount of $- T_0$, the new initial data are $u_0(x) \equiv u(x,T_0)$ and $v_0(x) \equiv \equiv v(x,T_0) = 0$. Problem (1) decouples into a MBP for $u$, $s (x < s(t))$ and an initial value problem for $v$ $(x > s)$:

$$\begin{cases} u_t - u_{xx} = 1, \quad v = 0 & (0 < x < s(t),\ 0 < t < \lambda), \\ -u_x + u + 1 = 0 & (x = 0,\ 0 < t < \lambda), \\ u = u_x = 0 & (x = s(t),\ 0 < t < \lambda), \\ u = u_0,\ v = 0 & (0 < x < 1,\ t = 0), \\ s = 1 & (t = 0), \end{cases} \tag{2}$$

$$u = 0,\ \lambda v_t = 1 \qquad (s(t) < x < 1,\ 0 < t < \lambda),$$

$$v = 0 \qquad (x = s(t),\ 0 < t < \lambda).$$

The front tracking algorithms built upon a common strategy for implicit MBPs like (2): to change over to an explicit MBP for a Baiocchi type transform $p$ of $u$, more precisely

$$p(x,t) = u_t(x,t) \quad (0 < x < s(t),\ t > 0) \quad \text{or}$$

$$u(x,t) = u_0(x) + \int_0^t p(x,t')dt'$$

with initial datum $p_0(x) \equiv p(x,0) = (u_0)_{xx}(t) + 1$. We mention that the maximum principle implies $0 \leqslant p_0(x) \leqslant p_0(1) \equiv P < 1$, and calculation confirms $P \approx 0.370$.

## 3. ALGORITHM.

We introduce the space-time grid $\omega\times\omega_\tau$ with

$$\omega = \left\{x = ih,\ i = 1,\dots,N-1,\ Nh = 1\right\} \quad \text{and}$$

$$\omega_\tau = \left\{t = n\tau < \lambda,\ \tau > 0,\ n = 1,\ 2,\dots\right\}.$$

Let $H$ denote the linear space of all grid functions on $\bar{\omega} = \omega\cup\{0,1\}$ with the scalar product

$$(y,w) = 0.5hy(0)w(0) + 0.5hy(1)w(1) + \Sigma_{x\in\omega}hy(x)w(x)$$

and the norm $\|y\| = \sqrt{(y,y)}$. Further we make use of

$$\|y\|_\infty = \max_{x\in\bar{\omega}} |y(x)|.$$

By $\delta(x,x')$ $(x \in \bar{\omega})$ we denote the discrete delta function for a fixed $x' \in \bar{\omega}$: $\delta(x,x') = 0$ if $x \neq x'$ and $(\delta(x,x'),1) = 1$. The difference quotients are symbolized by

$$y_x(x) = \frac{y(x+h) - y(x)}{h}, \quad y_{\bar{x}}(x) = y_x(x-h),$$

$$y_{\bar{x}x}(x) = \frac{y_{\bar{x}}(x+h) - y_{\bar{x}}(x)}{h}, \quad y_{\bar{t}}(t) = \frac{y(t) - y(\check{t})}{\tau}$$

with $\check{t} = t - \tau$. Summation and maximization over $t \in \omega_\tau$ is abbreviated as $\Sigma$ and $max$, respectively. The linear operator $A\colon H \to H$, defined by

$$(Ay)(0) = -2(y_x(0) - y(0))/h, \quad (Ay)(x) = -y_{\bar{x}x}(x) \quad (x \in \omega),$$

$$(Ay)(1) = 2y_{\bar{x}}(1)/h$$

is related to problem (1). Obviously, $A$ is self-adjoint and positive.

The discrete analogues of $p$, $u$ and $v$ are denoted by $w$, $y$ and $v$ respectively, the discrete moving boundary by $\xi$ and a "nearest" grid point by $x^0$. The small-time behaviour of $s$ requires the choice $\tau = ch^2$ with $c \leqslant (P^{-1} - 1)/2$. For the practical implementation of the algorithm a variable time step is recommended, that can be successively increased as the interface velocity is slowing down. Here is the complete algorithm followed by its output for $\lambda = 1$ and $h = 0.01$:

At time $t = 0$ set $\xi = x^0 = 1$ and $w = p_0$, $v = 0$, $y = u_0$ $(x \in \bar{\omega})$.

Let $\check{\xi}$, $\check{x}^0$, $\check{w}$, $\check{v}$, $\check{y}$ (the values at time $\check{t}$) be known. At time $t = \check{t} + \tau$ determine

$\xi$ as $\xi = 1 - hp_0(1-h)$ $(t=\tau)$ or via $k = \check{\xi} + \tau\check{w}_{\bar{x}}(\check{x}_0)$ $(t>\tau)$:

$$\xi = k \text{ if } \check{x}^0 - k \leqslant 0, \; \xi = k + h^2\check{w}_{\bar{x}}(\check{x}^0) \text{ if } \check{x}^0 - k > 0,$$

$x^0 \in \bar{\omega}$ from $-h < x^0 - \xi \leqslant 0$,

$w$ $(x \in \bar{\omega})$ as solution of $w_{\bar{t}} + Aw = 0$ $(x<x^0)$, $w = 0$ $(x \geqslant x^0)$,

$v(x \in \bar{\omega})$ from $v = 0$ $(x<x^0)$, $\lambda v_{\bar{t}} = 1$ $(x \geq x^0)$ and

$y(x \in \bar{\omega})$ from $y_{\bar{t}} = w$.

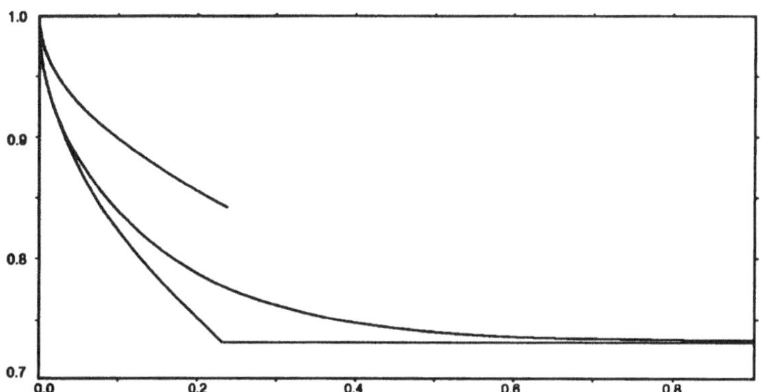


In the figure above the moving boundary of problem (2) is displayed vs.time, together with two of its bounds, namely the functions $max(1 - 0.55710\sqrt{t}, \sqrt{3} - 1)$ $(t > 0)$ and $1 - 0.324\sqrt{t}$ $(0<t<0.239)$ (cf. Ughi 1984)

## 4. ERROR ESTIMATES FOR THE FIELD VARIABLE.

From now on $C$ denotes a generic positive constant not depending on $h$ and $\tau$. First we point out that $y$, $\xi$ and $x^0$ have the following properties:

$$y_{\bar{t}} + Ay = 1 - \delta(x,0), \quad y < 0,$$

$$y_x > 0 \quad (0 \leq x < x^0),$$

$$-h/\tau < \xi_{\bar{t}} < 0$$

and

$$-\frac{h}{\tau} \leq (x^0)_{\bar{t}} \leq 0, \quad 0 < y_{\bar{x}}(x^0) < 1.5h, \quad -1.5h < y(x^0) \leq 0. \qquad (3)$$

In view of $y(x^0) \neq 0$ we introduce an adjusted approximate solution $\tilde{y}$:

$$\tilde{y}(x,t) = y(x,t) - y(x^0(t),t) \quad (x \leq x^0(t)),$$

$$\tilde{y}(x,t) = 0 \quad (x > x^0(t)).$$

Simple calculation shows that in $H$ the grid functions

$\tilde{y}(t),\ v(t)$ satisfy the operator-difference scheme

$$(\tilde{y} + \lambda\varphi)_{\bar{t}} + A\tilde{y} - 1 + \delta(x,0) = \psi_{\bar{t}} + \eta\delta(x,0) = \zeta\delta(x,x^0),$$

$$v \in \theta(\tilde{y}) \quad (t \in \omega_\tau),$$
$$\tilde{y} = u_0, \quad v = 0 \ (t = 0)$$

with

$$\eta(t) = -y(x^0,t), \quad \zeta(t) = y_{\bar{x}}(x^0,t) - h(x^0)_{\bar{t}}\, y_{\bar{x}}(\check{x}^0,\check{t}) \tag{4}$$

and

$$\psi(x,t) = -y(x^0,t) \qquad (x < x^0),$$

$$\psi(x,t) = -y(x^0(t^0(x) - \tau), t^0(x) - \tau), \quad (x \geqslant x^0),$$

where

$$t^0(x) = min\left\{ t \in \omega_\tau : x^0(t) = x \right\}.$$

Let $u_h$, $v_h$ be the solution of the finite difference analogue of problem (1):

$$(u_h + \lambda v_h)_{\bar{t}} + Au_h - 1 + \delta(x,0) = 0, \quad v_h \in \theta(u_h) \quad (t \in \omega_\tau),$$

$$u_h = u_0, \quad v_h = 0 \qquad (t = 0).$$

The following error estimate holds

$$\sum \tau\|(\tilde{y} - u_h)(t)\|^2 \leqslant C \sum \tau\left[\|\psi(t)\|^2 + |\eta(t)|^2 + |\zeta(t)|^2\right].$$

It is seen from (3) that the right hand side is of order $O(h^2)$, hence

**Estimate 1** $\left[\sum \tau\|(\tilde{y} - u_h)(t)\|^2\right]^{1/2} \leqslant Ch$

is valid. Moreover, (3) yields

**Estimate 2** $max\ \|(y - \tilde{y})(t)\|_\infty \leqslant Ch.$

All that remains is the analysis of $u_h - u$. If no special information on the behaviour of near the phase change region of problem (1) is available, then $(\Sigma\,\tau\|(u_h - u)(t)\|^2)^{1/2}$ is known to be at least of order $O(\sqrt{h})$, both for finite difference based schemes (Streit 1989a) and finite element methods, and both for methods with or without preliminary regularization of $\theta(u)$ (cf. Nochetto & Verdi 1988 and the references quoted therein).

We may derive a stronger estimate for $y$. Bearing the oxygen diffusion MBP in mind, we realize that the solution $u$, $s$ of (2) satisfies another fixed domain problem:

$$\begin{cases} u_t - u_{xx} + f = 0, \quad f \in \theta(u) - 1 & (x \in \Omega,\ 0 < t < \lambda), \\ -u_x + u + 1 = 0 & (x = 0,\ 0 < t < \lambda), \\ u_x = 0 & (x = 1,\ 0 < t < \lambda), \\ u = u_0 & (x \in \Omega,\ t = 0). \end{cases} \tag{5}$$

Defining $\varphi(x,t) = -1 \ (x < x^0)$, $\varphi(x,t) = 0 \ (x \geqslant x^0)$ as discrete analogue of $f$ yields the operator-difference scheme

$$\tilde{y}_{\bar{t}} + A\tilde{y} + \varphi + \delta(x,0) = \phi + \eta\delta(x,0) + \zeta\delta(x,x^0),$$

$$\varphi \in \theta(\tilde{y}) - 1 \quad (t \in \omega_\tau),$$

$$\tilde{y} = u_0 \qquad (t = 0)$$

with $\phi(x,t) = -(y(x^0,t)_{\bar{t}} \ (x<x^0)$, $\phi(x,t) = 0 \ (x \geqslant x^0)$ and $\eta$, $\xi$ from (4). By means of

$$(y(x^0,t)_{\bar{t}} = (x^0)_{\bar{t}}\, y_{\bar{x}}(\check{x}^0,\check{t})$$

and (3) we conclude

$$|\phi| \leqslant Ch|(x^0)_{\bar{t}}| = -CH(x^0)_{\bar{t}}$$

and thus,

$$\sum \tau\|\phi(t)\| \leqslant -Ch\sum \tau(x^0)_{\bar{t}} \leqslant Ch.$$

Now let $u_h$, $f_h$ be the solution of the finite difference analogue of problem (5):

$$(u_h)_{\bar{t}} + Au_h + f_h + \delta(x,0) = 0, \quad f_h \in \theta(u_h) - 1 \quad (t \in \omega_\tau),$$

$$u_h = u_0 \qquad (t = 0).$$

This time the crucial estimate is

$$max\|(\tilde{y} - u_h)(t)\|^2 + \sum \tau\|(\tilde{y} - u_h)(t)\|_A^2 \leqslant$$

$$\leqslant C\left[\left(\sum \tau\|\phi(t)|\right)^2 + \sum \tau(|\eta(t)|^2 + |\zeta(t)|^2)\right],$$

hence

**Estimate 3** $max\ \|(\tilde{y} - u_h)(t)\| + \left[\sum \tau\|(\tilde{y} - u_h)(t)\|_A^2\right]^{1/2} \leqslant Ch$

holds. The analysis of $u_h - u$ yields

$$max\ \|(u_h - u)(t)\| + \left[\sum \tau\|(u_h - u)(t)\|_A^2\right]^{1/2} \leqslant Ch$$

as long as $\tau = O(h^2)$, i.e. the total error $y - u$ is of order $O(h)$ (cf.Streit 1991).

## 5.MOVING BOUNDARY ERROR.

From the discrete maximum principle we obtain $0 \leqslant w \leqslant P$ $(0<x\leqslant x^0,\ 0<t<\lambda)$. Similarly, we have $0\leqslant p\leqslant P$ $(0<x<s(t),\ 0<t<\lambda)$. If $x^0(t) \leqslant s(t)$ at some time level $t \in \omega_\tau$, then we recall the Taylor series expansion of $u$:

$$-u(x^0) = -u(s) + (s - x^0)u_x(s) + \int_{x^0}^{s}\int_{x}^{s}(-u_{xx}(x'))dx'\ dx. \tag{6}$$

In view of

$$u(s) = u_x(s) = 0,$$

$$-u_{xx}(x') = 1 - u_t(x') = 1 - p(x') \geqslant 1 - P \equiv 2c_0 > 0$$

we derive

$$c_0|x^0 - s|^2 = 2c_0\int_{x^0}^{s}\int_{x}^{s}dx'\,dx \leqslant -u(x^0) \leqslant |y(x^0) - u(x^0)| + Ch,$$

i.e.

$$|(x^0 - s)(t)|^2 \leqslant C\|(y - u)(t)\|_\infty + Ch.$$

If $s(t) < x^0(t)$ for $t \in \omega_\tau$, then we get the same relation from the finite difference analogue of (6). The imbedding inequality $\|\tilde{y} - u\|_\infty \leqslant C\|\tilde{y} - u\|_A$, together with the result on

$\left(\sum \tau\|(\tilde{y} - u)(t)\|_A^2\right)^{1/2}$, finally yields

**Estimate 4.** $$\left(\sum \tau|(\xi - s)(t)|^4\right)^{1/4} \leqslant C\sqrt{h}.$$

## 6. RESULTS FOR RELATED PROBLEMS.

We give a brief review of other MBPs, for which the good performance of our approach was demonstrated.

First we announce that the algorithm for problem (1) beyond the time $T_0 + \lambda$ can be found in Streit (1989b) and is supplemented by numerical results in Streit & Andra (1990). Estimate 1 holds true as long as $\tau = O(h^2)$.

As mentioned earlier, problem (2) is very similar to the classical oxygen diffusion MBP (Crank & Gupta, 1972) that models the process of emptying an oxygen-filled region. For this problem and a "reversed" one describing the filling of an oxygen-free region, algorithms requiring $\tau = O(h^2)$ and $\tau = O(h^3)$ respectively, are given in Streit (1991). The error analysis yields the validity of Estimates 2, 3 and 4. Examples of nonmonotone moving boundaries are also discussed.

The classical two-phase Stefan problem with initial and boundary temperatures from $W_2^1$ and a moving boundary strictly inside $\Omega$ is the subject of Streit (1989a). Using backward time discretization

**Estimate 5** $$\left(\sum\left[\tau\|(y - u)(t)\|^2 + |(\xi - s)(t)|^2\right]\right)^{1/2} \leqslant Ch$$

is proved if $\tau = O(h^{3/2})$. To give an impression of the method, we apply it to problem (1) that turns into a two-phase problem with moving boundary conditions $u(s) = 0$, $\lambda s_t = u_x(s+0) - u_x(s-0)$, after the mushy region has vanished. This time let $k$ be the discrete moving boundary velocity and $\tilde{x} \in \bar{\omega}$ another approximation of $s$. Then the time stepping procedure is as follows:

Let $\check{\xi}$, $\check{x}^0$, $\check{y}$, $\check{k}$ be known. At time $t = \check{t} + \tau$ determine $\xi$ from $\xi_{\bar{t}} = \check{k}$ and $\tilde{x} \in \bar{\omega}$ from $-h/2 \leqslant \tilde{x} - \xi < h/2$.

If $\tilde{x} = \check{x}^0$ then set $x^0 = \check{x}^0$, determine $y(\ x \in \bar{\omega})$ as solution of

$$y_{\bar{t}} + Ay - 1 + \delta(x,0) = 0 \quad (x \neq x^0),\ y = 0 \quad (x = x^0), \tag{7}$$

and find $k$ from

$$-\lambda k + h(y_{\bar{t}} + Ay - 1)(x^0) = 0. \tag{8}$$

If $\tilde{x} \neq \check{x}^0$ then determine $y^{(1)}$, $y^{(2)}$ as solutions of (7), where $x^0$ is replaced by $x^{0(1)} = min(\tilde{x},\check{x}^0)$, $x^{0(2)} = max(\tilde{x},\check{x}^0)$ respectively, and next correspondingly define $k^{(1)}$ and $k^{(2)}$ by (8).

If $k^{(1)} \leqslant 0$ and $k^{(2)} \leqslant 0$ then set

$x^0 = x^{0(1)},\ y = y^{(1)},\ k = k^{(1)}$.

If $k^{(1)} > 0$ and $k^{(2)} \geqslant 0$ then set

$x^0 = x^{0(2)},\ y = y^{(2)},\ k = k^{(2)}$.

Otherwise set $x_0 = \check{x}^0$, $k = 0$ and determine $y$ as solution of

$$y_{\bar{t}} + Ay - 1 + \delta(x,0) = 0 \quad (x \in \bar{\omega}).$$

In view of monotonicity of $s$ an adequate behaviour of $\xi$ and thus simplifications in the algorithm can be expected.

Estimate 5 was also verified for an explicit finite difference method (Streit 1989c). For the lower regularity problem of freezing the half space (Neumann solution) convergence rates of $O(h^{3/5})$ for $u$ and $O(h^{1/4})$ for $s$ can be shown.

## REFERENCES


[1] Atthey D.R. A finite difference scheme for melting problems. J.Inst. Math. Appl. 13 (1974), 353-366.

[2] Crank J. & Gupta R.S. A moving boundary problem arising from the diffusion of oxygen in absorbing tissue. J.Inst.Math. Appl., 10 (1972) , 19 - 33.

[3] Meirmanov A.M. The Stefan problem, Nauka, Novosibirsk (1986) (in Russian).

[4] Nochetto R.H. & Verdi C. Approximation of degenerate parabolic problems using numerical integration. SIAM J. Numer. Anal. 25 (1988), 784-814.

[5] Primicerio M., Mushy regions in phase change problems. In: Gorenflo R. & Hoffmann K.-H. (ed.) Applied nonlinear functional analysis: Variational methods and ill-posed problems. P.Lang, Frankfurt a. M. (1983), 251-270.

[6] Streit U. An efficient enthalpy-type method for the Stefan problem. IMA J. Numer. Anal., 9 (1989), 353-372.

[7] Streit U. An efficient algorithm for determining the mushy region in a Stefan problem. Wiss. Z. Techn. Univ. Karl-Marx-Stadt 31 (1989), 265-269.

[8] Streit U. Error estimates for the explicit finite difference method solving the Stefan problem. Z. Anal. Anwendungen 8 (1989), 57-68.

[9] Streit U. & Andra H. Numerical results for a Stefan problem with mushy region. Wiss. Z. Techn. Univ. Karl-Marx-Stadt, 32 (1990), 7-11.

[10] Streit U. One-dimensional implicit moving boundary problems. A new front tracking approach, Numer. Funct. Anal. Optim. (1991) (to appear).

[11] Ughi M. A melting problem with a mushy region: qualitative properties. IMA J. Appl. Math. 33 (1984), 135-152.

# ON CAUCHY PROBLEM FOR LONG WAVE EQUATIONS

V.M.Teshukov
Lavrentyev Institute of Hydrodynamics
Novosibirsk 630090, RUSSIA

The Cauchy problem for the system of integro-differential equations modelling free-boundary vortex flows is considered. For the initial data satisfying the hyperbolicity conditions, the solvability of the Cauchy problem, local in time, is proved.

Key words: vortex flows, integro-differential equations, hyperbolicity.

## 1. FORMULATION OF THE PROBLEM.

The solution of the free-boundary problem

$$
\begin{aligned}
&u_t + uu_x + vu_y + \rho^{-1}p_x = 0,\\
&v_t + uv_x + vv_y + \rho^{-1}p_y = -g \quad (0 \leqslant y \leqslant h(x,t)),\\
&u_x + v_y = 0, \qquad\qquad (1)\\
&h_t + u(x,h,t)h_x = v(x,h,t), \quad p(x,h,t) = 0,\\
&v(x,0,t) = 0, \quad u(x,y,0) = u_0(x,y), \; h(x,0) = h_0(x)
\end{aligned}
$$

describes a plane-parallel unsteady wave motion of ideal incompressible fluid. Here $\vec{u} = (u,v)$ is the velocity vector, $p$ is the pressure, $\rho$ is the density ($\rho = const.$), $g$ is the gravitational acceleration, $x$, $y$ are the Cartesian coordinates on a plane, $t$ is time, $y = h(x,t)$ is the equation of free boundary, $y = 0$ corresponds to an even bottom. Given functions $u_0(x,y)$, $h_0(x)$ define an initial velocity field and an initial position of free surface.

After introducing new variables

$$t' = \varepsilon^{1/2}t, \; x' = x, \; y' = \varepsilon^{-1}y, \; u' = \varepsilon^{-1/2}u, \; v' = \varepsilon^{-3/2}v, \quad p' = \varepsilon^{-1}p$$

and passing to the limit $\varepsilon \to 0$ in the transformed equations, we obtain a mathematical model of long-wave approximation. In the limit only the second equation of (1) changes; it takes the form

$$\rho^{-1}p'_{y'} = -g \tag{2}$$

(a hydrostatical law of pressure distribution). Hence, the long-wave approximation gives the problem of form (1), but with the only difference consisting in replacement of the second equation in (1) by (2).

This problem may be reduced to Cauchy problem in fixed region (strip $0 \leqslant \lambda \leqslant 1$, $\lambda$ is the analogy of the Lagrangian coordinate) by the change of variables

$$x' = \bar{x},\ t' = \bar{t},\ y' = \Phi(\bar{x},\lambda,\bar{t}),$$

where $\Phi(\bar{x},\lambda,\bar{t})$ is the solution of problem

$$\Phi_{\bar{t}} + \left[ \int_0^{\bar{t}} u(\bar{x},y,\bar{t})dy \right]_{\bar{x}} = 0, \quad \Phi(\bar{x},\lambda,0) = \lambda h_0(\bar{x}).$$

As a result, the initial problem is equivalent to the Cauchy problem for the system of integro-differential equations $(H(x,\lambda,t) = g\Phi_\lambda(x,\lambda,t),\ x = \bar{x},\ \ t = \bar{t})$

$$u_t + uu_x + \int_0^1 H'_x\, d\nu = 0, \quad u(x,\lambda,0) = U_0(x,\lambda) = u_0(x,\lambda h_0(x)),$$

$$H_t + (uH)_x = 0, \quad H(x,\lambda,0) = H_0(x,\lambda) = gh_0(x). \tag{3}$$

If $u_{0y} \equiv 0$ (vortex-free initial velocity field) then $u_\lambda \equiv 0$, $H_\lambda \equiv 0$ and equations (3) transform into classical equations of shallow water theory. We shall consider vortex flows $(u_{0y} \neq 0)$.

The generalized concept of hyperbolicity of the system of first-order equations applicable to integro-differential equations have been formulated in [ 1 ], and in [2] it was shown that the conditions

$$u_\lambda > 0, \quad \chi^{\pm}(u) \neq 0, \quad \Delta arg \left( \frac{\chi^{+}(u)}{\chi^{-}(u)} \right) = 0 \tag{4}$$

are sufficient for hyperbolicity of equations (3). Here $\Delta$ denotes the increment of the function on the interval $\lambda \in [0,1]$,

$$\chi^{\pm}(u) \equiv (u-u_0)(u-u_1)\left[ 1 - g \int_0^1 \left(\frac{1}{\omega'}\right)_\nu \frac{dv}{u' - u} \mp \pi i \left(\frac{1}{\omega}\right)_\lambda \frac{1}{u_\lambda}\right] +$$

$$+ \omega_1^{-1}(u_0-u) - \omega_0^{-1}(u_1-u),$$

($\omega = u_\lambda/H$; $u'$, $u$ are the abbreviated designations of $u(x,v,t)$, $u(x,\lambda,t)$, indices "$0$", "$1$" designate values of functions at $\lambda = 0$, $\lambda = 1$. System of equations (3) has the family of characteristics corresponding to continuous characteristic spectrum

$$x = x^\lambda(t) \quad (\lambda \in (0,1)):$$

$$\frac{dx^\lambda}{dt} = u(x^\lambda,\lambda,t),$$

and characteristics $x = x^i(t)$, corresponding to discrete characteristic spectrum:

$$\frac{dx^i}{dt} = k_i(x^i,t).$$

The characteristic numbers $k_i$ are defined as roots of the equation

$$\int_0^1 \frac{H'\ dv}{(u' - k_i)^2} = 1. \tag{5}$$

If conditions (4) are fulfilled, this equation has only two real roots

$$k_1(x,t) < \min_\lambda u(x,\lambda,t), \quad k_2(x,t) > \max_\lambda u(x,\lambda,t).$$

and the system (3) is equivalent to the relations on the characteristics

$$R_t + uR_x = 0, \quad \omega_t + u\omega_x = 0,$$
$$r_{it} + k_i r_{ix} = 0 \quad (i = 1, 2), \tag{6}$$

where

$$R = u - \int_0^1 \frac{H'\,d\nu}{u' - u}, \quad r_i = k_i - \int_0^1 \frac{H'\,d\nu}{u' - k_i}. \tag{7}$$

The function $H$ may be expressed via $\omega$ and $u_\lambda$: $H = u_\lambda \omega^{-1}$, therefore further $H$ will be eliminated with the help of this equality.

## 2. A PRIORI ESTIMATES.

To obtain a priori estimates of the solution of Cauchy problem (3) the analogue of energy integral method is used. The system of equations (3) is reduced to a symmetric form by introducing of Riemann invariants $R$, $\omega$, $r_i$. This enables us to use such scheme of obtaining the estimates of the solution and its derivatives in Sobolev classes of functions as that used in the case of quasilinear hyperbolic systems of differential equations. Some differences are associated with the nonlocality of transformations of (3) into (6) and with the singularities of $R(x,\lambda,t)$ when $\lambda = 0$ and $\lambda = 1$.

To obtain estimates of the solution we need the formulae relating the derivatives of $u$ with the derivatives of Riemann invariants .After differentiating (7) we obtain the relations

$$\begin{aligned} u_x - \int_0^1 \frac{1}{\omega}\frac{\partial}{\partial \nu}\left[\frac{u'_x - u_x}{u' - u}\right] d\nu &= R_x + \int_0^1 \left[\frac{1}{\omega}\right]_x \frac{u'_\nu\,d\nu}{u' - u} = F_1, \\ - \int_0^1 \frac{1}{\omega}\frac{\partial}{\partial \nu}\left[\frac{u'_x}{u' - k_i}\right] d\nu &= r_{ix} + \int_0^1 \left[\frac{1}{\omega}\right]_x \frac{u'_\nu\,d\nu}{u' - k_i}. \end{aligned} \tag{8}$$

If we represent the function $u$ in the form

$$u_x = w_1 + \sum_{i=1}^{2} \alpha_i^1 (u - k_i)^{-1}, \tag{9}$$

where $w_1(x,\lambda,t)$ vanishes when $\lambda = 0$, $\lambda = 1$ and $\alpha_i^1$ depends only on $x$, $t$, then we obtain singular integral equation for determining $w_1(x,\lambda,t)$:

$$R_u w_1 + \int_{u_0}^{u_1} \left[\frac{1}{\omega'}\right]_{u'} \frac{w_1' \, du'}{u' - u} = F_1, \tag{10}$$

where

$$R_u = R_\lambda u_\lambda^{-1}, \quad (\omega'^{-1})_{u'} = (\omega^{-1})_\nu u_\nu^{-1}.$$

With the help of (8) we find $\alpha_i^1$ in an explicit form

$$\alpha_i^1 = \left[ 2\int_0^1 \frac{1}{\omega'} \frac{u_\nu' \, d\nu}{(u' - k_i)^3} \right]^{-1} \left[ r_{ix} + \int_0^1 \left[\frac{1}{\omega'}\right]_x \frac{u_\nu' \, d\nu}{u' - k_i} - \int_0^1 \left[\frac{1}{\omega'}\right]_\nu \frac{w_1' \, d\nu}{u' - k_i} \right]. \tag{11}$$

The hyperbolicity conditions (4) guarantee univalent solvability of singular integral equation (10) in the class of functions satisfying Holder condition at the internal points of the interval $[u_0, u_1]$ and bounded at its ends [3].

The equation (10) may be solved explicitly

$$w_1 = \frac{fF_1 \cdot fR_u}{|\chi^+|^2} - \frac{f}{g} \int_{u_0}^{u_1} \left[\frac{1}{\omega'}\right]_{u'} \frac{f' F_1' \, du'}{|\chi^+|^2 g' (u' - u)}, \tag{12}$$

where

$$f = (u - u_0)(u - u_1), \quad g = (u - k_1)(u - k_2).$$

Representation (9), (11) and (12) of function $u_x$ allows estimating this function via norms of the derivatives of Riemann invariants $\omega$, $R$, $r_i$. Analogous representations may be obtained for $u_{xx}$, $u_{xxx}$.

In $R^3(x,\lambda,t)$ let us consider the domain $Q_t$ bounded by the planes $t = 0$, $t = const$; $\lambda = 0$, $\lambda = 1$, and by surfaces $\Gamma_i$ prescribed by the equations $x = x^i(t)$:

$$\frac{dx^i}{dt} = k_i(x^i, t); \quad x^1(0) = b, \quad x^2(0) = a \quad (b > a).$$

The cross-section of $Q_t$ by the plane $t = \tau$ is a rectangle

$$\Omega_\tau = \left\{(x,\lambda):\ x \in [x^2(\tau),x^1(\tau)],\ \lambda \in [0,1]\right\}.$$

By the symbol $H^s$ we shall denote the Sobolev space of functions which are square integrable over some region together with the derivatives up to the $s$-th order, $s > 0$. The symbol $\|h\|_{H^s}(\tau)$ denotes the norm of the function $h$ in $H^s(\Omega_\tau)$, $\|h\|_q(\tau)$ denotes the norm in $L_q(\Omega_\tau)$, $\|h\|_C(\tau)$ is the norm in $C(\Omega_\tau)$. Analogous norms for the functions depending only on $x$, $t$ ($x \in [x^2(\tau),x^1(\tau)]$) are designated as $|h|_{H^s}(\tau)$, $|h|_q(\tau)$, $|h|_C(\tau)$.

**LEMMA.** Let initial data satisfy conditions (4) and norms

$$\|u\|_{H^3}(0),\ \|\omega\|_{H^3}(0),\ \|\omega^{-1}\|_C(0),\ \|u_\lambda^{-1}\|_C(0),\ |r_i|_{H^3}(0)\quad (i = 1,\ 2)$$

be bounded. Then there exist $t_0 > 0$ and positive functions $C_i(t)$such that inequalities

$$\|u\|_{H^3}(t) \leqslant C_1(t),\ \|\omega\|_{H^3}(t) \leqslant C_2(t),$$

$$\|\omega^{-1}\|_C(t) \leqslant C_3(t),\quad \|u_\lambda^{-1}\|_C \leqslant C_4(t),$$

$$|r_i|_{H^3}(t) \leqslant C_5(t)$$

hold at the interval $0 < t < t_0$.

The functions $C_i(t)$ and value $t_0$ depend on the above-mentioned norms of the initial data.

**Scetch of proof.** Consider the equations for the squares of the derivatives of the functions $u_\lambda$, $u_\lambda^{-1}$, $\omega$, $\omega^{-1}$, $r_i$ $(i = 1,\ 2)$. Integration of these equations over $Q_t$ allows to obtain the inequality

$$\sum_{i=1}^{6} z_i^2(t) \leqslant \sum_{i=1}^{6} z_i^2(0) + C_1\int_0^t\Bigg[\Big(\|u_x\|_2 + \|u_{xx}\|_2 + \|u_{xxx}\|_2 + \sum_{i=1}^{2}\Big(|k_{ix}|_2 + |k_{ixx}|_2 + |k_{ixxx}|_2\Big) + \sum_{i=1}^{6} z_i\Bigg](\tau)\sum_{i=1}^{6} z_i^2(\tau)\,d\tau. \tag{13}$$

Here $C_1$ is positive constant, $z_1 = \|u_\lambda\|_{H^2}(t)$, $z_2 = \|u_\lambda^{-1}\|_{H^2}(t)$, $z_3 = \|\omega\|_{H^3}(t)$, $z_4 = \|\omega^{-1}\|_{H^3}(t)$, $z_5 = |r_1|_{H^3}(t)$, $z_6 = |r_2|_{H^3}(t)$.

The inequality

$$\sum_{i=7}^{9} z_i^2(t) \leqslant \sum_{i=7}^{9} z_i^2(0) + C_2\int_0^t\Bigg(\|u_x\|_2 + \|u_{xx}\|_2 + \|u_{xxx}\|_2 + \sum_{i=1}^{2} z_i\Bigg)\Bigg(\sum_{i=7}^{9} z_i + \|fR_x\|_C + \|fR_{xx}\|_4\Bigg)\sum_{i=7}^{9} z_i(\tau)\, d\tau. \tag{14}$$

may be obtained analogously. Here

$$z_7 = \|fR_x\|_2(t), \quad z_8 = \|fR_{xx}\|_2(t), \quad z_9 = \|fR_{xxx}\|_2(t).$$

Representation (9) of $u_x$ as an operator over derivatives of the functions $r_i$, $\omega^{-1}$ and $fR_x$ as well as analogous representations for $u_{xxx}$, allow to obtain the final inequality

$$Z(t) \leqslant Z(0) + \int_0^t \Phi(Z(\tau))\, d\tau \tag{15}$$

for

$$Z(t) = \sum_{i=1}^{9} z_i^2(t)$$

with a monotonically increasing positive function $\Phi(Z)$. Bounded solution of this inequality provides the required estimates on interval $[0, t_0]$ only if conditions (4) hold. To complete the proof we estimate the differences between the expressions (4) and their initial values and provide fulfillment of conditions (4) by the choice of $t_0$.

The existence theorem for the Cauchy problem (3) is proved when initial data satisfy conditions of Lemma with uniformly bounded norms and uniformly fulfilled conditions (4) (uniformly with respect to $a$ when $|b - a| = const$).

**THEOREM.** There exists $t_0 > 0$ such that $0 < t < t_0$ problem

(3) has the solution at the interval $0 < t < t_0$ with the norms

$$\|u\|_{H^3}(\tau),\quad \|\omega\|_{H^3}(\tau),\quad \|\omega^{-1}\|_C(\tau),\quad \|u_\lambda^{-1}\|_C(\tau),\quad |r_i|_{H^3}(\tau)$$

uniformly bounded with respect to $a$.

When a priori estimates of the solution are obtained, the existence theorem is proved by a standard scheme.

## REFERENCES.

[1] Teshukov V.M. On the hyperbolicity of long-wave equations. Soviet Math. Dokl. 32, N 2 (1985), 469-473.

[2] Teshukov V.M. Long wave approximation for vortex free-boundary flows . In: Numerical Methods for Free Boundary Problems. ISNM 99, Birkhauser Verlag (Basel), 1991.

[3] Muskhelishvili N.I. Singular integral equations. Nauka Publishers, Moscow, 1968.

# ON FIXED POINT (TRIAL) METHODS FOR FREE BOUNDARY PROBLEMS

*Timo Tiihonen*
*Department of Mathematics*
*University of Jyvaskyla*
*Seminaarinkatu 15, 40100 Jyvaskyla, FINLAND*

*Jari Järvinen*
*Centre for Scientific Computing*
*Box 40, 02101 Espoo, FINLAND*

In this note we consider the trial methods for solving steady state free boundary problems. For two test examples (electrochemical machining and continuous casting) we discuss the convergence of a fixed point method. Moreover, using the techniques of shape optimization we introduce a modification of the method, which gives us superlinear convergence rate. This is also confirmed numerically.

Keywords: free boundary problems, trial method, shape optimization.

## 1. A TRIAL METHOD FOR ELECTROCHEMICAL MACHINING PROBLEM

Consider the following problem: given $\Omega_0$ with boundary $\Gamma$, find the domain $\Omega = \Omega_1 \backslash \Omega_0$ and the free boundary $\Sigma = \partial\Omega_1$ such that

$$\begin{cases} -\Delta u = 0 & \text{in } \Omega, \\ u = 1 & \text{on } \Gamma, \\ u = 0 & \text{on } \Sigma, \\ \dfrac{\partial u}{\partial n} = -\lambda & \text{on } \Sigma, \quad \lambda > 0 \text{ given.} \end{cases} \tag{1.1}$$

One possible motivation is the following: electrical potential $u$ acts between the tool $\Omega_0$ and the work piece $R^n \backslash \Omega_1$. The work piece melts into the electrolyte $\Omega$ if the flux at $\Sigma$ is greater than $\lambda > 0$. The free boundary problem gives the steady state of the process. For more reference and the existence theory see [Alt, Caffarelli].

There are different strategies that can be used to find $\Sigma$. In the methods that are based on optimization we take out one of the two conditions imposed on $\Sigma$. Then the problem (1.1) becomes well posed and we can formulate an optimization problem where the goal is to find $\Sigma$ so that the removed boundary condition is also satisfied as well as possible. For example, if we remove the Dirichlet condition on $\Sigma$, we have a cost function of the type $\int_\Sigma u^2\, ds$ to be minimized with respect to $\Sigma$. If a zero value is found for the cost, we have reached the free boundary.

The trial methods, which we consider in this note are of fixed point type. Here we also ignore one of the conditions on $\Sigma$ and solve the resulting boundary value problem. Next, using the information obtained from the solution we try to construct a new $\Sigma$ so that the ignored condition would also be satisfied.

Let us give an example of a trial method for problem (1.1).

0. Give some initial guess $\Sigma$ such that $\Omega_0 \subset \Omega_1$.

1. Solve $-\Delta u = 0$, $u|_\Gamma = 1$, $\dfrac{\partial u}{\partial n} = -\lambda$ on $\Sigma$.

2. Set $\Sigma = \Sigma + \lambda^{-1} u|_\Sigma$.

3. Repeat from 1 until $u|_\Sigma$ is small enough.

We notice immediately that the method has a fixed point whenever $u|_\Sigma \equiv 0$. Thus if the method converges, it gives a solution of the free boundary problem. But when does the method converge?

As above, the method is not rigorously defined. In the second step $\Sigma \subset R^n$ whereas $u$ is real valued. To give a precise meaning to step 2 we can either modify if to the form $\Sigma = \Sigma + \lambda^{-1} un$, where $n$ is the unit normal of $\Sigma$ or we can assume that $\Sigma$ can be identified with some function . (For example, if $\Omega_1$ is starshaped, $\Sigma$ can be identified with a $2\pi$ periodic function. The same is true for $u|_\Sigma$). Let us assume that step 2 defines a mapping from some function space $V$ into itself, $F\colon V \to V$, $F(\Sigma) = \Sigma + \lambda^{-1}u|_\Sigma$.

Now, if $\|dF(\Sigma)/d\Sigma\|_{L(V,V)} = q < 1$ at the solution, the method converges at least locally with convergence rate $q$. So let us begin with the computation of $dF(\Sigma)/d\Sigma$.

Let $\sigma \in V$ be arbitrary. Then the directional derivative of $F$ is

$$\frac{d}{d\Sigma} F(\Sigma,\sigma) = \frac{d}{dt} F(\Sigma + t\sigma)\Big|_{t=0}.$$

Before proceeding, we introduce the notation of the shape derivative [Zolesio]: we denote by $u' = u'(x) = \frac{\partial}{\partial t}u_{\Sigma+t\sigma}(x)$ the shape derivative of $u_\Sigma$ to direction $\sigma$. By $u_\Sigma$ we denote the solution at step 1 when the boundary is $\Sigma$. It is known that $u'$ depends only on the normal component of the perturbation $\sigma$.

Now suppose that step 2 is of the form $\Sigma = \Sigma + \lambda^{-1}u|_\Sigma n_\Sigma$ and that the perturbation $\sigma$ has only the normal component. Then direct computation gives

$$\frac{d}{d\Sigma} F(\Sigma,\sigma) = \sigma + \frac{1}{\lambda}\left[\frac{\partial u}{\partial n}\Big|_\Sigma \sigma + u(\Sigma)\frac{d}{d\Sigma} n_\Sigma + u'(\Sigma)\right].$$

If we evaluate this at the solution, where $u(\Sigma) = 0$ and $\frac{\partial u}{\partial n}\Big|_\Sigma = -\lambda$, we get

$$\frac{d}{d\Sigma} F(\Sigma,\sigma) = \lambda^{-1}u'(\Sigma).$$

Thus if $\|\lambda^{-1}u'(\Sigma)\|_V \leqslant q < 1$ for all $\sigma$, $\|\sigma\|_V \leqslant 1$ the method converges. If this is not the case, or if the convergence is very slow, a natural question arises: is it possible to obtain (faster) convergence by modifying the algorithm? This leads us to the following idea presented in [Garabedian], see also [Cuvelier, Schulkes].

## 2. A SECOND ORDER FIXED POINT METHOD.

On $\Sigma$ we have two boundary conditions, $u = 0$ and $\partial u/\partial n = -\lambda$. However, it is possible to write a whole family of equivalent boundary conditions: $u = 0$ and $\alpha u + \frac{\partial u}{\partial n} = -\lambda$, where $\alpha$ can be chosen arbitrarily. Now, if we change the Neumann condition in step 1 to the Newton condition, we still get a method which has the free boundary as a fixed point, but in addition we have a free parameter $\alpha$ which can possibly be used to speed up the convergence.

Before we analyse the situation further we recall some tools of 'shape calculus'. Let us be given a domain $\Omega \subset \mathcal{O}$, a vector field $V$ defined in $\Omega$ and a function $f = f(t,x)$ defined in $[0,\varepsilon]\times\mathcal{O}$. Let us denote $f' = \partial f/\partial t$. Further, let

$$\Omega_t = \Omega + tV = \left\{ x + tV(x) \;\middle|\; x \in \Omega \right\}$$

(resp. $\Gamma_t = \Gamma + tV$). Then we have the following formulas

$$\frac{d}{dt}\int_{\Omega_t} f(t,x)\,dx\,\Big|_{t=0} = \int_{\Omega} f'(0,x)\,dx + \int_{\partial\Omega} f(0,s)\,\langle V,n\rangle\,ds,$$

$$\frac{d}{dt}\int_{\Gamma_t} f(t,s)\,ds\,\Big|_{t=0} = \int_{\Gamma} f'(0,s)\,ds + \int_{\Gamma}\left(\frac{\partial f}{\partial n} + fH\right)\langle V,n\rangle\,ds,$$

where $H$ is the mean curvature of $\Gamma$. The precise assumptions on regularity and so on can be found in [Zolesio]. The formulas are also well defined and valid when $f$ is defined only in $\left\{(t,x) \mid x \in \Omega_t\right\}$.

Now we are ready to formulate the equation for $u'$. The step 1 with $\alpha \neq 0$ can be written in the weak form as

$$\int_{\Omega} \nabla u\,\nabla\varphi\,dx + \int_{\Sigma} \alpha u\varphi\,ds = \int_{\Sigma} -\lambda\varphi\,ds, \quad \varphi \in H^1(\Omega), \quad \varphi|_{\Gamma} = 0.$$

Now, suppose we perturb $\Sigma$ (and $\Omega$) to some direction $V$. Differentiating in that direction we get

$$\int_{\Omega} \nabla u' \ \nabla\varphi \ dx + \int_{\Sigma} \alpha u' \varphi \ ds = - \int_{\Sigma} \nabla u \ \nabla\varphi \ \langle V,n\rangle \ ds -$$

$$- \int_{\Sigma} \alpha\left( \frac{\partial u}{\partial n} + uH \right)\varphi \ \langle V,n\rangle \ ds - \int_{\Sigma} \lambda H\varphi \ \langle V,n\rangle \ ds.$$

Here we have used the facts that $\varphi'$ and $\frac{\partial\varphi}{\partial n}|_{\Sigma}$ can be chosen to vanish and $\lambda' = \frac{\partial\lambda}{\partial n}|_{\Sigma} = 0$. If $\Sigma$ is the solution of the free boundary problem, we have that $u = 0$ and $\frac{\partial u}{\partial n} = -\lambda$ on $\Sigma$. Thus after dividing $\nabla u \ \nabla\varphi$ into tangential and normal components we get

$$\int_{\Omega} \nabla u' \ \nabla\varphi \ dx + \int_{\Sigma} \alpha u' \varphi \ ds = \int_{\Sigma} \alpha\lambda\varphi \ \langle V,n\rangle \ ds - \int_{\Sigma} \lambda H\varphi \ \langle V,n\rangle \ ds.$$

Thus, if we choose $\alpha = H$, we get $u' \equiv 0$.

We have arrived to the following fixed point algorithm

0. Choose initial guess $\Sigma$.
1. Solve $-\Delta u = 0$, $u|_{\Gamma} = 1$, $H_{\Sigma}u + \frac{\partial u}{\partial n} = -\lambda$ on $\Sigma$.
2. Set $\Sigma = \Sigma + \lambda^{-1}u|_{\Sigma}n_{\Sigma}$.
3. Repeat from 1.

The formal calculus made above suggests that the algorithm converges quadratically. The analysis can be made more rigorous using the following observtions. Let us assume that we work in $R^2$. Then it is known that the free boundary is $C^{\infty}$, see [Alt, Caffarelli]. Assuming that $\Omega_0$ is convex, we obtain that the free boundary is also convex. Thus, in a small neighbourhood of the solution, $H$ is nonnegative. The step 1 is thus uniquely solvable and $u|_{\Gamma} \in C^{\infty}(\Sigma)$. So the step 2 defines a mapping from $C^{\infty}_{per}(0,2\pi)$ into itself. Moreover this mapping can be made contractive with constant $q$ for any $q > 0$ by restricting the neighbourhood around the solution, as we can get $H_\Sigma$ to approximate the curvature at the solution.

It should be noticed that the above calculus was made by assuming that the correction in step 2 is made in the normal direction. If, for example, 'radial' correction is used, the step 2 should be modified to $\Sigma = \Sigma + \frac{1}{\lambda n_r} u\big|_\Sigma$, where $n_r$ is the radial component of the unit normal.

As a numerical example we consider the above problem in the axisymmetric case. Then the relevant equations can be written as

$$-u_{rr} - \frac{1}{r} u_r = 0,$$

$$u(1) = 1,$$

$$u(s) = 0,$$

$$\frac{\partial u(s)}{\partial r} = -\lambda,$$

where $s > 1$ denotes the free surface. The mean curvature $H$ at $s$ is $s^{-1}$. If we denote $F(s) = s + \lambda^{-1}u(s)$, we get by simple calculation that $|\partial F(s)/\partial s| \leqslant 1$ if $1 < s \leqslant e$. Thus, the original fixed point method converges only for $\lambda > e^{-1}$. The second order method converges also for smaller $\lambda$. In Figures 2.1 and 2.2 we present the iteration history of $s$ for the original (marked with crosses) and the second order method (marked with circles) for two different values of $\lambda$. The initial guess for the free boundary $s$ was 1.5 in the both cases.

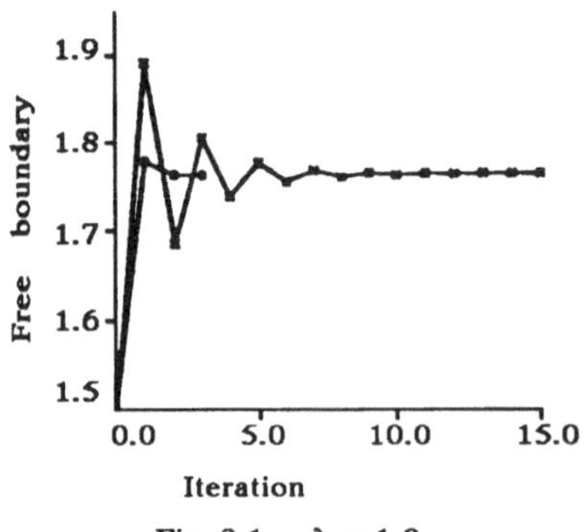


Fig. 2.1 $\lambda = 1.0$

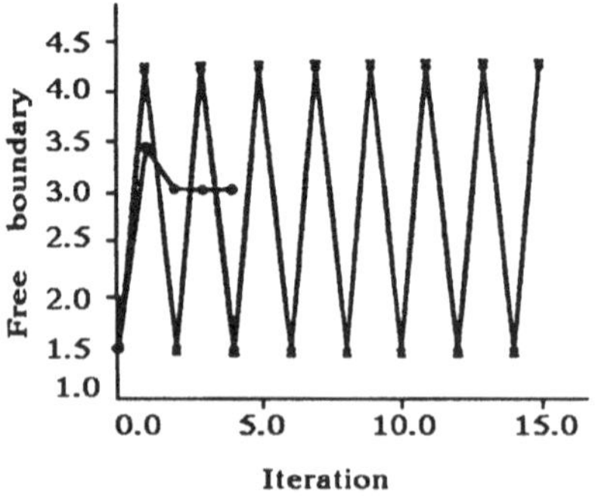


Fig. 2.2 $\lambda = 0.3$

## 3. STEADY STATE CONTINUOUS CASTING.

As the second example of a stationary free boundary problem we consider the determination of the steady state configuration of a continuous casting process. That is, we study the two-phase Stefan problem in a system where the material flows with constant velocity and the free boundary remains stationary in the laboratory frame.

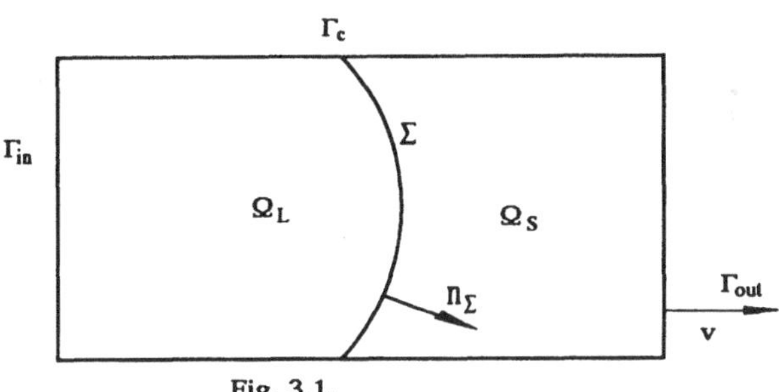


Fig. 3.1.

Let the situation be as in Figure 3.1: the free boundary $\Sigma$ separates the domain into two parts $\Omega_L$ (liquid) and $\Omega_S$ (solid). We suppose that the material flows from $\Omega_L$ to $\Omega_S$ with velocity $v$. By $n_\Sigma$ we denote the normal of $\Sigma$, directed to $\Omega_S$. For simplicity we scale the heat conductivity to be 1 (in both phases), as well as the volumetric heat capacity. The phase change temperature is normalized to 0. Then the relevant equations are

$$-\nabla u_S + v\cdot\nabla u_S = 0 \quad \text{in} \quad \Omega_S,$$

$$-\nabla u_L + v\cdot\nabla u_L = 0 \quad \text{in} \quad \Omega_L,$$

and the boundary conditions

$$u_L = u_{LO} > 0 \quad \text{on} \quad \Gamma_{in},$$

$$u_S = u_{SO} < 0 \quad \text{on} \quad \Gamma_{out},$$

$$\frac{\partial u}{\partial n} = g \quad \text{on} \quad \Gamma_c,$$

$$u_L = u_S \quad \text{on} \quad \Sigma,$$

$$-\frac{\partial u_L}{\partial n_\Sigma} + \frac{\partial u_S}{\partial n_\Sigma} = -Lv\cdot n_\Sigma \quad \text{on} \quad \Sigma.$$

Let us now try a fixed point method of the following type:

0. Give initial guess for $\Sigma$

1. Solve $u_S$ in $\Omega_S$ using the condition $u_S = 0$ on $\Sigma$.

2. Solve $u_L$ in $\Omega_L$ using the condition $\dfrac{\partial u_L}{\partial n_\Sigma} = \dfrac{\partial u_S}{\partial n_\Sigma} + Lv\cdot n_\Sigma$.

3. Set $\Sigma = \Sigma - \dfrac{u_L}{\dfrac{\partial u_S}{\partial n_\Sigma} + Lv\cdot n_\Sigma}$.

4. Repeat from 1.

In step 3 we assume that $\Sigma$ "grows" when it moves downstream with respect to flow $v$. Of course it is possible to change the roles of $\Omega_S$ and $\Omega_L$ and solve a Dirichlet problem in $\Omega_L$ and the Neumann problem in $\Omega_S$.

We are interested in finding a fast method. Thus, as before, we study the effect of changing the Neumann condition in step 2 to

$$\alpha u_L + \frac{\partial u_L}{\partial n_\Sigma} = \frac{\partial u_S}{\partial n_\Sigma} + Lv\cdot n_\Sigma.$$

In the weak form the step 3 now reads

$$\int_{\Omega_L} \nabla u_L\ \nabla\varphi\ dx + \int_{\Omega_L} v\cdot\nabla u_L\varphi\ dx + \int_\Sigma \alpha u_L\varphi\ ds =$$

$$= \int_\Sigma \left(\frac{\partial u_S}{\partial n_\Sigma} + Lv\cdot n_\Sigma\right)\varphi\ ds + \int_{\Gamma_c(\Sigma)} g\varphi\ ds.$$

Now, assume for simplicity that $g = 0$ near $\Sigma$. Then we get by differentiating in direction V with respect to $\Sigma$ (as before we can choose $\varphi' = \partial\varphi/\partial n\,\big|_\Sigma = 0$)

$$\int_{\Omega_L} \nabla u_L'\ \nabla\varphi\ dx + \int_{\Omega_L} v\cdot\nabla u_L'\varphi\ dx + \int_\Sigma \alpha u_L'\ \varphi\ \ ds\ =$$

$$= -\int_{\Sigma} \nabla_{\Gamma} u_L \nabla_{\Gamma} \varphi \, \langle V, n_{\Sigma} \rangle \, ds - \int_{\Sigma} v \cdot \nabla u_L \varphi \, \langle V, n_{\Sigma} \rangle \, ds -$$

$$- \int_{\Sigma} \alpha \left[ \frac{\partial u_L}{\partial n_{\Sigma}} + u_L H \right] \varphi \, \langle V, n_{\Sigma} \rangle \, ds + \int_{\Sigma} L(v \cdot n_{\Sigma}) \, H\varphi \, \langle V, n_{\Sigma} \rangle \, ds +$$

$$+ \frac{\partial}{\partial \Sigma} \left[ \int_{\Sigma} \frac{\partial u_S}{\partial n_{\Sigma}} \varphi \, ds \right] =$$

$$= \int_{\Sigma} \left[ -(v \cdot n_{\Sigma}) \frac{\partial u_L}{\partial n_{\Sigma}} - \alpha \frac{\partial u_L}{\partial n_{\Sigma}} + L(v \cdot n_{\Sigma}) H \right] \varphi \, \langle V, n_{\Sigma} \rangle \, ds +$$

$$+ \frac{\partial}{\partial \Sigma} \left[ \int_{\Sigma} \frac{\partial u_S}{\partial n_{\Sigma}} \varphi \, ds \right].$$

at the solution, when $u_L = 0$ on $\Sigma$.

The last term on the right corresponds to the change in the flux into $\Omega_S$ which is due to the change of geometry of $\Omega_S$.

Our next step is to try to find a formula of the type

$$\frac{\partial}{\partial \Sigma} \left[ \int_{\Sigma} \frac{\partial u_S}{\partial n_{\Sigma}} \varphi \, ds \right] = \langle g, \varphi \langle V, n \rangle \rangle_{D'(\Sigma) \times D(\Sigma)},$$

where $D(\Sigma)$ denotes the space of test functions on $\Sigma$ and $D'(\Sigma)$ is the space of distributions on $\Sigma$. It is known from the structure theorem of [Zolesio] that such a representation exists. In general $g$ can be a genuine distribution, but in our case, assuming sufficient regularity, $g$ reduces to a function. Using the well-known adjoint state technique we arrive to the following result:

$u_L' \equiv 0$ in $\Omega_L$ if $\alpha$ satisfies

$$\alpha \frac{\partial u_L}{\partial n_{\Sigma}} = -(v \cdot n_{\Sigma}) \frac{\partial u_L}{\partial n_{\Sigma}} + L(v \cdot n_{\Sigma}) H - \frac{\partial p}{\partial n_{\Sigma}} \frac{\partial u_S}{\partial n_{\Sigma}},$$

where $p$ is the solution of the adjoint equation in $\Omega_S$

$$-\Delta p - \nabla\cdot(vp) = 0 \quad \text{in} \quad \Omega_S,$$
$$p = 0 \quad \text{on} \quad \Gamma_{out},$$
$$\frac{\partial p}{\partial n} = 0 \quad \text{on} \quad \Gamma_c,$$
$$u_L = u_S \quad \text{on} \quad \Sigma,$$
$$p = 1 \quad \text{on} \quad \Sigma.$$

We can summarize the above into following algorithm

0. Give initial guess for $\Sigma$.
1. Solve $u_S$ and $\Omega_S$ using the condition $u_S = 0$ on $\Sigma$.
2. Solve $p$ in $\Omega_S$ from the adjoint equation.
3. Compute

$$\alpha = -(v\cdot n_\Sigma) + \frac{L(v\cdot n_\Sigma)H - \dfrac{\partial p}{\partial n_\Sigma}\dfrac{\partial u_S}{\partial n_\Sigma}}{\dfrac{\partial u_S}{\partial n_\Sigma} + L(v\cdot n_\Sigma)}$$

4. Solve $u_L$ in $\Omega_L$ using the condition

$$\alpha u_L + \frac{\partial u_L}{\partial n_\Sigma} = \frac{\partial u_S}{\partial n_\Sigma} + Lv\cdot n_\Sigma.$$

5. Set $\Sigma = \Sigma - \dfrac{u_L}{\partial u_L/\partial n_\Sigma}$.

6. Repeat from 1.

The algorithm has some weak points. For example, the denominator in steps 3 and 5 $\partial u_L/\partial n_\Sigma$ is typically strictly negative at the solution. However, away from the solution a zero value may be reached. Also the parameter $\alpha$ may be negative leading to problems in step 4.

If the free boundary is not $C^\infty$ the algorithm does not work without some kind of regularization as at each step the correction term has only the regularity of $\partial u_L/\partial n_\Sigma$ which normally is less smooth as $\Sigma$.

The method has been tested for a 1-dimensional problem which is an idealization of the Czochralski method for growing silicon crystals. The dimensional parameters, where the phase change temperature was normalized to 0, were chosen to be $\rho = 2300\ kg/m^3$, $c=0.7034\ kJ/kg^0C$, $k=0.149\ kW/m^0C$, $L=165.0\ kJ/kg$, $v=0.6\ mm/min$, $u_{in} = 1.0^0C$, $u_0 = 0.0^0C$ and $u_{out} = -1.0^0C$. Our 1-dimensional domain was $[-0.1,\ 0.1]$.

In Table 3.1 the iteration history is shown in the case where the problem was solved using analytic solution for the boundary value problems. In Table 3.2 we show the error in the free boundary due to FE-discretization for different grid sizes. The accuracy is very good.

Table 3.1

| N | $\Sigma$ | $u(\Sigma)$ |
|---|---|---|
| 1 | 4.1752436911054716E-02 | -0.1161924328163856 |
| 2 | 3.9725124110066330E-02 | -1.7078710860503415E-02 |
| 3 | 3.9629816803940680E-02 | -6.9283019728305639E-04 |
| 4 | 3.9629619584925786E-02 | -1.4233396130880482E-06 |
| 5 | 3.9629619584084128E-02 | -6.0741967011779252E-12 |

Table 3.2

| Grid size | $\Sigma$ | Error |
|---|---|---|
| 2 | 0.0396296118243561 | 7.7597280129237944E-09 |
| 4 | 0.0396296176768944 | 1.9071897533473603E-09 |
| 8 | 0.0396296191239731 | 4.6011098750464718E-10 |
| 16 | 0.0396296194776923 | 1.0639182590477247E-10 |
| 32 | 0.0396296195620684 | 2.2015715639422950E-11 |

## REFERENCES.

[1] H.W.Alt, L.A.Caffarelli. Existence and regularity for a minimum problem with free boundary, J.Reine Angew. Math.325 (1981), pp.105-144.

[2] C.Cuvelier, R.M.S.M.Schulkes. Some numerical methods for the computation of capillary free boundaries governed by the Navier-Stokes equations, SIAM Review 32 (1990), pp.355-423.

[3] P.R.Garabedian. The mathematical theory of three dimensional cavities and jets, Bull Amer.Math.Soc. 62 (1956), pp.219-235.

[4] J.P.Zolesio. Identification de domaines par deformations, These d'etal, Univ.Nice (1979).

International Series of Numerical Mathematics, Vol. 106, © 1992 Birkhäuser Verlag Basel 

# NONLINEAR THEORY OF DYNAMICS OF A VISCOUS FLUID WITH A FREE BOUNDARY IN THE PROCESS OF A SOLID BODY WETTING

*O.V.Voinov*
*The Institute of Multiphase Systems Mechanics*
*Tyumen 625026, RUSSIA*

Hydrodynamic aspects of wetting a solid surface and, in particular, a flow near a moving line of contact of three phases are considered. The asymptotic theory is developed so as to derive a closed-form solution.

Key words: viscous fluid, free boundary, hydrodynamics of wetting.

A flow analysis for the case of a viscous liquid with a free boundary where the contact line moves is complicated by the necessity to use special closing conditions for a low-scale region at the contact line. This is caused by the fact that imposing the no-slip condition on a viscous flow within a corner results in a singular stress field.

There are three methods for closing the equations [1,2]. In the first method a small region near the boundary line where hydrodynamic laws fail is subject to exclusion from the domain of definition of the boundary-value problem. The hydrodynamic problem may also be regularized by changing its formulation for a narrow region near the boundary line. Accordingly, the pure hydrodynamic analysis of the free-boundary movement may be completed with two other techniques also taking into account

2) the effects of far-field molecule interaction forces and

3) the "viscosity-velocity gradient" relation which is of importance for certain fluids.

The main problems to be considered are the capillary phenomena paradoxical from the viewpoint of the hydrodynamics.

Mentioned as the first must be spontaneous yielding of a small droplet of a well wetting liquid on a smooth solid. Here, the wetting angle $\alpha_0$ is time-dependent. Despite the state of nonequilibrium the droplet shape is close to a spherical segment.

The second principal phenomenon is dependence of the viscous fluid free surface shape on the speed $v$ at which a narrow capillary is filled. In this case, as well as the previous one, the free surface can be nearly spherical. The contact angle $\alpha_0$ at the interface may be zero in the event of equilibrium wetting but movement causes this angle to be non-zero (namely, positive).

The basic formulation of the problem in the hydrodynamic wetting theory includes determination of the dynamic contact angle, i.e. the slope of the tangent of the free surface $S$. The viscous flow is assumed to be stationary two-dimensional, at low Reynolds numbers. The domain of interest is also bounded by circular arcs $L_0$ and $L_m$ of major radius and of major radius and of minor radius, respectively; the arcs are taken in a cross-section normal both to the solid and the free surface. The solid $(y = 0)$ moves at the velocity of $-v$. Fluid-solid interaction area is characterized by the no-slip condition $u_x = -v$, $u_y = 0$ ($u$ is the velocity of the fluid). Over the boundary $S$ between regions 1 and 2, the no-penetration condition should be imposed, $u_n = 0$; as well, the tangential stress fields should be compatible, $P_{\tau_1} = P_{\tau_2}$ (zero if region 2 is filled with gas). The normal stresses undergo a jump due to capillary forces:

$$P_{n1} - P_{n2} = -\sigma \frac{d \cos \alpha}{dh},$$

here $h$ is an ordinate of a point on the free surface, $n$ is the external normal to region 1, $\sigma$ is the capillary tension coefficient.

Let a free surface point $y = h$ be put in correspondence with a circular ark $L$ whose center is at the point where the tangent intersects the axis $y = 0$, the point being on the arc. Then a portion of the free boundary can be said to possess certain domain of the flow and to be characterized by the scale of the distance from the portion to the solid.

The speed $u$ over the arcs $L_0$ and $L_m$ is assumed to be known and continuous, including the field at the points $y = 0$. Let us require that ratios of these speeds to the solid surface velocity

$v$ are limited values when passage to limits is performed. We can also consider some model values of the speed over these arcs and make use of the velocity field inherent in corner flows.

We should consider simultaneously two passages to limits,

$$\frac{\mu v}{\sigma} \to 0, \quad \frac{h_0}{h_m} \to \infty,$$

where $\mu$ is the fluid viscosity coefficient, values of $h_0$ and $h_m$ are related to the boundary arcs. The curvature $q$ at a point $h$ of the free surface must decrease:

$$q(h_0) = -\left.\frac{d \cos \alpha}{dh}\right|_{h_0} \to 0.$$

The least distance $h_m$ from the free boundary to the solid is associated with an angle

$$\alpha = \alpha_m, \quad h = h_m.$$

The $h_m$ has a lower limit of a few molecule dimensions $a$, independent of another limitations. Therefore, in using the first method for closing the mathematical problem, the $h_m$ may be taken as low as $2a$ to $3a$ if no other limitations are imposed. However, if the molecular forces and the viscosity-velocity gradient dependence are taken into account, the angle $\alpha_m$ can turn out to be other than the equilibrium value and the distance $h_m$ may increase.

An asymptotic solution is sought at conditions $h_m \ll h \ll h_0$. With this, the velocity field within a domain of characteristic length $h$ is, in the main approximation, independent of conditions on the boundary arcs $L_m$ and $L_0$. A tangent line slope angle $\alpha$ may slowly vary as a function of the height $h$:

$$h \left| \frac{d\alpha}{dh} \right| \ll \alpha.$$

The main approximation is based on the fact that the flow at $\mu v/\alpha \ll 1$ is insignificantly differing from the corner flow. The jump in normal stresses at the boundary is

$$p_{n1} - p_{n2} = \frac{\mu v}{h} \mathcal{F}_1(\alpha), \quad \mathcal{F}_1(\alpha) = \frac{2 \sin^2\alpha}{\alpha - \sin \alpha \cos \alpha}$$

and the angle $\alpha$ satisfies the differential equation

$$\sin \alpha \frac{d\alpha}{dh} = \frac{\mu v}{\sigma h} \mathcal{F}_1(\alpha).$$

In the next order with respect to the parameter $\mu v/\sigma$, account is made for the boundary being locally differing from a straight line, a minor curvature is evaluated on the basis of the first approximation etc. As a result, the jump in stresses may be represented by an asymptotic series in $\mu v/\sigma$:

$$p_{n1} - p_{n2} = \frac{\mu v}{h} \sum_{k=1}^{n} \left( \frac{\mu v}{\sigma} \right)^{k-1} \mathcal{F}_k(\alpha).$$

The corresponding expression for the angle $\alpha$ has been written for $N = 2$ in [3].

Of main interest here is the first approximation which, after employing the condition at $h = h_m$, provides a relation

$$\frac{1}{2} \int \left[ \frac{\alpha}{\sin \alpha} - \cos \alpha \right] d\alpha = \frac{\mu v}{\sigma} \ln \frac{h}{h_m}.$$

This equation can be simplified over a wide range $(\alpha < 5\pi/6,\ \alpha_m \leqslant 1)$ to give the formula

$$\alpha = \left[ \alpha_m^3 + \frac{9\mu v}{\sigma} \ln \frac{h}{h_m} \right]^{1/3}.$$

Estimation indicates that the approximate closure condition at $h = h_m$ is allowed to be used if an error of the order of $(\ln (h/h_m))^{-1}$ is tolerable. Such error is produced by either the value of $h_m$ being specified roughly or the solution being with deviation for $h \sim h_m$; the latter is observed in the case of $\alpha_m = 0$.

The equations derived do help to disclose a crise in wetting. Say, increasing the parameter $\mu v/\sigma$ results in the angle $\alpha$ increasing up to the maximum possible value, i.e. $\pi$. If $\alpha_m > 0$, the critical maximum magnitude of the speed $v < 0$ (at which the surface becomes free from the fluid) is described by the equation

$$\frac{9\mu|v|}{\sigma} \ln \frac{h}{h_m} = \alpha_m^3.$$

The crise is in the dynamic wetting angle being of no sense, the free boundary not being stationary, and a film formation being possible.

Considering the case of two fluids in a capillary requires an additional parameter to be introduced - the ratio of viscosity coefficients $\mu_* = \mu_2/\mu_1$. The method remains the same as for the liquid-gas pair for which $\mu_* = 0$. The first approximation to the solution [4] is

$$\int_{\alpha_m}^{\alpha} f(\alpha,\mu_*)\, d\alpha = \frac{2v\mu_1}{\sigma} \ln \frac{h}{h_m},$$

$$f(\alpha,\mu_*) = \frac{(\alpha-\cos\alpha\ \sin\alpha)(\varphi^2-\sin^2\alpha) + \mu_*(\varphi-\cos\varphi\ \sin\varphi)(\alpha^2-\sin^2\alpha)}{\sin\alpha[\pi-(1-\mu_*)(\alpha-\sin\alpha)][\pi-(1-\mu_*)\ (\alpha+\sin\alpha)]},$$

$$\varphi = \pi - \alpha.$$

critical speeds associated with the angle limiting values ($\pi$ or $0$) depend on $\mu_*$, their values being subject to numerical determination from the latter equation. Asymptotic behavior (at $\mu_* \to 0$) for $v < 0$ is the same as for a liquid with gas, and for $v > 0$ is describable by the relation

$$\frac{9\mu_1 v}{\sigma} \ln \frac{h}{h_m} = \frac{3}{2} \pi \ln \frac{1}{\mu_*} - 4.059 - \alpha_m^3.$$

The second method of closing the theory in the case of perfect wetting ($\alpha_m = 0$) is based on introduction of the Van-der-Waals forces in the thin layer approximation. The equations for a steady flow in a film include an additional pressure $A/6\pi\ h^3$, where $A < 0$ is the Hamaker parameter. An approximate analytical solution [2] to the corresponding boundary-value problem is of the following form when $h \gg h'_m$:

$$\alpha = \left[ \frac{9\mu v}{\sigma} \ln \frac{h}{h'_m} \right]^{1/3}, \quad h'_m = \left[ \frac{\sigma}{3\mu v} \right]^{1/3} \left[ \frac{|A|}{2\pi\sigma} \right]^{1/2}.$$

In $h \ll h'_m$, the profile of the layer is significantly depending on molecule interaction forces. Estimate indicates that the $h'_m$ at the contact angle $\alpha_0$ of an order of 1 radian is as low as a molecule size. Therefore, the Van-der-Waals force is of no meaning at $\alpha_0 \geqslant 1$, although the solution is formally valid when $h \to 0$. The molecule interaction force is to be accounted at extremely low velocities $v$ when the angle $\alpha_0 \to 0$. Influence of this force becomes notable only if $h'_m$ considerably exceeds the molecule size. It should be concluded that these results validate the closure based on "truncating" the solution at the height $h_m$ of several molecule sizes, with exception for extremely low angles. Any method for closing the theory must be validated by experiment. From this point of view, experiments with yielding droplets seem to be useful.

The nonstationary problem on spontaneous yielding of a droplet on a solid is solved on the basis of the asymptotic theory for a flow in the vicinity of the contact of three phases. In the main approximation, the radius variation rate is a function of the dynamic wetting angle: $v \sim \alpha^3$. The theory predicts $\mu v/\sigma \ll 1$, and the droplet shape when considered at the scale of the contact area diameter is a sphere segment. As time $t$ goes on, the diameter varies as $t^{1/10}$ [1], this law being confirmed by a large amount of experimental data (for example, see [5]).

The third version for closing the asymptotic theory in an area of small size (the "rheological" version) was considered to cover polymeric liquids which can be nonlinear viscous at high velocity gradients (the higher the gradient, the less the viscosity coefficient).

The formula for the dynamic wetting angle can be generalized to the case of weak dependence of viscosity on the velocity gradient. The angle is described (in [1]) as a function of an integral in $h$, which converges at the lower limit, thus reflecting convergence of the energy dissipation integral for the case of a nonlinear viscous liquid.

Analogous expressions have been derived by using a thin-layer approximation within low-scale domains. The viscosity coefficient was assumed to be a power function of the velocity gradient $\varepsilon = |\partial u/\partial y|$ if the latter exceeds a certain critical value $\varepsilon_*$ ($\mu = \mu(\varepsilon_*/\varepsilon)^n$, $\varepsilon > \varepsilon_*$, $n > 0$). The characteristic length of the domain is $h = 2v/\varepsilon_*$. It is principally important that this length $h_*$ at "macroscopic" angles $\alpha_0 \lesssim 1$ is as low as a molecule size; this is evidenced by experiment [6]. So the "rheological" closure is not physically meaningful, only the first method should be valued as feasible. The "rheological" closure may be used at relatively great angles $\alpha_0 > \pi/2$. It is interesting that the results obtained with such closure principle have low difference from those produced by using the first method.

Yielding of a small droplet has been analyzed in [1] with due account for nonlinearity of the viscosity-velocity gradient relation on the basis of the experimental data available. The results differ insignificantly from those with the first closure method employed.

The asymptotic theory based on the dynamic wetting angle made it possible to determine the shape of an interface boundary which moves in a capillary. The results obtained are in good agreement with experimental data on filling a narrow capillary with a viscous liquid [6'7]. Agreement holds over the angle range of $0$ to $\pi$ when the capillary constant vary by a factor of several powers of ten.

## REFERENCES.

[1] Voinov O.V. Hydrodynamics of wetting, Izv.,Akad.Nauk SSSR, Mekh.Zhid. i Gasa, 5, (1976), 76-84.

[2] Voinov O.V. On the interface inclination in moving liquid layers, Zh. Prikl.Mekh. i Tekh.Fez., 2, (1977), 92-99.

[3] Voinov O.V. The free surface asymptotic for a viscous liquid in a creeping motion and the dependence of the contact wetting

angle on the speed, Dokl.Akad.Nauk SSSR, 243, 6, (1978),1422-1425..

[4] Voinov O.V. Hydrodynamic wetting theory, Proceedings of the 1st All-Union Seminar on "Hydrodynamics of applying the polymer coast", Pereslavl: preliminary publication of the Thermophysics Institute (SO AN SSSR), No.179,(1988).

[5] Welygan D.G., Burns Ch.M. Spreading of viscous, well wetting liquids on plane surfaces, J.Adhes.,10, (1979), 123-138.

[6] Hoffman R.L. A study of the advancing interface, J.Colloid and Interface Sci., 50, (1975), 228.

[7] Zheleznyi B.V. Experimental study of the contact angle dynamic hysteresis, Dokl.Akad.Nauk SSSR, 207, 3, (1972), 647-650.

Zeitfracht Medien GmbH
Ferdinand-Jühlke-Straße 7
99095 Erfurt, Deutschland
produktsicherheit@kolibri360.de